Preface

As the world enters the AI era, auditors face new challenges and opportunities that require them to use the right tools and technologies to stay ahead. With over 20 years of experience in audit and business data analytic, I have gained a deep understanding of the importance of transitioning from traditional audit operations to smart auditing that involves proactive warning or prediction using AI. Smart auditing enables businesses to obtain valuable insights and identify potential risks that traditional auditing methods may not detect. By using AI to analyze large amounts of structured and un-structured data, auditors can provide more accurate and insightful audits, as well as assist with compliance and risk management.

JCAATs is a new audit software for smart auditing. It utilizes AI language Python and can run on Windows and MAC operating systems. It offers traditional computer-aided audit tools (CAATs) data analysis functions along with AI functions such as text mining, machine learning, and data crawling, resulting in smarter audit analysis. The software allows for the analysis of large amounts of data and an open data architecture that enables interfacing with various databases, cloud data sources, and different file types, making data collection and integration more convenient and faster. Additionally, the multiple language and visual user interface makes generating Python audit programs simple and easy, even for auditors not familiar with Python language. Integrating with open-source Python program resources enables more scalability and openness for audit programs, eliminating the limitations of only a few software programs.

This textbook explains the use of technology such as big data analysis, text mining, and machine learning in auditing through practical cases. Readers will gain an understanding of data analysis and smart audit and their advancements. JCAATs, which includes data fusion technology and an OPEN DATA connector, helps auditors quickly obtain heterogeneous data for audit operations, enhancing effectiveness and efficiency. The textbook includes exercise data for practicing with JCAATs to fully experience intelligent auditing practices. It's suitable for professionals like accountants, auditors, legal and compliance personnel, risk management, and information security, as well as managers at all levels, college teachers, and students with data analysis needs.

Sherry Huang ICCP, CEAP, CFAP, CIA, CCSA
Jacksoft Ltd., Taipei, Taiwan
2023/03/15

Readers Guide

For Lecturers

To cultivate AI programming skills in business schools and other institutions, the International Computer Auditing Education Association (ICAEA) recommends starting with No Code. This approach allows students to develop audit applications to solve real-world problems without needing to write any code, giving them practical experience and enhancing their professional value. Once proficient in No Code, students can progress to Read Code training, where they gain an understanding of coding logic. Finally, Write Code training enables them to create their own AI applications. This approach prepares students for a future working environment centered on data analysis and smart auditing, where they can collaborate with AI auditing robots. As the AI era presents new challenges to the education of business professionals, colleges must adopt innovative and effective approaches to prepare their students for the workforce of the future.

This book is designed to be a comprehensive guide for learners to develop their skills in big data analysis, text mining, and machine learning using the JCAATs AI audit software. The book is structured to provide a progressive learning experience, with chapters covering different concepts and applications, as well as exercises and practice questions. The book also includes simulated independent audit case exercises to help learners apply what they have learned to real-world business environments and develop their problem-solving abilities. The operating manuals for each audit instruction enable teachers to deliver No Code courses that are more practical and aligned with real-world applications. Overall, this book aims to provide a powerful knowledge system for smart auditing and equip learners with the knowledge and skills needed for the modern workplace.

This book provides an educational version of the JCAATs - AI audit software for trial use, with a multi-language interface. This allows students to install the software on their personal computers for operational practice and learning. Since the JCAATs software is based on Python, it is easier to incorporate external resources into advanced Read Code and Write Code teaching, providing students with the necessary tools to develop their skills in big data analysis, text mining, and machine learning for the future auditing.

For students

As technology continues to advance, computer auditing plays an increasingly important role in ensuring the reliability and accuracy of financial statements and other key business information. Computer audit software enables auditors to analyze large amounts of data quickly and effectively, allowing them to identify potential areas of risk or fraud. It is essential for those entering the field of computer auditing to have a strong understanding of the latest computer audit software and to be able to use these tools effectively in practice.

This book offers an educational version of the software for trial use, which features a multi-language interface. Students can install the software on their personal computers for practice and learning, providing them with more opportunities for self-study. The book includes simulated independent audit case exercises, which go beyond theoretical knowledge transfer and enable students to integrate course content with practical applications more effectively. This approach makes learning more engaging and interesting.

Trail education software and exercise dataset

By scanning the QR code located in the image below, readers can access the teaching resource portal for this book. Once they enter the unique registration code, they can browse a variety of learning materials, download exercise datasets, and even apply for a trial use of the educational software version.

Login to download the test data and trail software

CAATs Professional Certificate

The **International Certified CAATs Practitioner (ICCP)** designation is a personal, professional certification to signify that you possess CAATs foundation knowledge and skills. Individuals qualifying for this designation must satisfy and substantiate the extensive skills and knowledge requirements established by the International Computer Auditing Education Association (ICAEA). It is the most common and fundamental certification for CAATs. The software currently used in the certification exams includes ACL, IDEA, JCAATs, etc.

- **ICCP Exam Method**

This closed-book, 120 minutes-long certification exams is divided into two components: a Knowledge Inventory and two short computer auditing case studies. The case study is CAATs free independent. You can use any CAATs, such as ACL, IDEA, JCAATs, Picalo, etc., to attend the exam.

1. **Multiple choice questions, worth 60 points in total -**

   Test the knowledge of the use of CAATs computer-aided audit tools and the basic concepts of JCAATs audit instructions.

2. **Example operation, 40 points in total -**

   The on-board test is conducted in the computer classroom, one person has one computer. Before the test, each computer will be pre-installed with JCAATs software, test data files and test questions. The scoring method is based on the answers you fill in and the planning process.

Part 2:   Exercise Solution + Practice

Chapter 3 Exercise solutions for Project 1

Chapter 4 Exercise solutions for Data 11

Chapter 5 Exercise solutions for Expressions 125

Chapter 6 Exercise solutions for Validation 153

Chapter 7 Exercise solutions for Analysis 243

Chapter 8 Exercise solutions for Script 287

Chapter 9 Exercise solutions for Text Mining 289

Chapter 10 Exercise solutions for Machine Learning 303

Chapter 11 Exercise solutions for Reporting 341

Chapter 12 Exercise solutions for Sampling 351

Chapter 13 Exercise solutions for Tools 359

Appendix A References

Appendix B Other Learning Resources

Python Based Computer-Assisted Audit Techniques (CAATs)

Data Analysis and Smart Audit

JCAATs - AI Audit Software In-Class Exercises

JCAATs - AI Audit Software

Chapter 3 - Exercise

Chapter 3 - Exercise

- Exercise 3.1: Please open the JCAATs AI audit software.
- Exercise 3.2: Please open the JCAATs sample project file.

Exercise 3.1
Please open the JCAATs AI audit software.

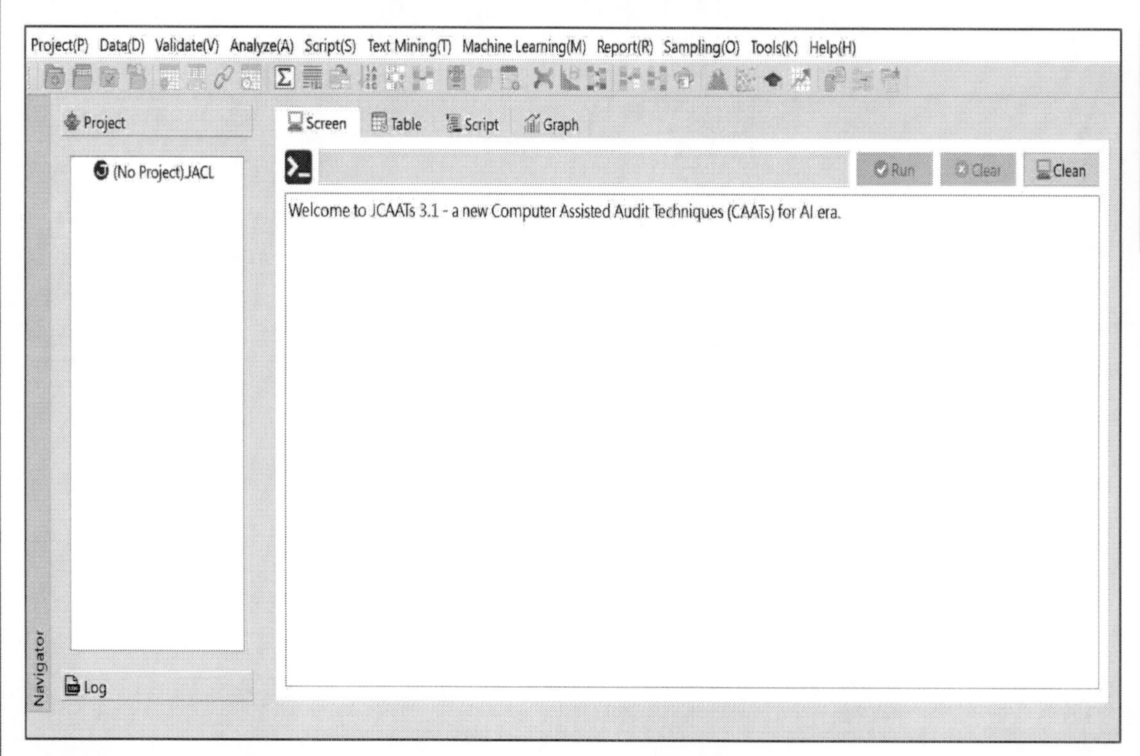

Exercise 3.2
Please open the JCAATs sample project file.

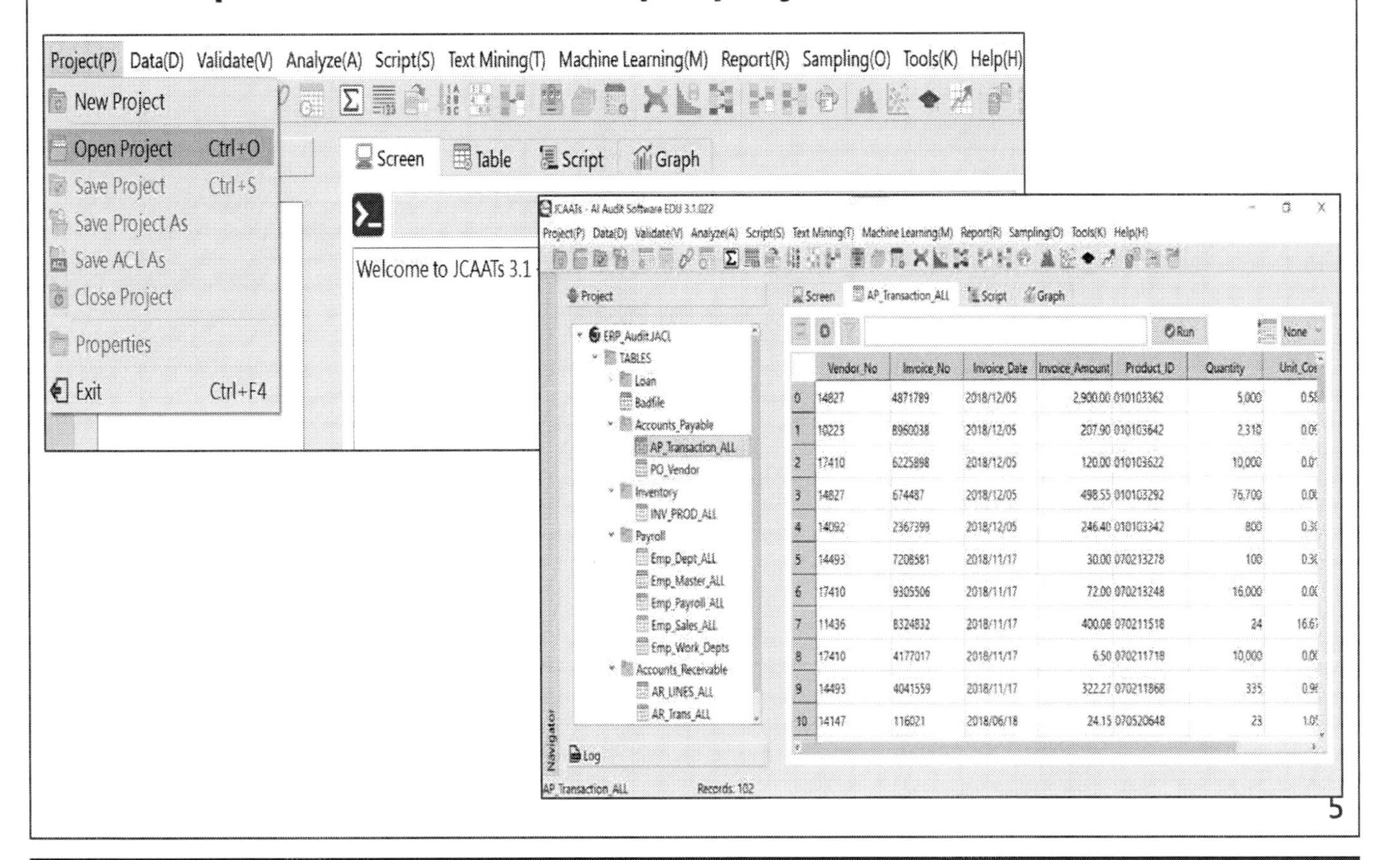

JCAATs-AI Audit Software Copyright © 2023 JACKSOFT.

Chapter 3 - Exercise

- Exercise 3.3: Please browse each data tables for analysis on the project navigator.
- Exercise 3.4: Please confirm whether the data table is read-only.
- Exercise 3.5: Please browse the operation log.

Exercise 3.3
Please browse each data tables for analysis on the project navigator.

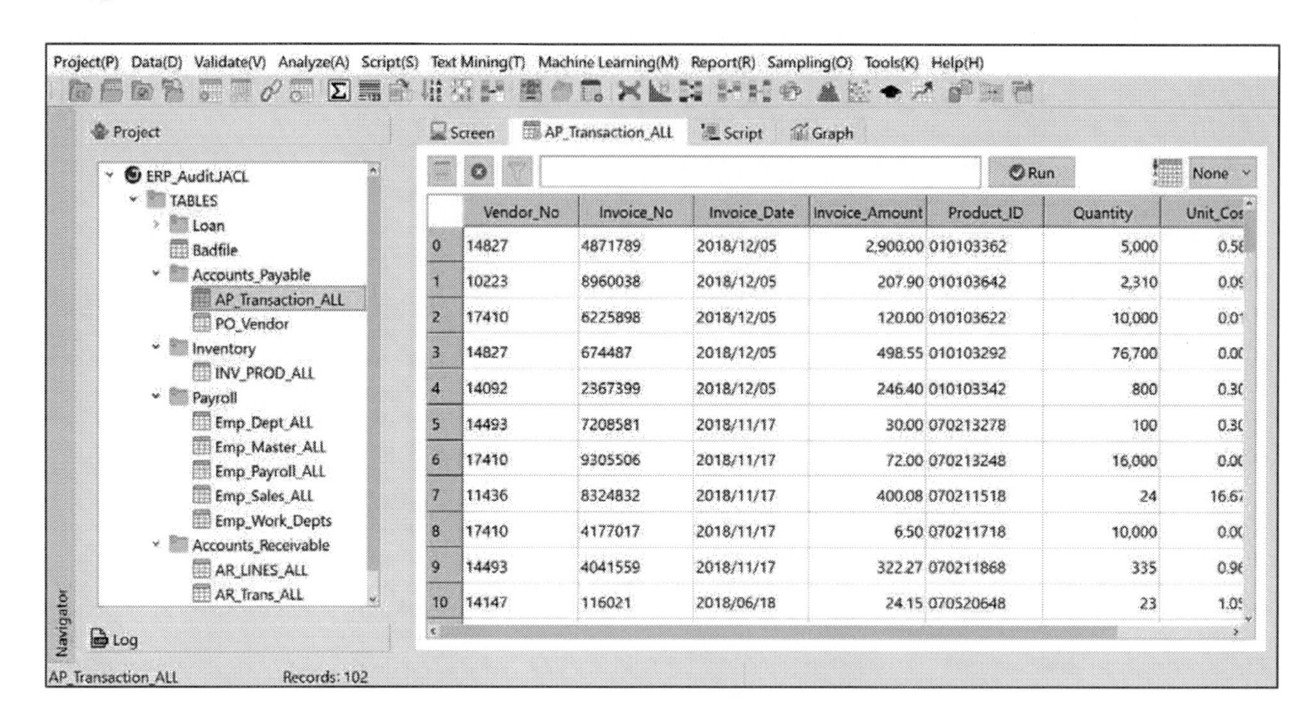

	Vendor_No	Invoice_No	Invoice_Date	Invoice_Amount	Product_ID	Quantity	Unit_Cos
0	14827	4871789	2018/12/05	2,900.00	010103362	5,000	0.58
1	10223	8960038	2018/12/05	207.90	010103642	2,310	0.09
2	17410	6225898	2018/12/05	120.00	010103622	10,000	0.01
3	14827	674487	2018/12/05	498.55	010103292	76,700	0.00
4	14092	2367399	2018/12/05	246.40	010103342	800	0.30
5	14493	7208581	2018/11/17	30.00	070213278	100	0.30
6	17410	9305506	2018/11/17	72.00	070213248	16,000	0.00
7	11436	8324832	2018/11/17	400.08	070211518	24	16.67
8	17410	4177017	2018/11/17	6.50	070211718	10,000	0.00
9	14493	4041559	2018/11/17	322.27	070211868	335	0.96
10	14147	116021	2018/06/18	24.15	070520648	23	1.05

Exercise 3.4
Please confirm whether the data table is read-only.

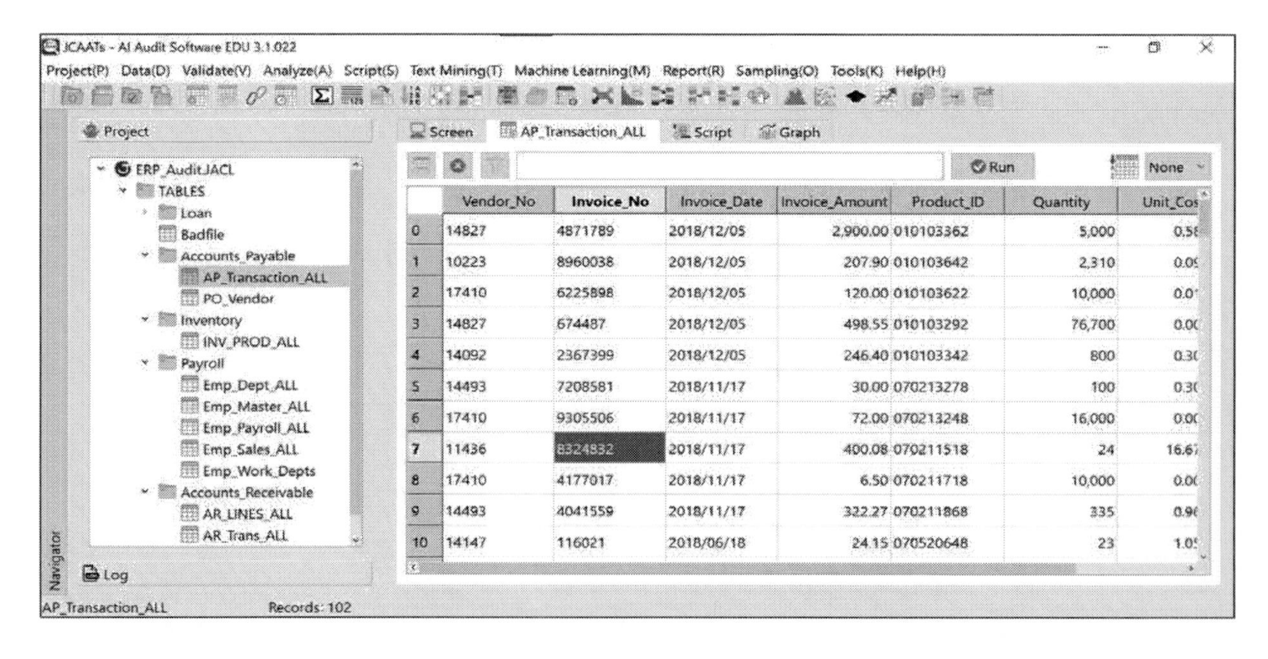

	Vendor_No	Invoice_No	Invoice_Date	Invoice_Amount	Product_ID	Quantity	Unit_Cos
0	14827	4871789	2018/12/05	2,900.00	010103362	5,000	0.58
1	10223	8960038	2018/12/05	207.90	010103642	2,310	0.09
2	17410	6225898	2018/12/05	120.00	010103622	10,000	0.01
3	14827	674487	2018/12/05	498.55	010103292	76,700	0.00
4	14092	2367399	2018/12/05	246.40	010103342	800	0.30
5	14493	7208581	2018/11/17	30.00	070213278	100	0.30
6	17410	9305506	2018/11/17	72.00	070213248	16,000	0.00
7	11436	8324832	2018/11/17	400.08	070211518	24	16.67
8	17410	4177017	2018/11/17	6.50	070211718	10,000	0.00
9	14493	4041559	2018/11/17	322.27	070211868	335	0.96
10	14147	116021	2018/06/18	24.15	070520648	23	1.05

Click on any field in the form to confirm whether it is read-only. (cannot be modified)

Exercise 3.5
Please browse the operation log.

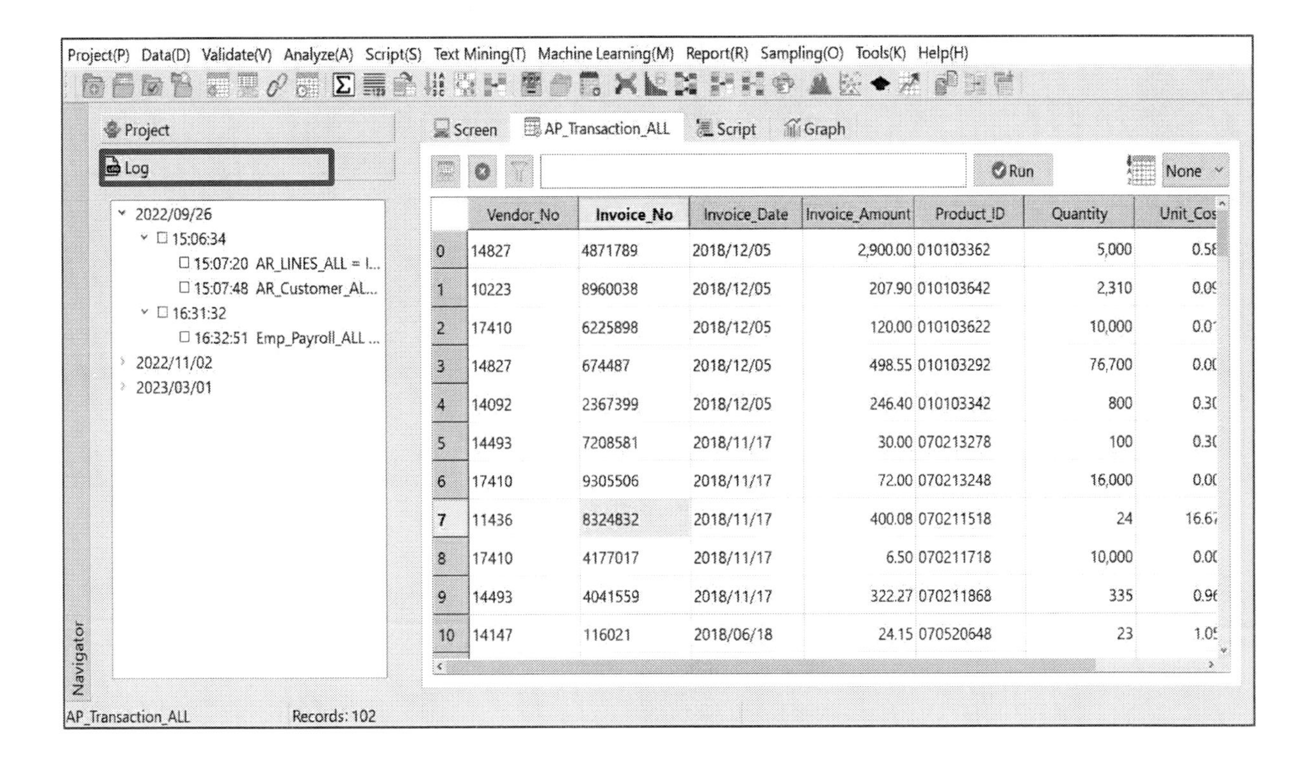

JCAATs Learning Note:

JCAATs - AI Audit Software

Chapter 4 - Exercise

Chapter 4 - Exercise

- Exercise 4.1: Please create a file folder on your own computer for storing the project and related files. Please do not use special symbols for naming conventions, and do not start with numbers.

- Exercise 4.2: Please obtain various files in different formats to facilitate the practice of data import.
  - **ACL project files**
  - **MS Excel**
  - **Delimited text files, fix length flat files**
  - **OpenDocument (ODS)**
  - **JSON**
  - **PDF, PDF-Table**
  - **XML.**

Exercise 4.1

Please create a file folder on your own computer for storing the project and related files. Please do not use special symbols for naming conventions, and do not start with numbers.

Create a new folder named JCAATs_CO

Exercise 4.2

Please obtain various files in different formats to facilitate the practice of data import.

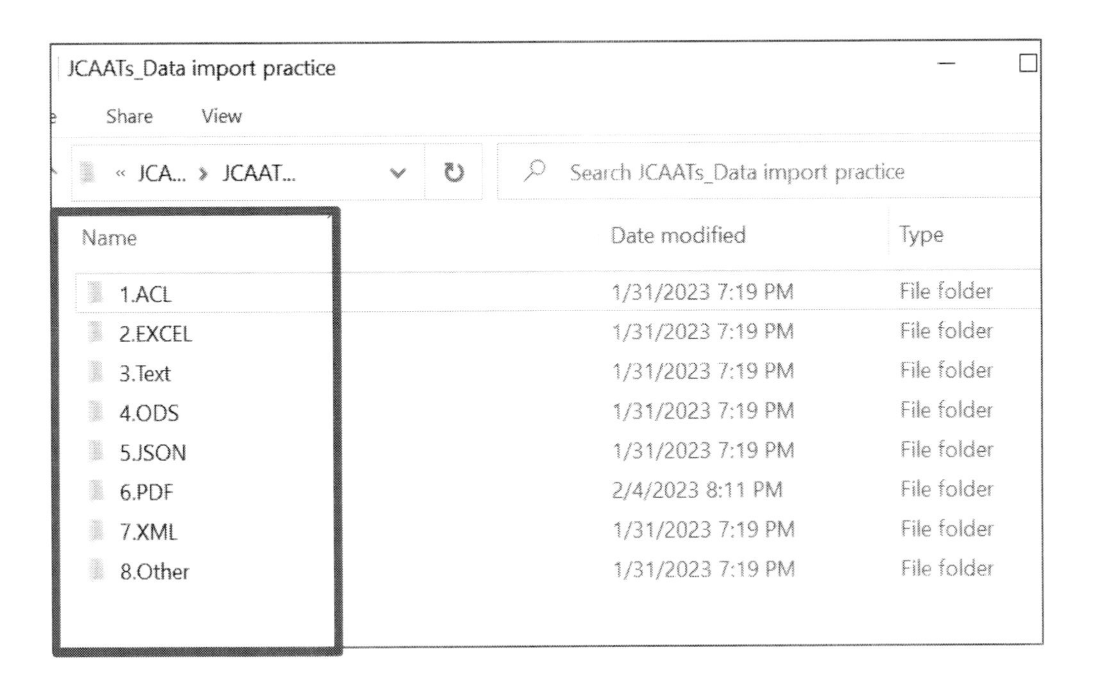

Chapter 4 - Exercise

- Exercise 4.3: Please create a smart project file, i.e. JCAATs project, under the previously established file folder, and name it according to specifications without using special symbols and without starting with a number.

- Exercise 4.4: Please practice importing the ACL project data file into the JCAAT project.

JCAATs - AI Audit Software

Please create a smart project file, i.e. JCAATs project, under the previously established file folder, and name it according to specifications without using special symbols and without starting with a number. (Project name: JCAATs_1)

JCAATs AI Audit Software: Add a New Project

1. Create a new folder
2. Click JCAATs-AI audit software
3. Click "Project > Select New Project"
4. Define a project name
5. Save

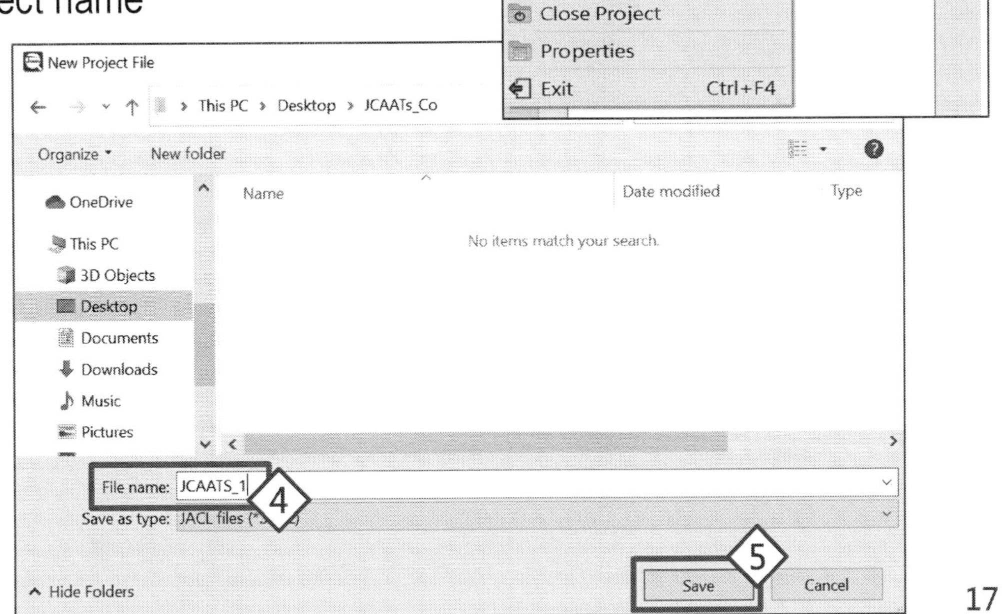

JCAATs - AI Audit Software

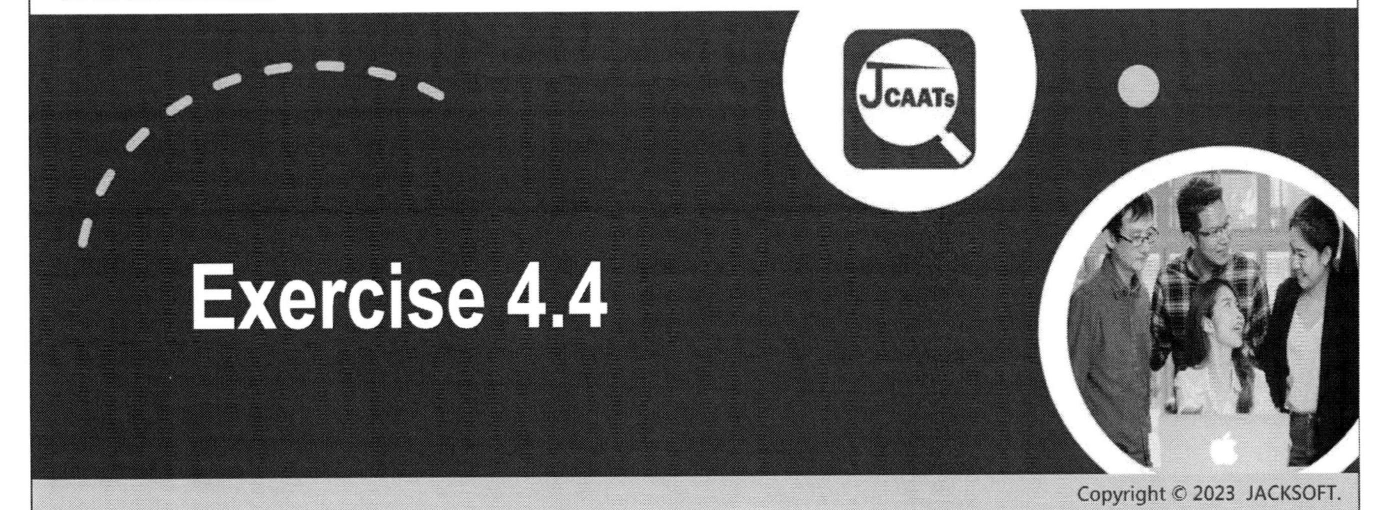

Please practice importing the ACL project data file into the JCAAT project.

Exercise 4.4
Please practice importing the ACL project data file into the JCAAT project.

Step1: Add a New Table-Select data source

- Select "Data > New Table".
- After selecting the data source platform as "File", click "Next".

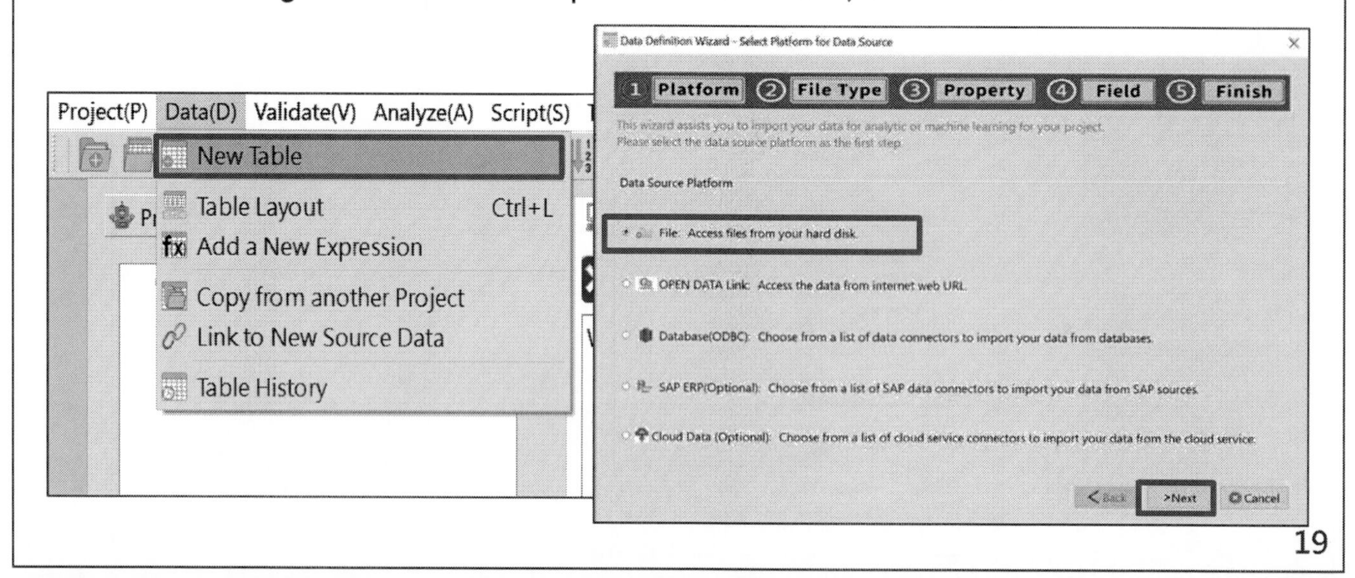

Exercise 4.4
Please practice importing the ACL project data file into the JCAAT project.

- After locating the ACL project file path, select "Open" to open the project.

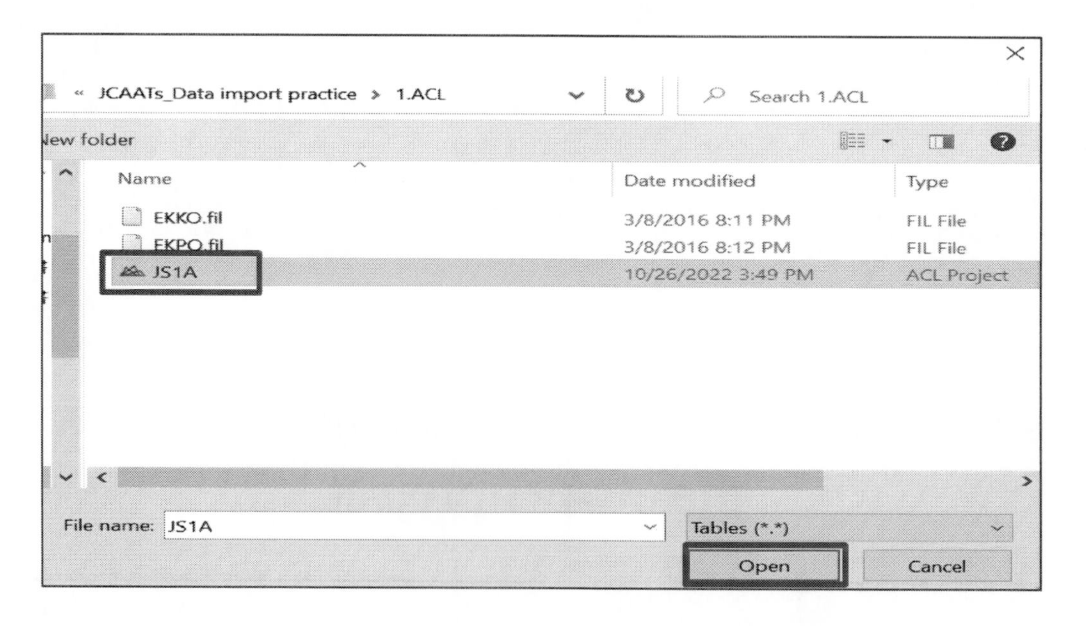

Exercise 4.4
Please practice importing the ACL project data file into the JCAAT project.

Step2: Select File Format

- JCAATs will automatically detect the file format to ensure accuracy. If there are no errors, select "Next ".

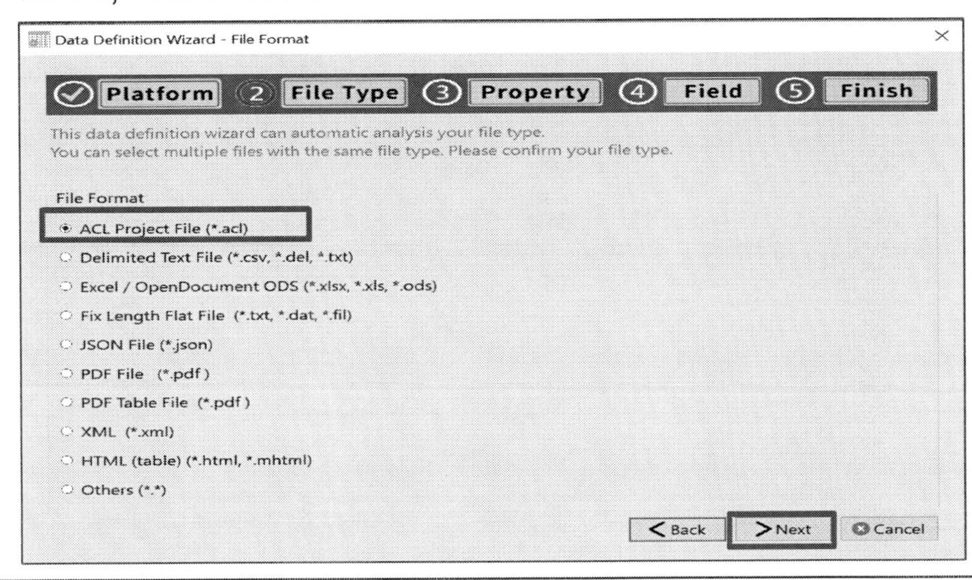

JCAATs-AI Audit Software

Exercise 4.4
Please practice importing the ACL project data file into the JCAAT project.

- **Step 3: Data features**
- JCAATs will automatically detect the character encoding method of the file and display the encoded result at the bottom. You can import multiple tables at once by clicking "Select All Tables" or by selecting the data tables that are needed and then clicking "Next".

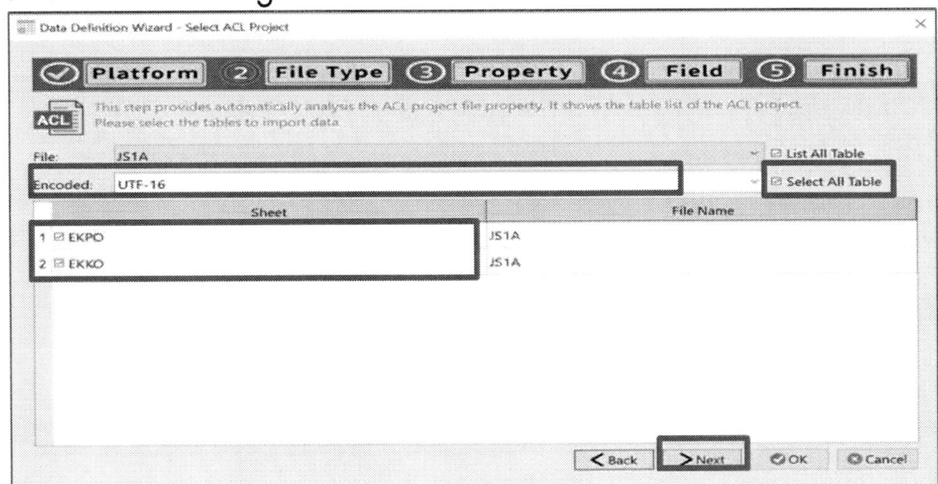

Exercise 4.4
Please practice importing the ACL project data file into the JCAAT project.

- After defining the data features, click "Next" to proceed.

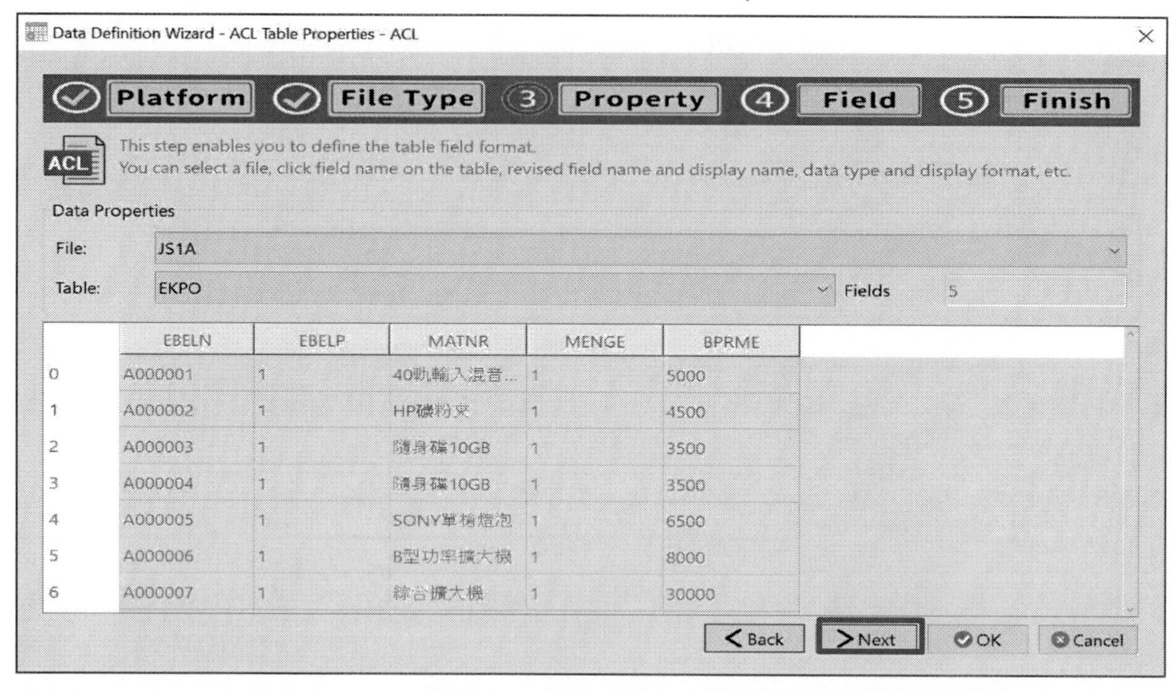

Exercise 4.4
Please practice importing the ACL project data file into the JCAAT project.

- **Step 4: Field Definition**
  After defining the field name, display name, data type, and data format, select "Next".

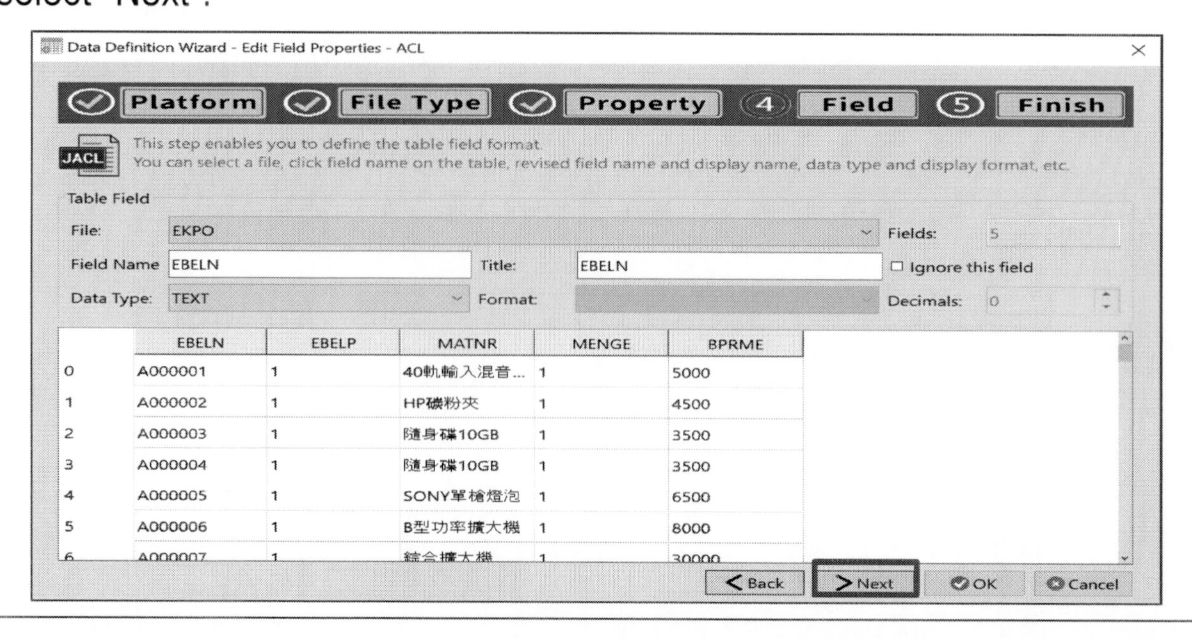

Exercise 4.4
Please practice importing the ACL project data file into the JCAAT project.

- **Step5: Confirm the data file path**
- The default data file path is the project folder, but it can be modified as needed. After ensuring that the path and information are correct, select "Done".

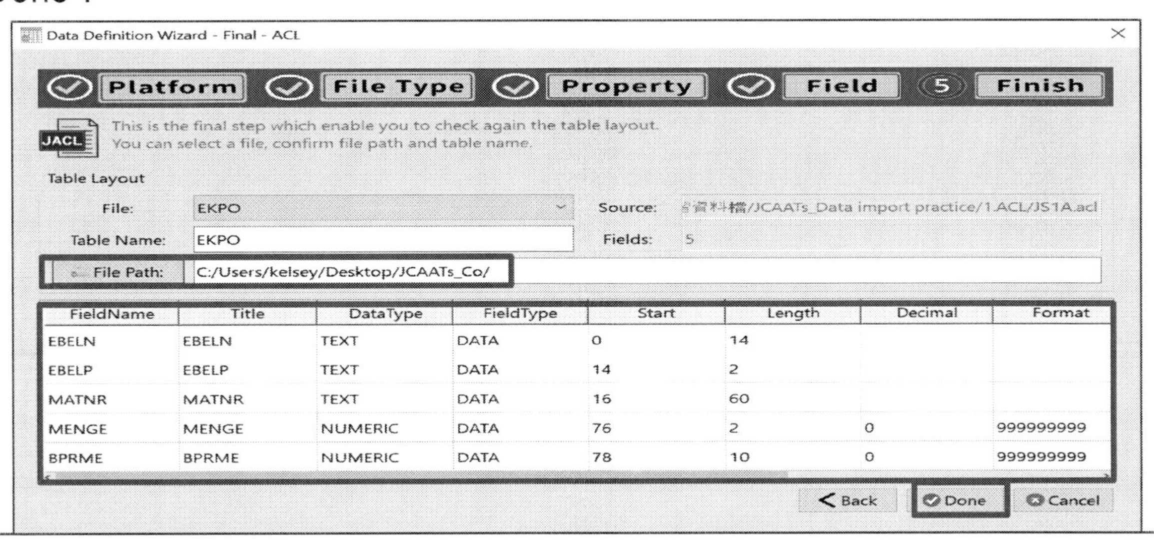

Exercise 4.4
Please practice importing the ACL project data file into the JCAAT project.

- After completing the steps to import the ACL project file using the Data Definition Wizard, make sure that the imported data is correct.

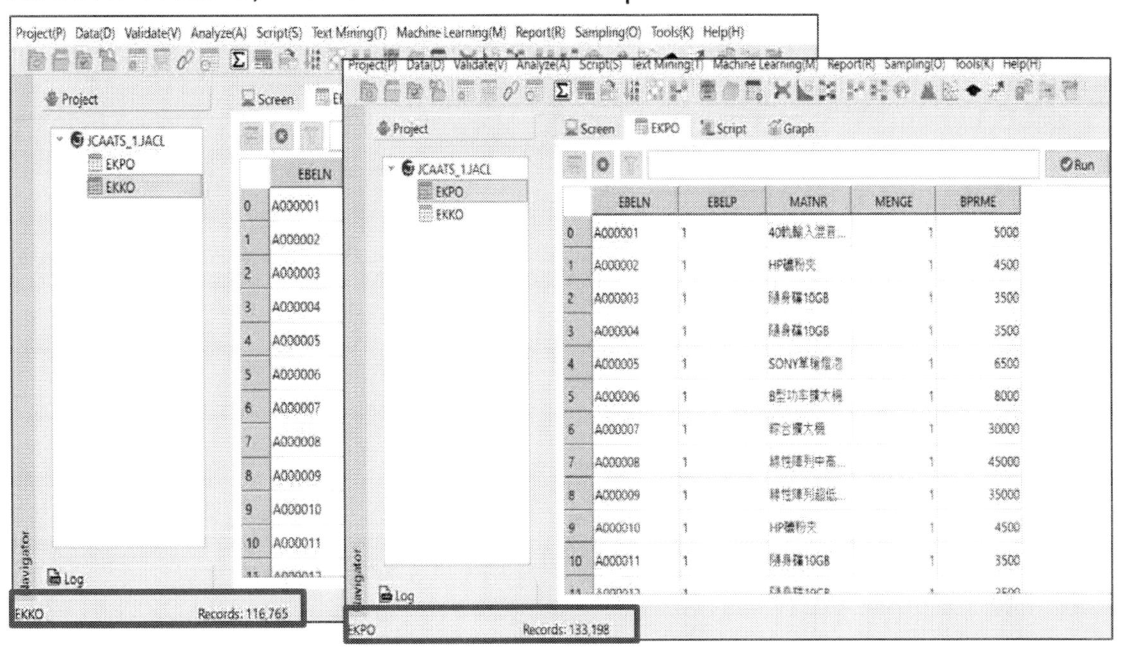

Chapter 4 - Exercise

- Exercise 4.5: Please practice importing an Excel file into the JCAATs project.
- Exercise 4.6: Please practice importing a delimited text file into the JCAATs project.
- Exercise 4.7: Please practice importing a fixed-length text file (Flat File) into the JCAATs project.

JCAATs - AI Audit Software

Please practice importing an Excel file into the JCAATs project.

Credit_cards_metaphor.xls

Exercise 4.5
Please practice importing an Excel file into the JCAATs project.

Step1: Add a New Table-Select data source

- Click "Data>New Table"
- After selecting the data source platform as "File", select "Next".

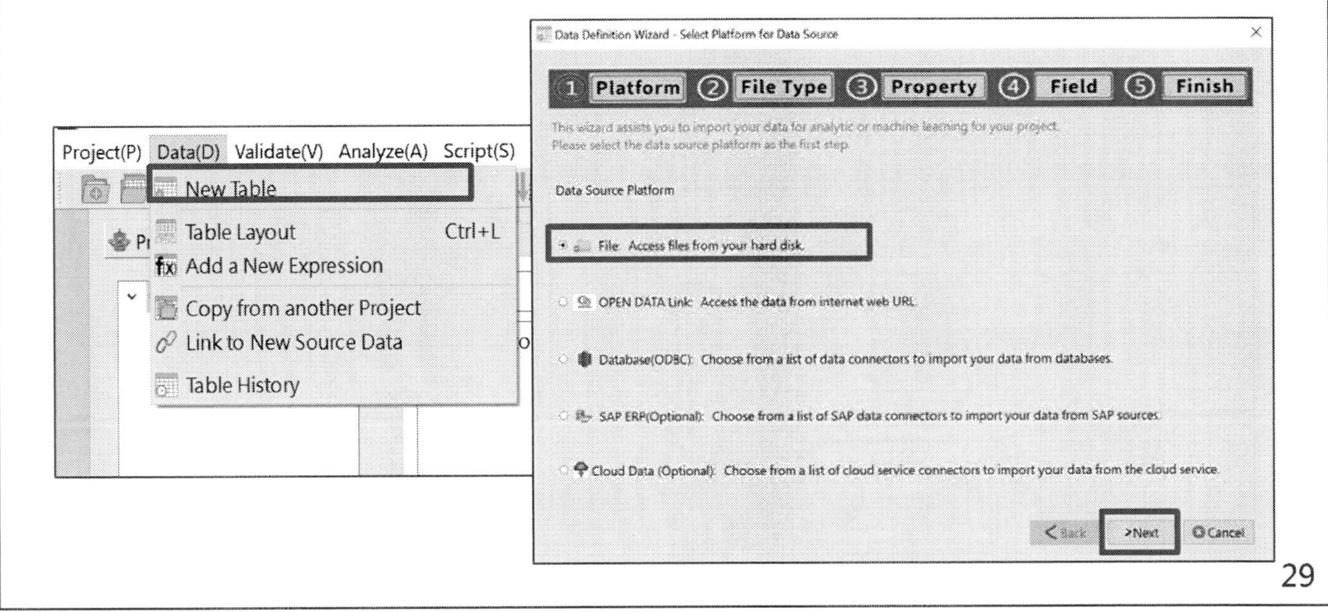

Exercise 4.5
Please practice importing an Excel file into the JCAATs project.

- Select the file we need to import: Credit_cards_metaphor.xls
- Select "Open "

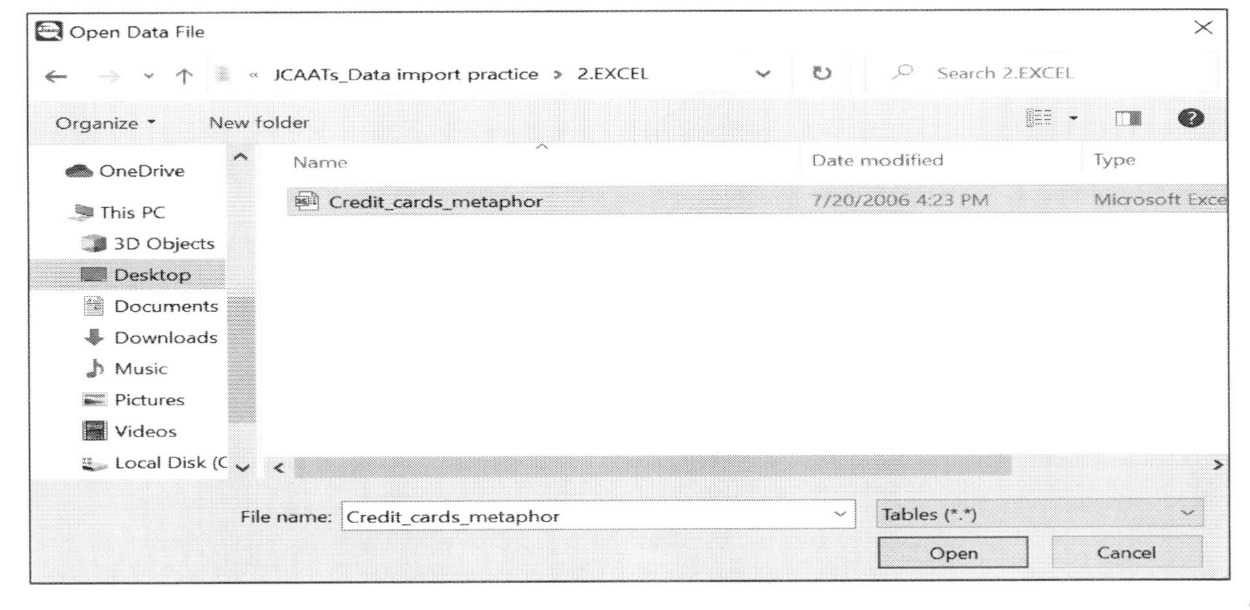

Exercise 4.5
Please practice importing an Excel file into the JCAATs project.
Step2: Select file type

JCAATs will automatically detect the file format to ensure accuracy. If there are no errors, select "Next ".

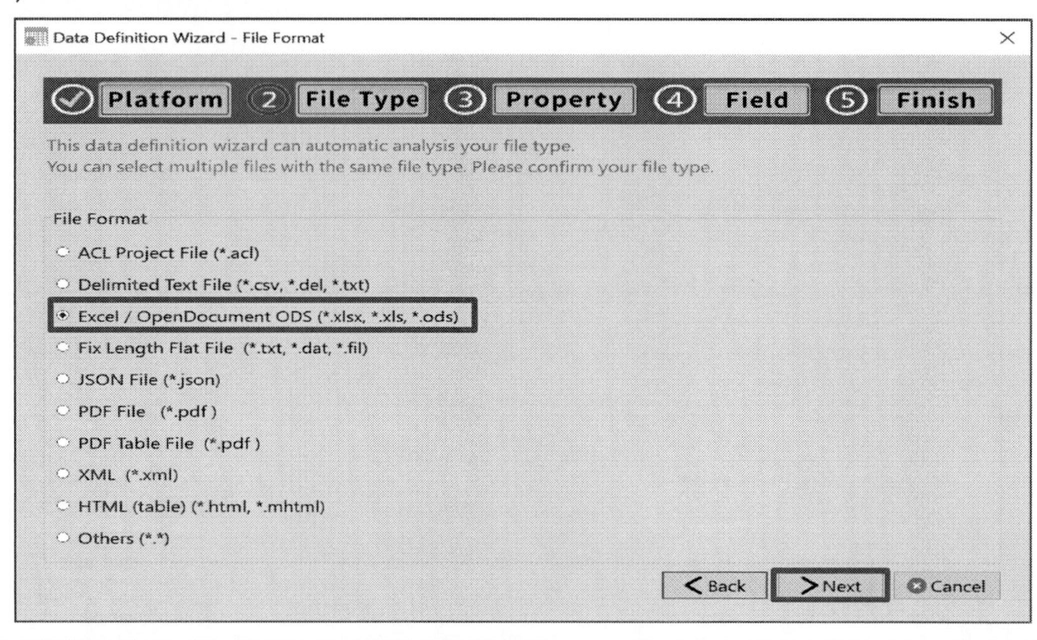

Exercise 4.5
Please practice importing an Excel file into the JCAATs project.

- JCAATs can import a single table or multiple tables at the same time. Click "Next" after selecting.

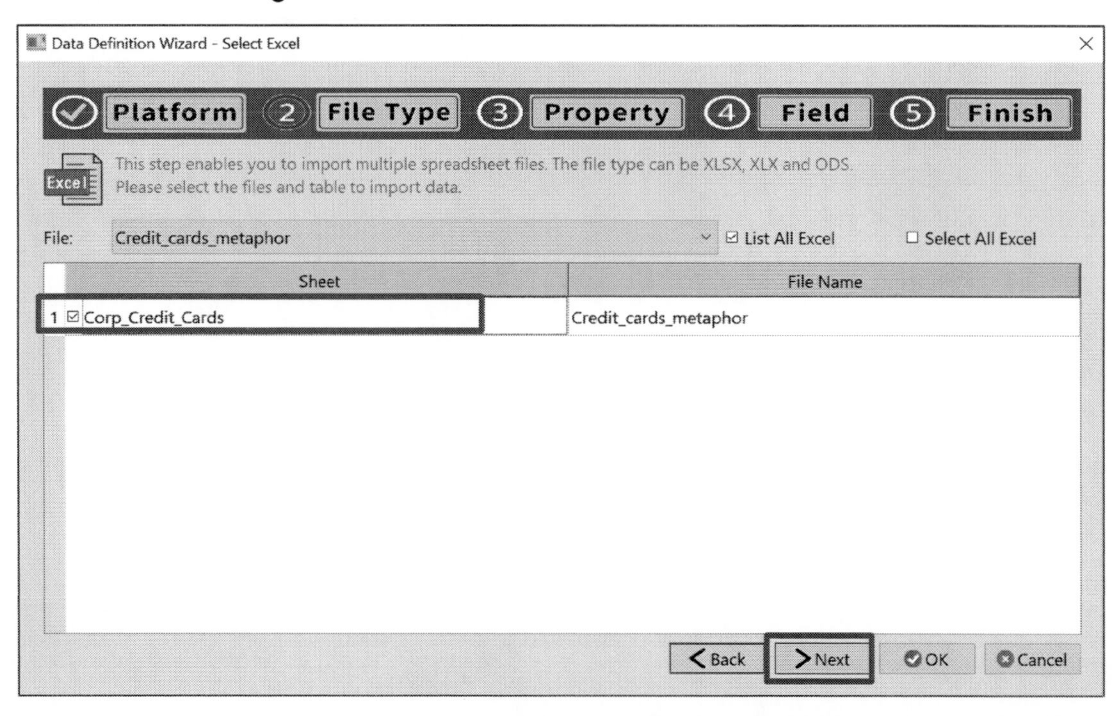

Exercise 4.5

Please practice importing an Excel file into the JCAATs project.

- **Step3: Confirm the Data Features**
- Ensure whether the first row is the field name or not, or specify the starting line. If there are no errors, select "Next".

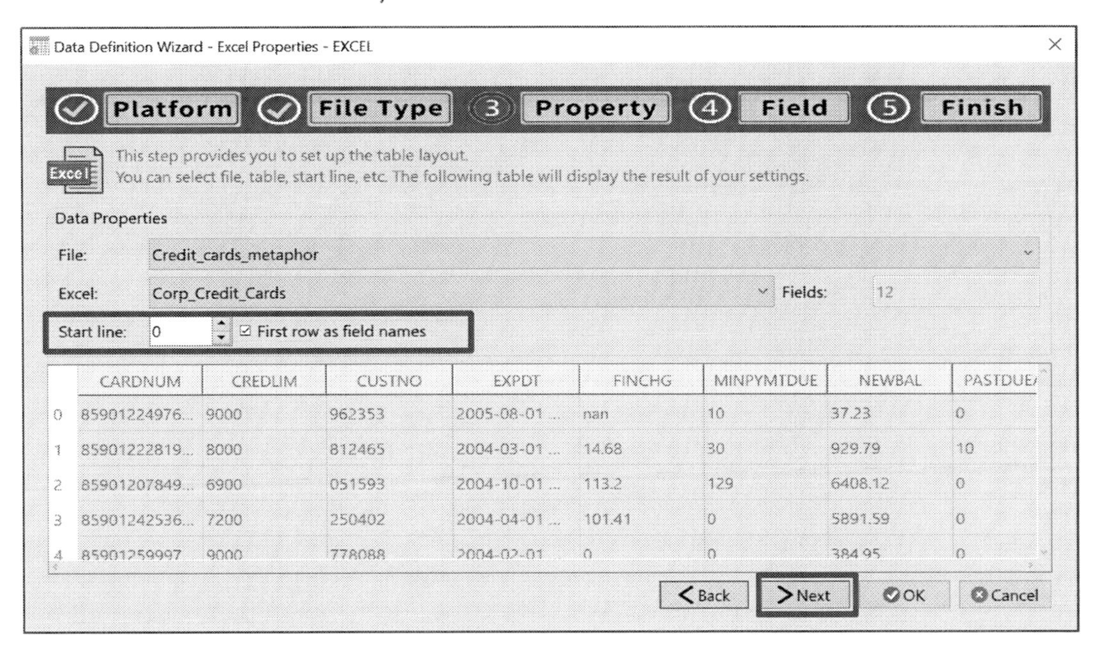

- **Step4: Define the fields**

  JCAATs can define the field name, display name, data type, and data format of each file one by one.

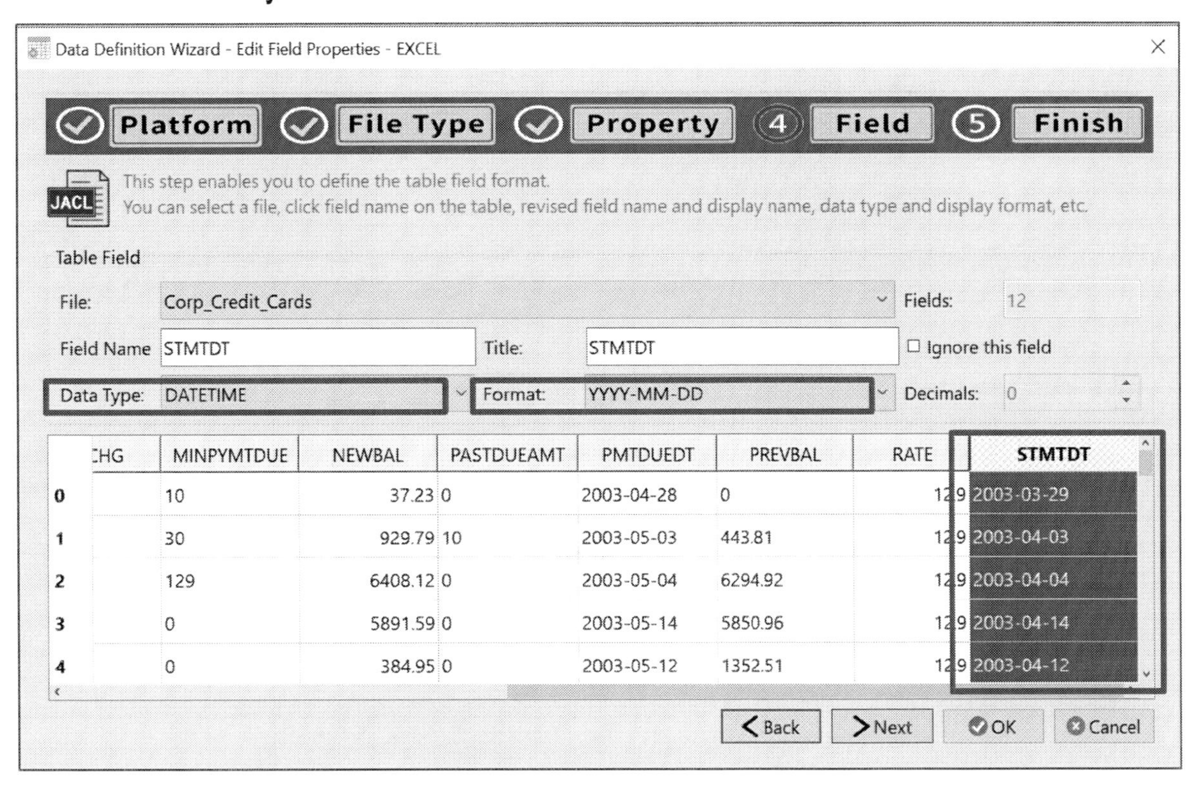

Exercise 4.5
Please practice importing an Excel file into the JCAATs project.

- After completing the settings, select "Next" to proceed.

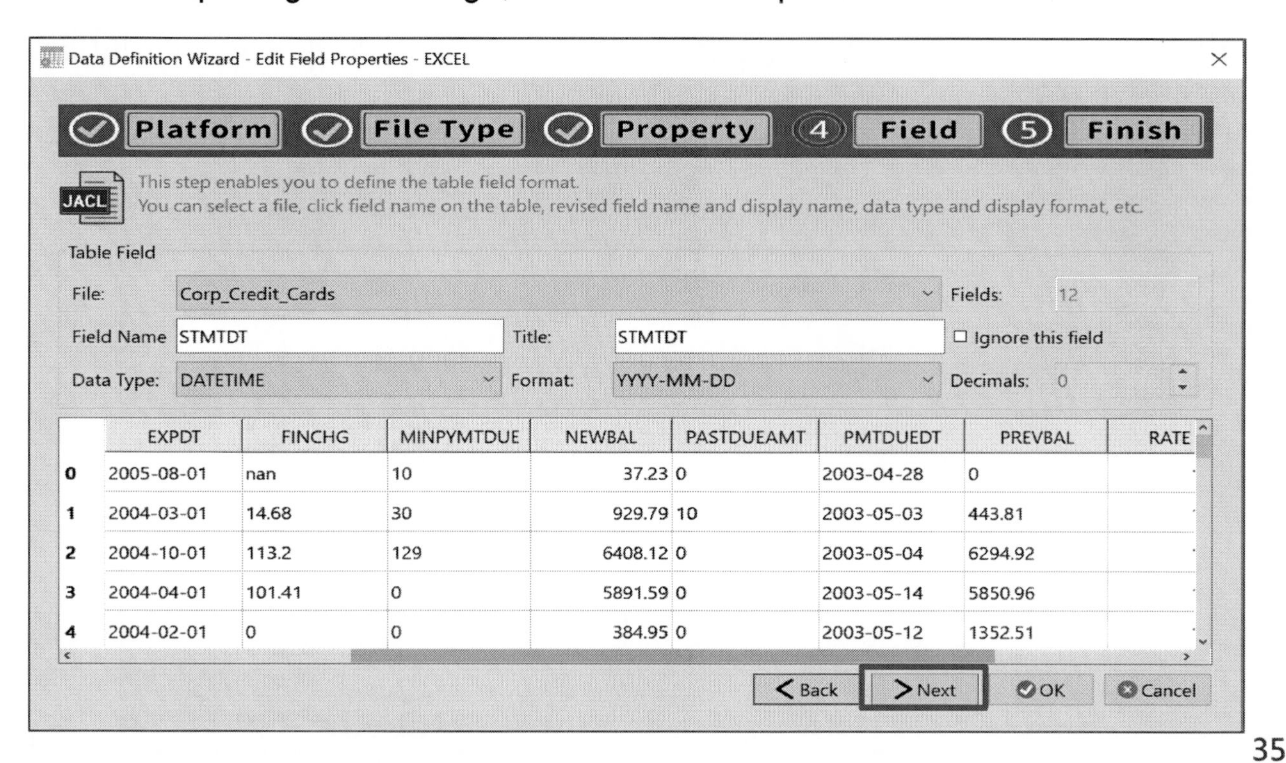

JCAATs-AI Audit Software
Copyright © 2023 JACKSOFT.

Exercise 4.5
Please practice importing an Excel file into the JCAATs project.

- **Step5: Confirm the data file path**
- The default data file path is the project folder, but it can be modified as needed. After ensuring that the path and information are correct, select "Done".

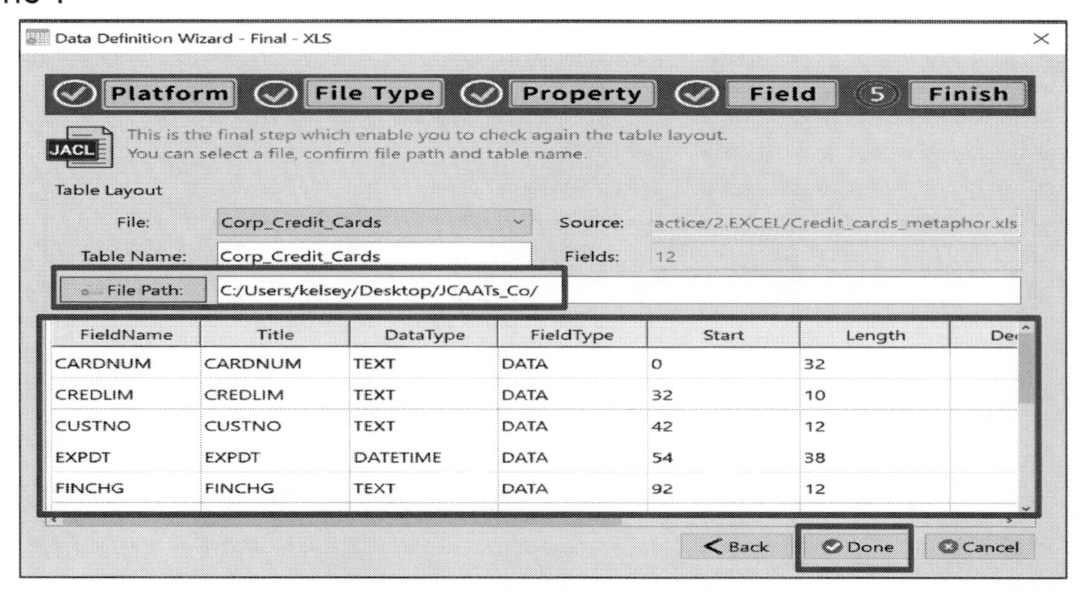

Exercise 4.5

Please practice importing an Excel file into the JCAATs project.

- Complete the import of the Excel file.
- Check whether the records and the information of imported files have been successfully imported or not.

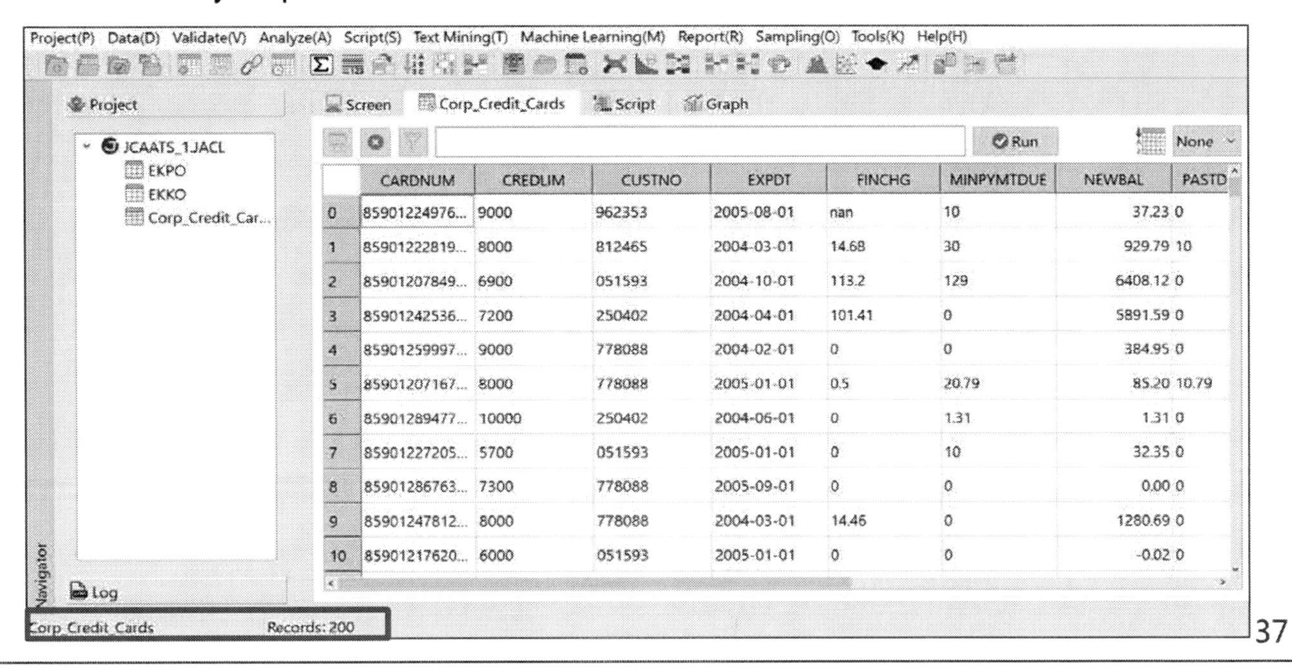

	CARDNUM	CREDLIM	CUSTNO	EXPDT	FINCHG	MINPYMTDUE	NEWBAL	PASTD
0	85901224976...	9000	962353	2005-08-01	nan	10	37.23	0
1	85901222819...	8000	812465	2004-03-01	14.68	30	929.79	10
2	85901207849...	6900	051593	2004-10-01	113.2	129	6408.12	0
3	85901242536...	7200	250402	2004-04-01	101.41	0	5891.59	0
4	85901259997...	9000	778088	2004-02-01	0	0	384.95	0
5	85901207167...	8000	778088	2005-01-01	0.5	20.79	85.20	10.79
6	85901289477...	10000	250402	2004-06-01	0	1.31	1.31	0
7	85901227205...	5700	051593	2005-01-01	0	10	32.35	0
8	85901286763...	7300	778088	2005-09-01	0	0	0.00	0
9	85901247812...	8000	778088	2004-03-01	14.46	0	1280.69	0
10	85901217620...	6000	051593	2005-01-01	0	0	-0.02	0

Exercise 4.5

Please practice importing an Excel file into the JCAATs project.

- Confirm whether the table structure is defined correctly. Modify the table structure if there are any errors.

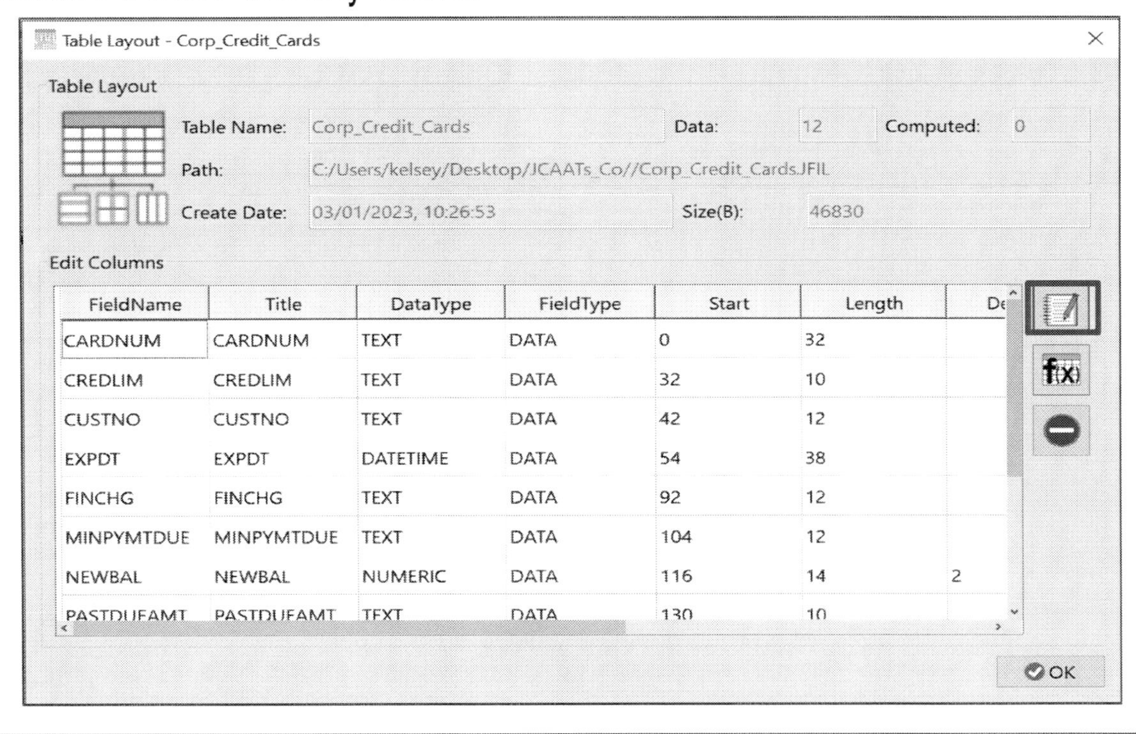

FieldName	Title	DataType	FieldType	Start	Length	De
CARDNUM	CARDNUM	TEXT	DATA	0	32	
CREDLIM	CREDLIM	TEXT	DATA	32	10	
CUSTNO	CUSTNO	TEXT	DATA	42	12	
EXPDT	EXPDT	DATETIME	DATA	54	38	
FINCHG	FINCHG	TEXT	DATA	92	12	
MINPYMTDUE	MINPYMTDUE	TEXT	DATA	104	12	
NEWBAL	NEWBAL	NUMERIC	DATA	116	14	2
PASTDUEAMT	PASTDUEAMT	TEXT	DATA	130	10	

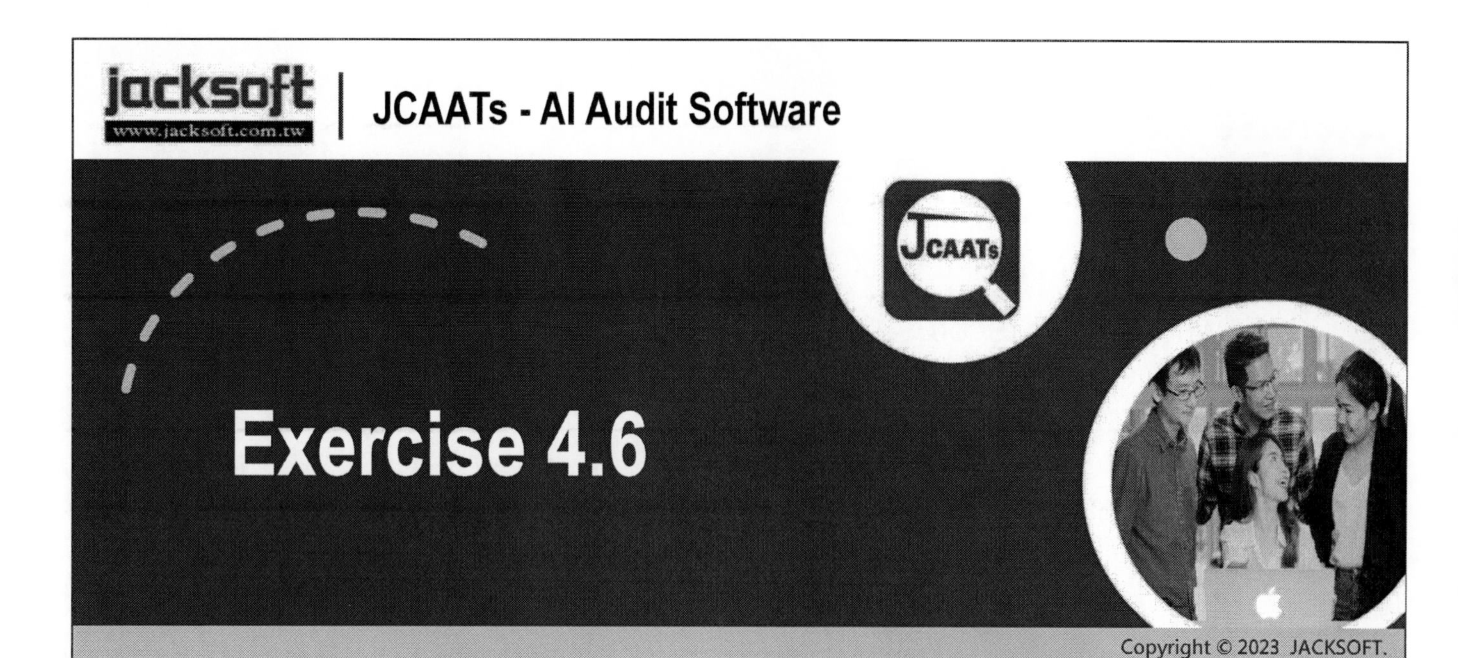

Please practice importing a delimited text file into the JCAATs project.

Employees.csv
Payment_0.CSV and Payment_1.CSV
Company_Departments.txt

Exercise 4.6.1
Please practice importing a delimited text file into the JCAATs project.

- JCAATs will detect the file format automatically.
- If there are no errors with the detected format, select "Next" to proceed.

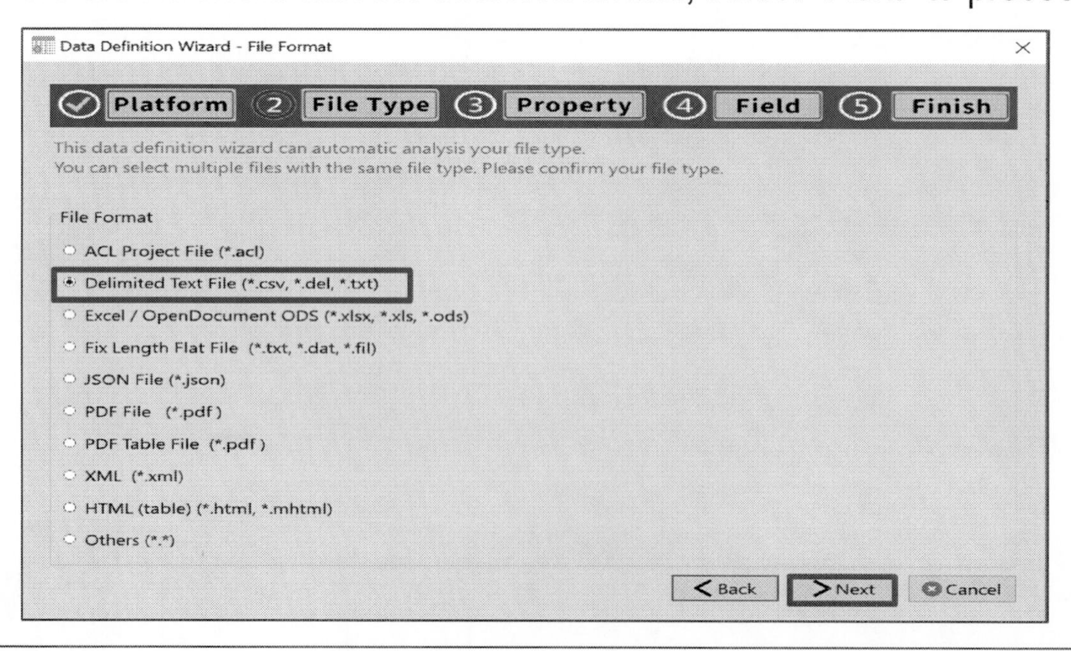

Exercise 4.6.1
Please practice importing a delimited text file into the JCAATs project.

- JCAATs will detect file character encoding automatically.
- Click "Next " to proceed.

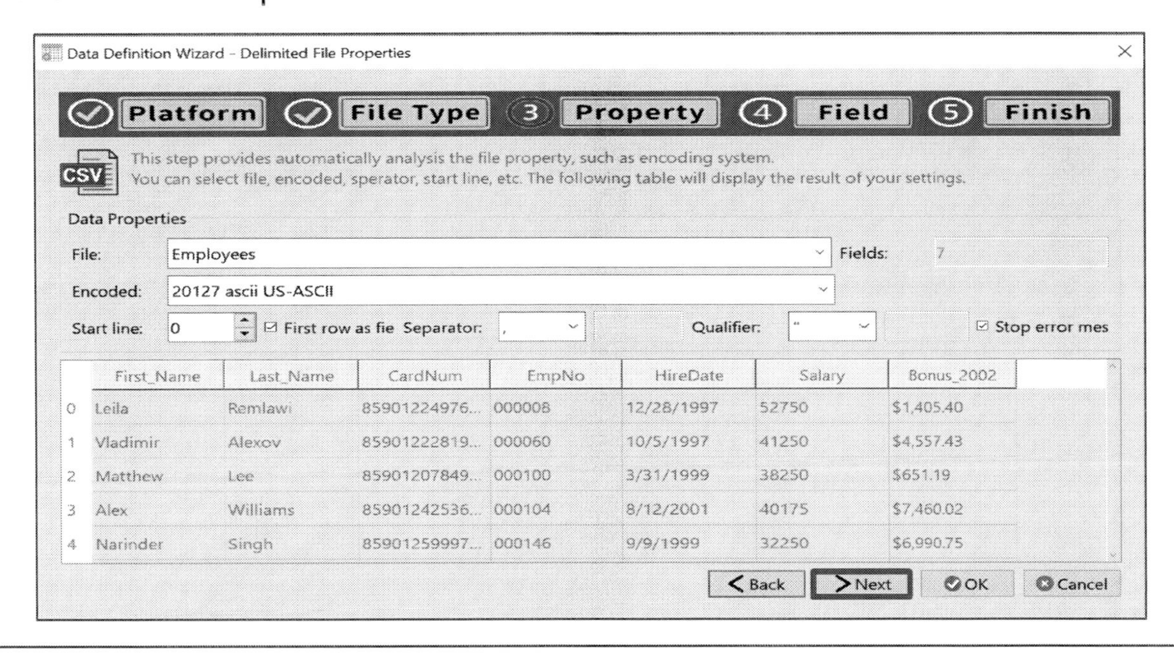

JCAATs-AI Audit Software Copyright © 2023 JACKSOFT.

Exercise 4.6.1
Please practice importing a delimited text file into the JCAATs project.

- Field Definition: JCAATs can define the field name, display name, data type and data format of each field. After completing the settings, select "Next" to proceed.

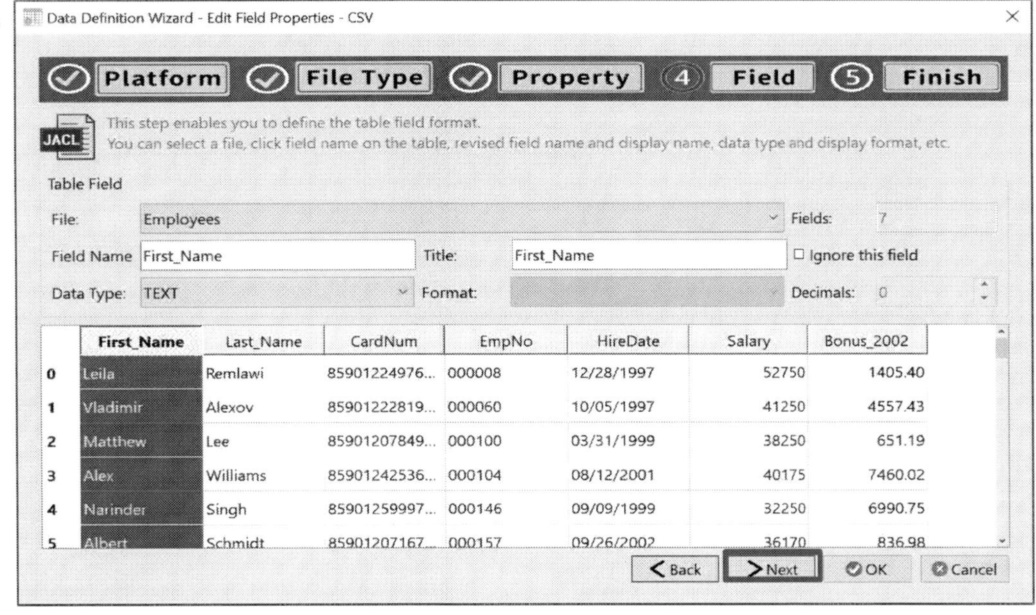

Exercise 4.6.1
Please practice importing a delimited text file into the JCAATs project.

- The default data file path is the project folder.
- After ensuring that the path and information are correct, select "Done " to complete the process.

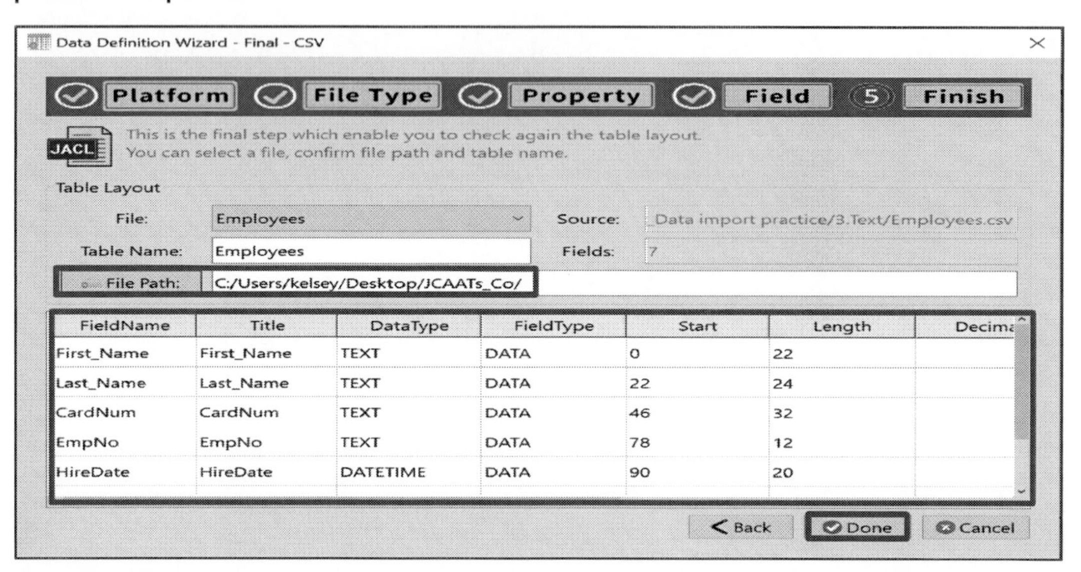

JCAATs-AI Audit Software
Copyright © 2023 JACKSOFT.

Exercise 4.6.1
Please practice importing a delimited text file into the JCAATs project.

- After the import progress is completed, we can see that the data table has been successfully imported.

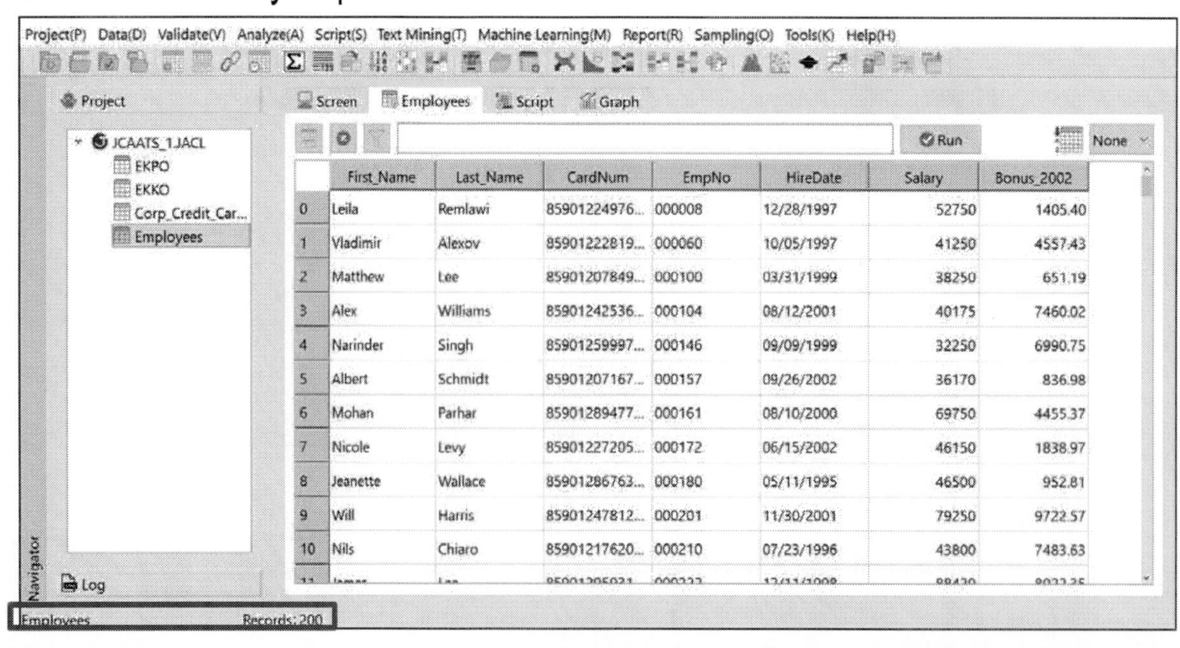

Exercise 4.6.2
Please practice importing a delimited text file into the JCAATs project.

Step1: Add a New Table-Select data source

- Select "Data" > "New Table.
- After selecting the data source platform as "File, " click "Next".

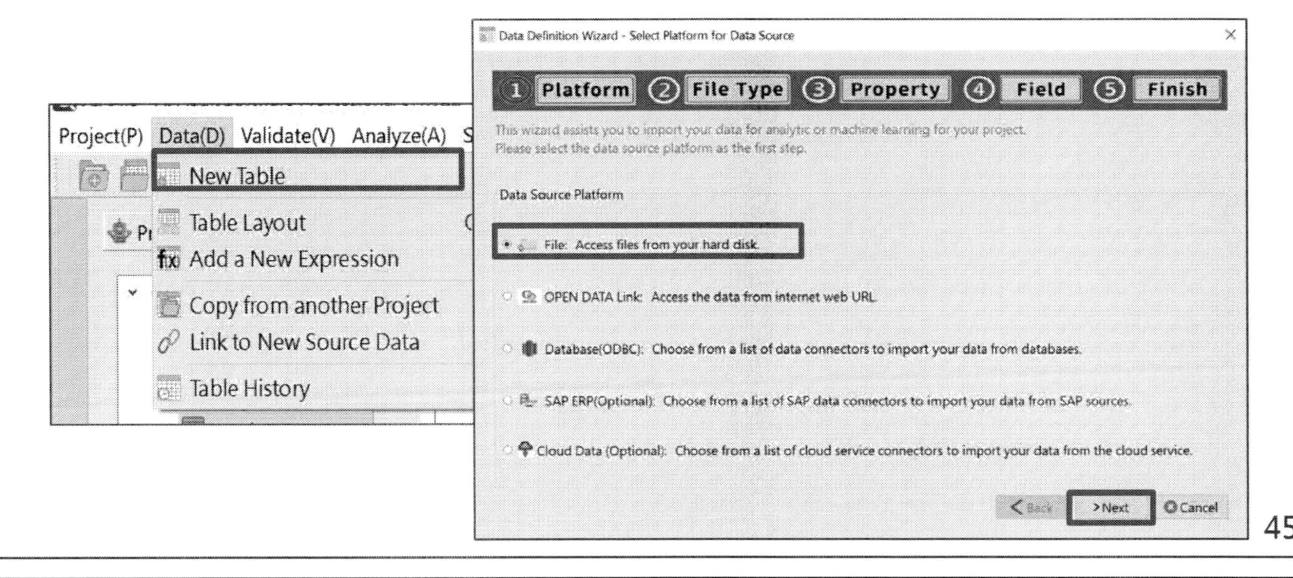

JCAATs-AI Audit Software Copyright © 2023 JACKSOFT.

Exercise 4.6.2
Please practice importing a delimited text file into the JCAATs project.

- Select Payment_0.CSV and Payment_1.CSV to import.
- Click "Open" to proceed.

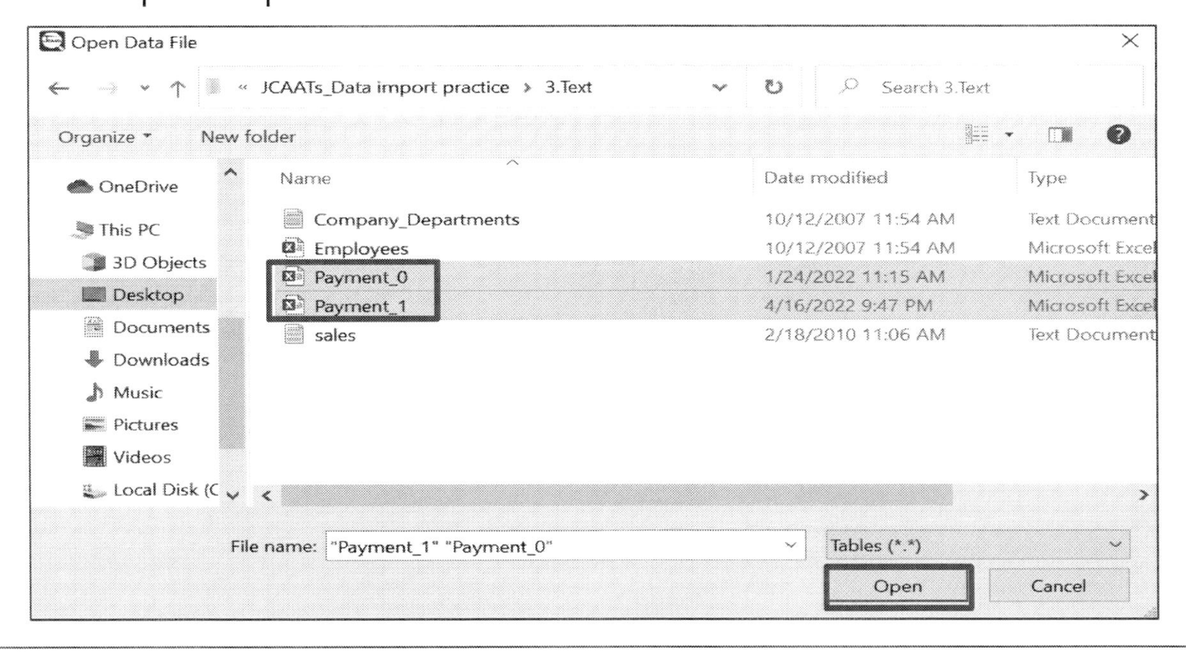

Exercise 4.6.2
Please practice importing a delimited text file into the JCAATs project.
Step2: Select File Format

JCAATs will detect the file format automatically.

If there are no errors with the detected format, select "Next" to proceed.

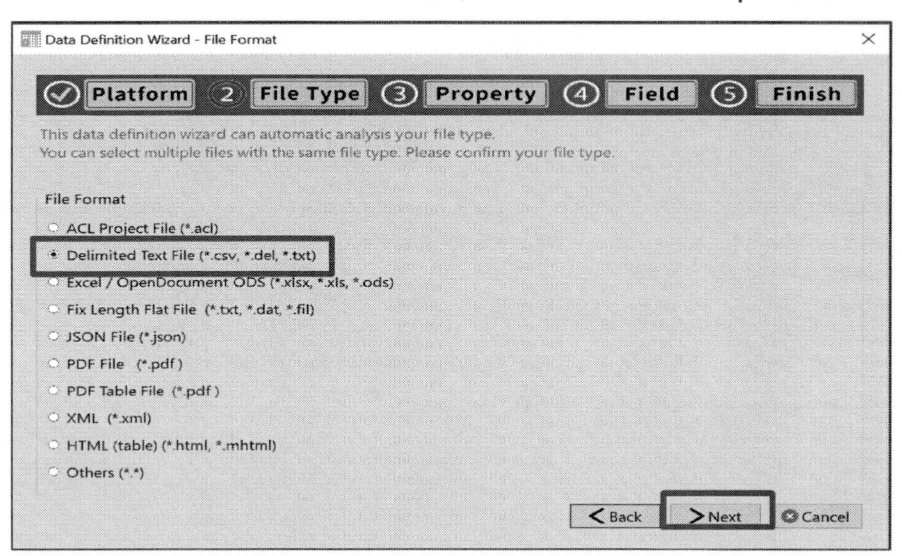

JCAATs-AI Audit Software
Copyright © 2023 JACKSOFT.

Exercise 4.6.2
Please practice importing a delimited text file into the JCAATs project.

- **Step3: Confirm the Data Features**
- Set the data file separator to "Tab" one by one, check the first line field name, set the starting line number to 0, and click "Next" after completing the settings.

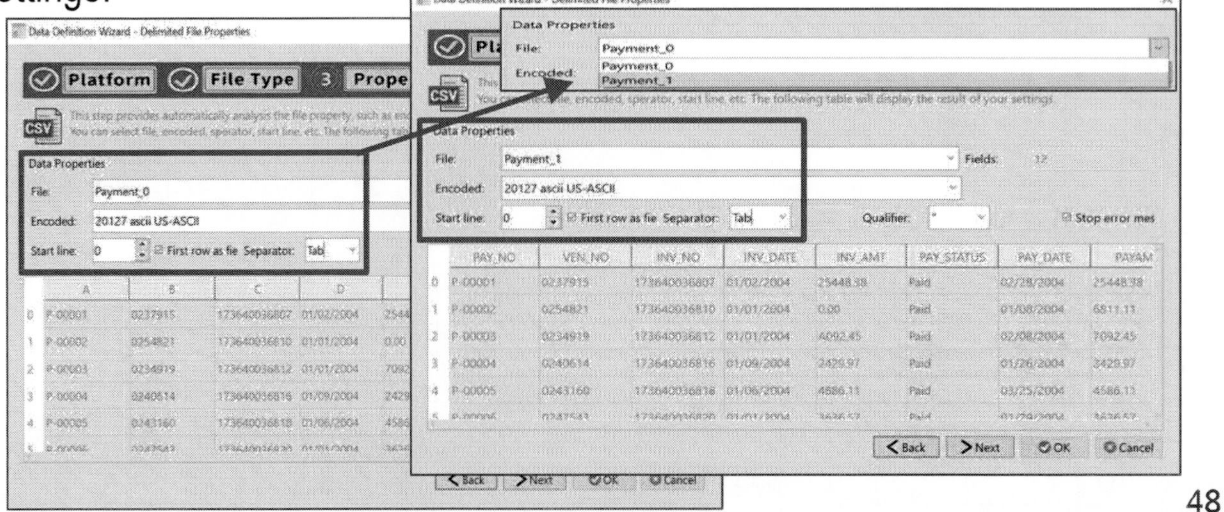

Exercise 4.6.2
Please practice importing a delimited text file into the JCAATs project.

- **Step4: Define the fields**
- Define the field name , display name , data type and data format of each data table field one by one. After setting, select "Next".

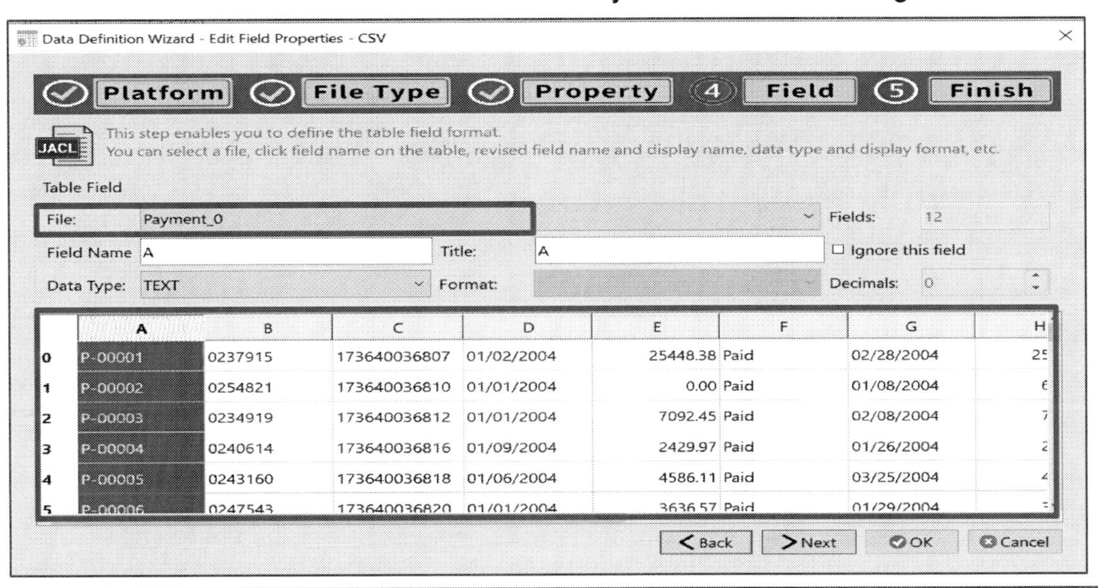

Exercise 4.6.2
Please practice importing a delimited text file into the JCAATs project.

- **Step5: Confirm the data file path**

The default data file path is the project folder, but it can be modified as needed. After ensuring that the path and information are correct, select "Done".

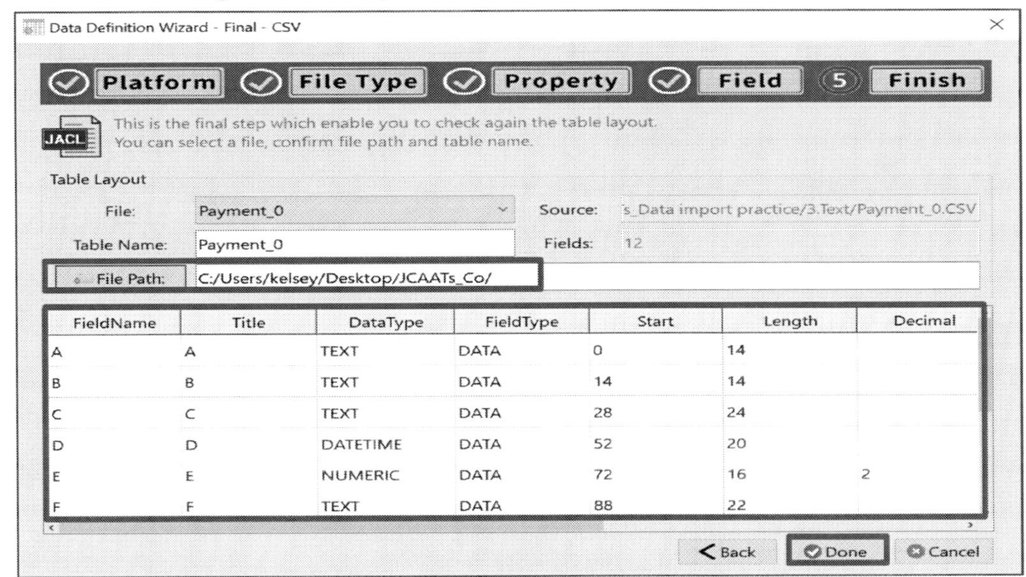

Exercise 4.6.2
Please practice importing a delimited text file into the JCAATs project.

- After the import progress is completed, we can see that there are two data tables successfully imported.

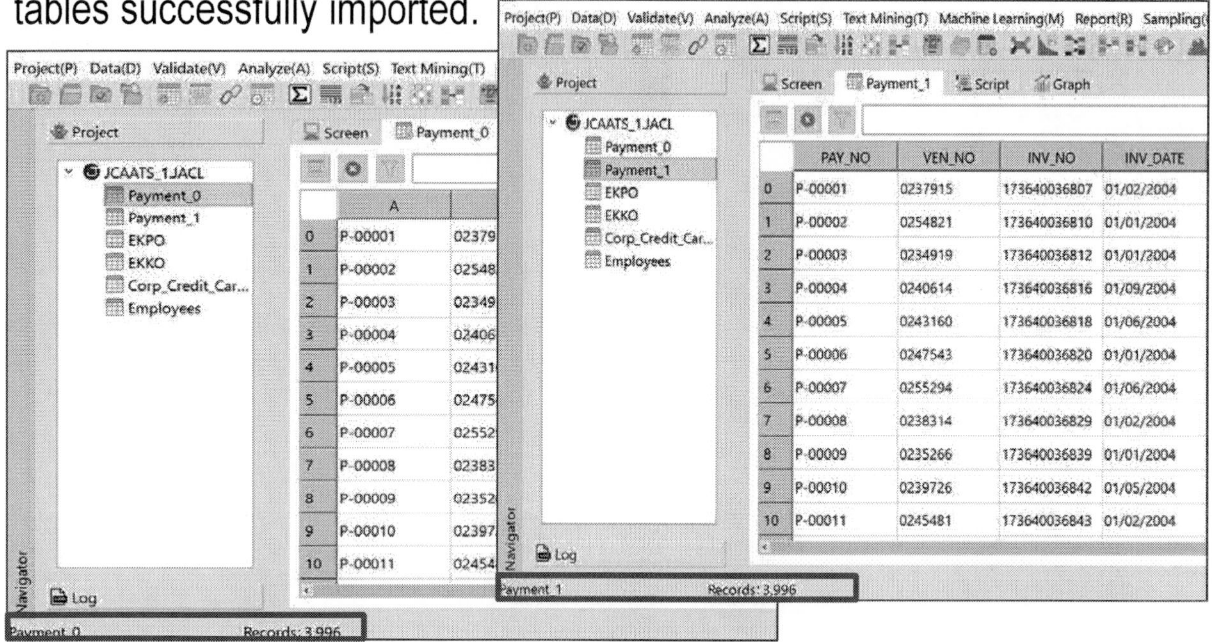

Exercise 4.6.3
Please practice importing a delimited text file into the JCAATs project.

- Select "Data > New Table".
- After selecting the data source platform as "File", click "Next".

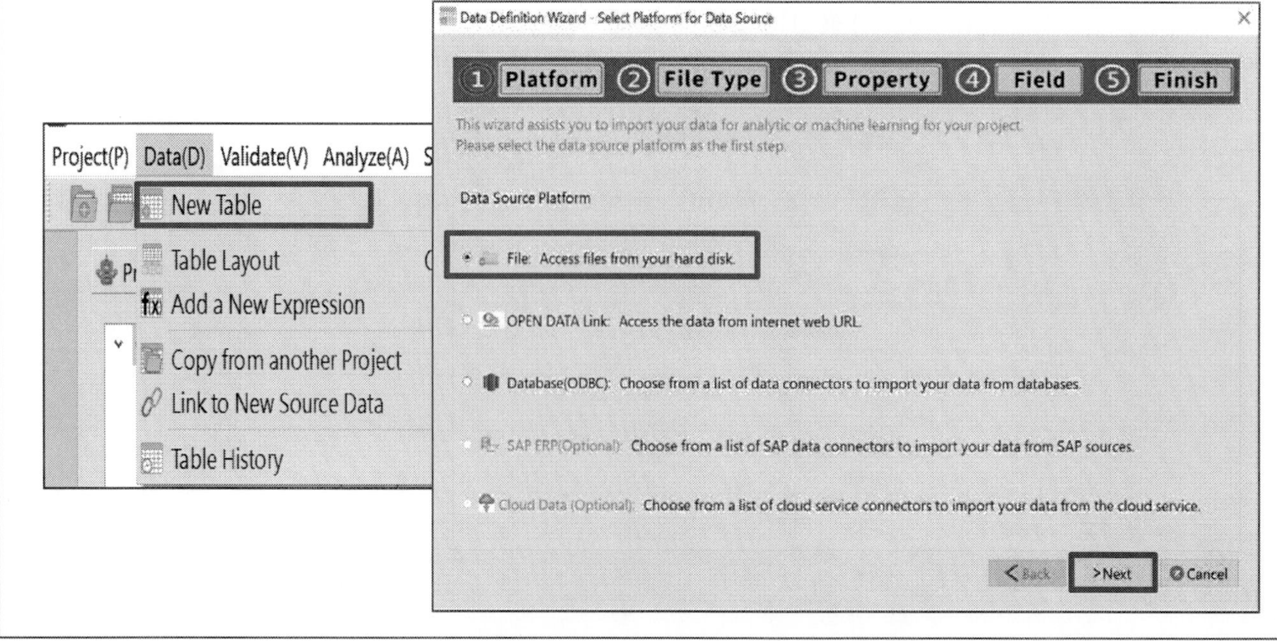

Exercise 4.6.3
Please practice importing a delimited text file into the JCAATs project.

- JCAATs will detect the file type automatically. We can change it to "Delimited Text File" to make the import process easier.
- Confirm that the file type is correct and click "Next" to proceed."

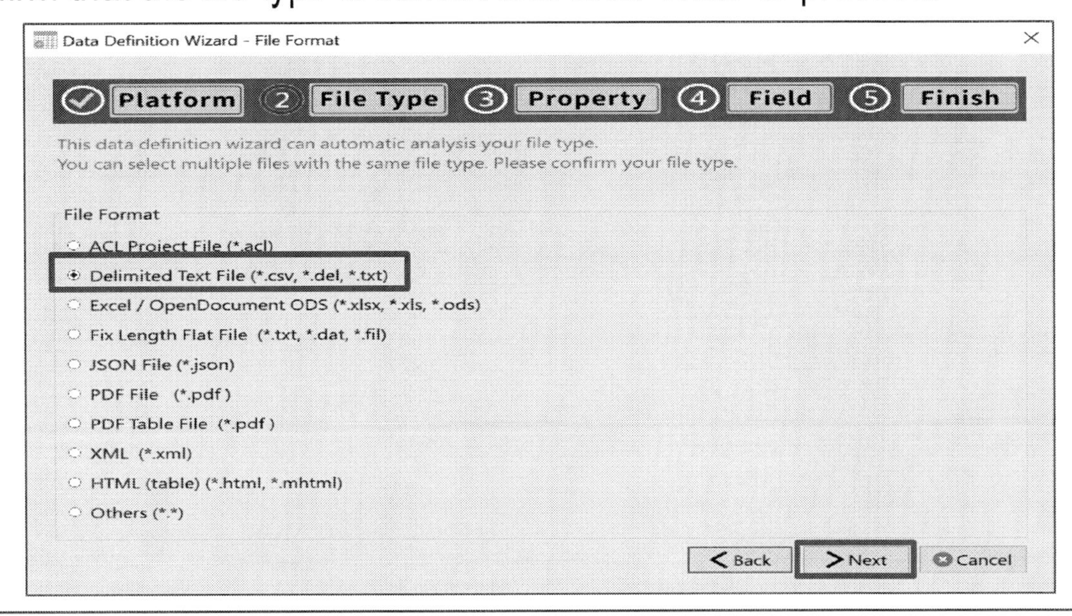

JCAATs-AI Audit Software

Exercise 4.6.3
Please practice importing a delimited text file into the JCAATs project.

- **Data Features:** JCAATs will identify the file encoding method automatically. Uncheck the 'First Row Field Name' box, select the separator as 'Tab,' and click 'Next' after completing the settings.

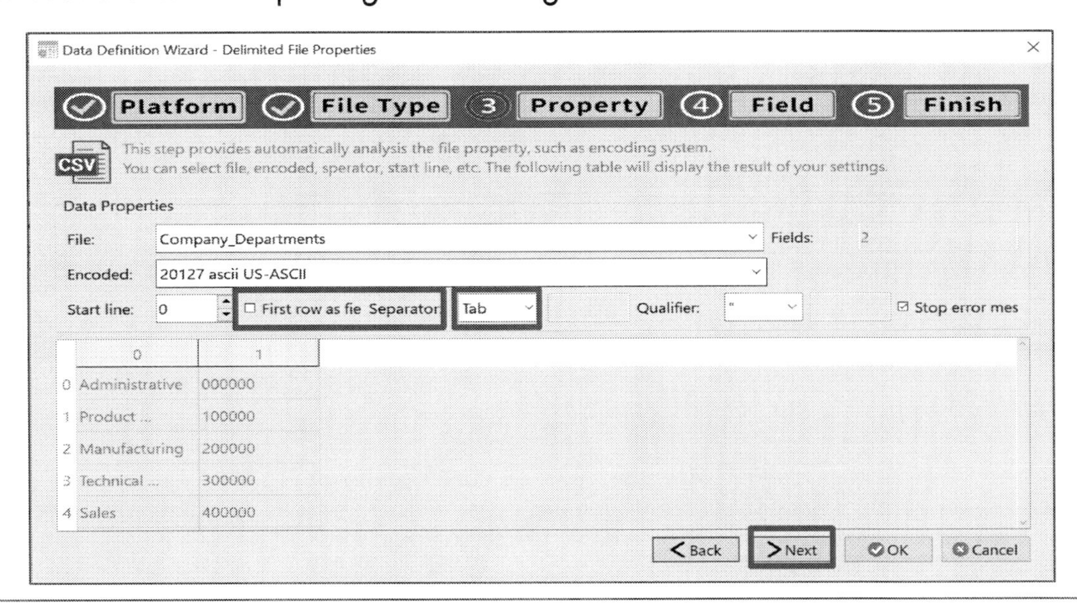

Exercise 4.6.3
Please practice importing a delimited text file into the JCAATs project.

- **Field Definition:** JCAATs can define the field name, display name, data type and data format of each field. After completing the settings, select "Next" to proceed.

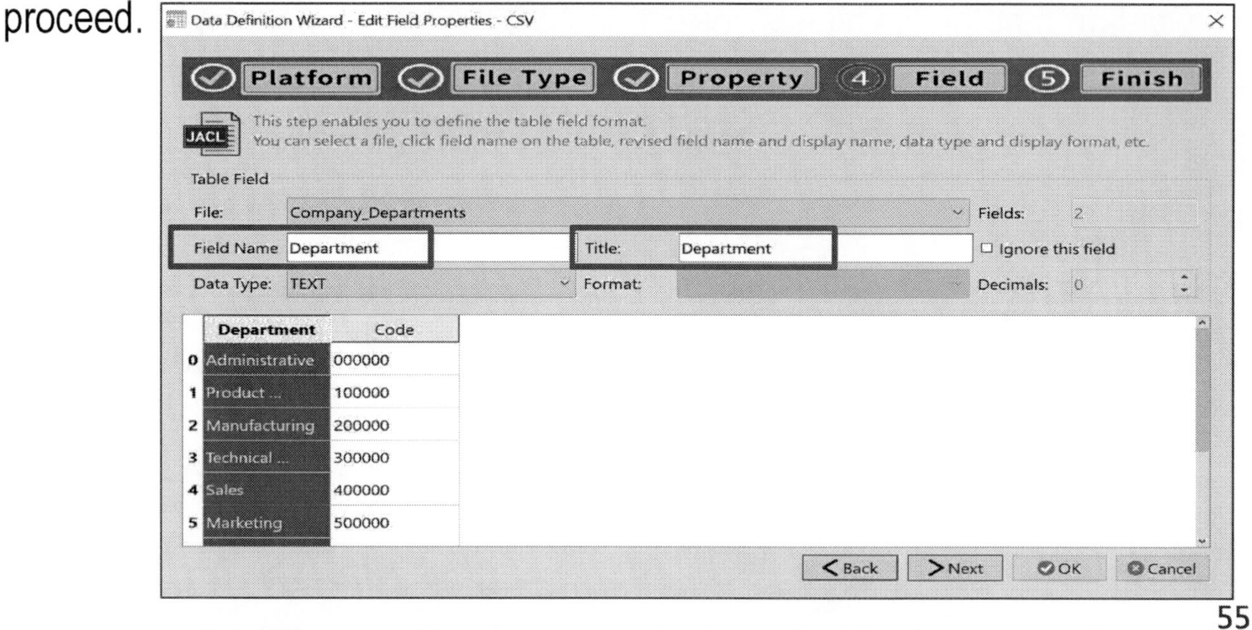

JCAATs-AI Audit Software
Copyright © 2023 JACKSOFT.

Exercise 4.6.3
Please practice importing a delimited text file into the JCAATs project.

- When the import progress is complete, we can see that the table has been successfully imported.

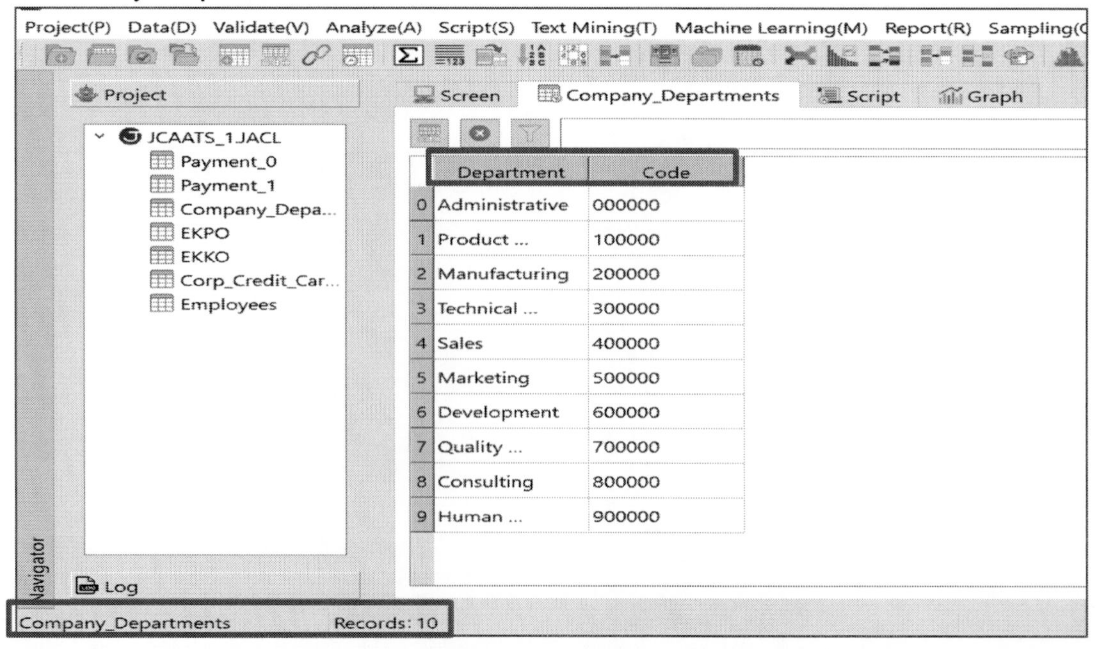

Exercise 4.7

Please practice importing a fixed-length text file (Flat File) into the JCAATs project.
sales.txt

Exercise 4.7
Please practice importing a fixed-length text file (Flat File) into the JCAATs project.

- Select "Data > New Table".
- After selecting the data source platform as "File", click "Next".

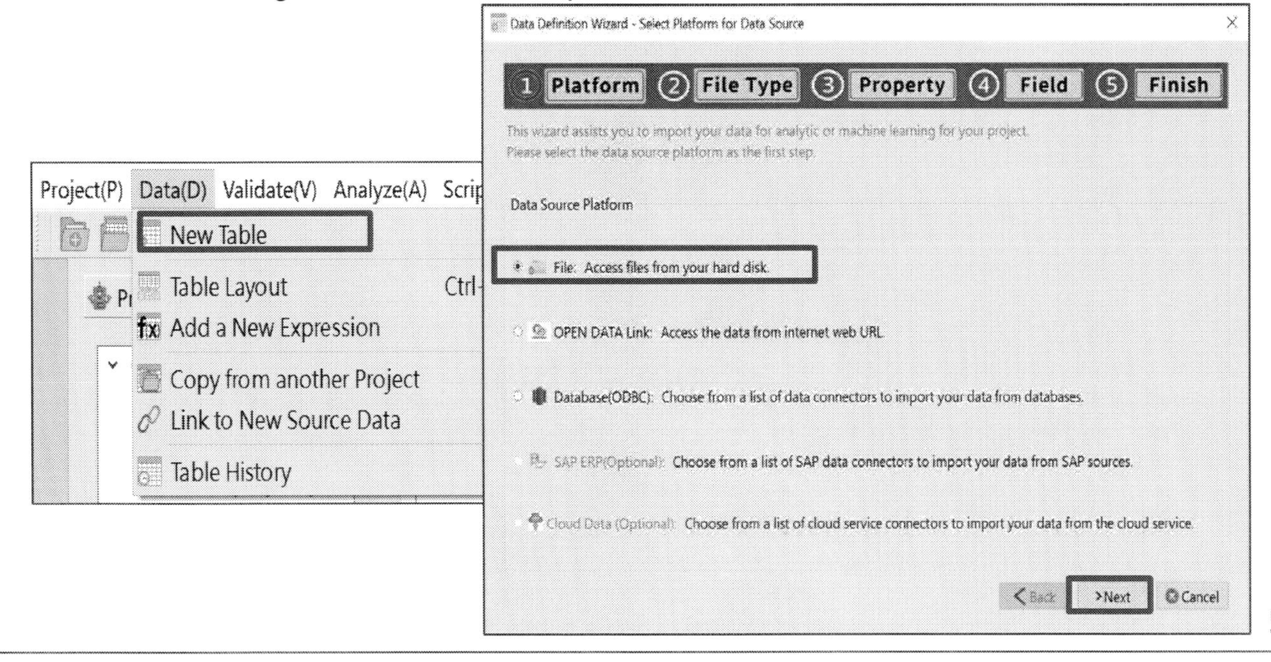

Exercise 4.7

Please practice importing a fixed-length text file (Flat File) into the JCAATs project.

- sales.txt will be detected automatically, confirm that the file type is correct, and click "Next" to proceed.

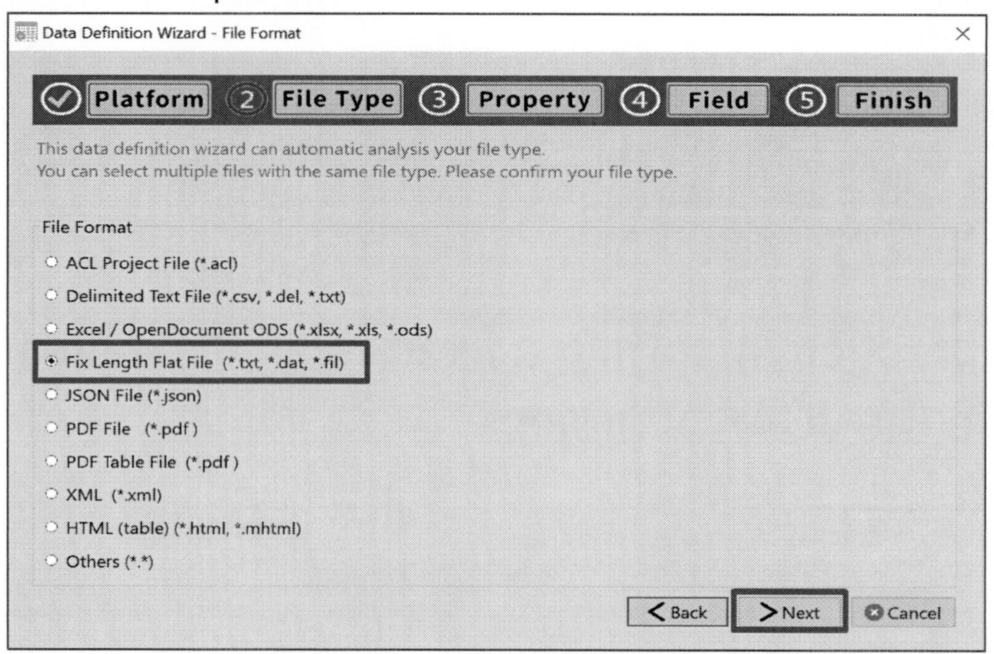

JCAATs-AI Audit Software
Copyright © 2023 JACKSOFT.

Exercise 4.7

Please practice importing a fixed-length text file (Flat File) into the JCAATs project.

- Click the Add Field Position button, ,and set the data length to "30".
- Select OK to proceed.

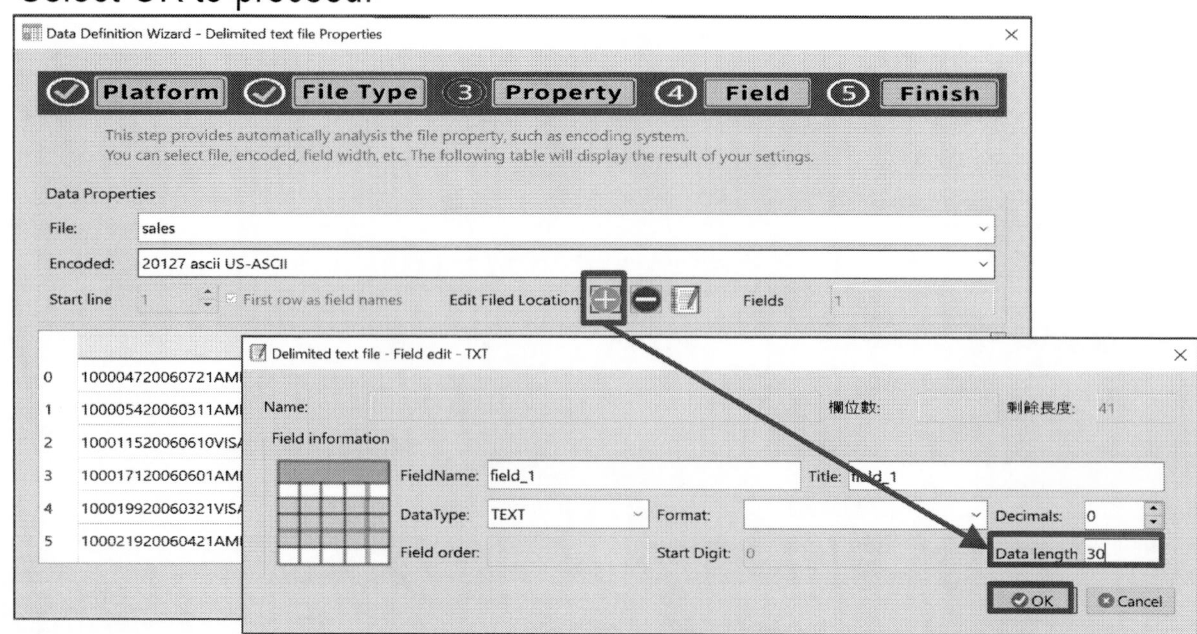

Exercise 4.7
Please practice importing a fixed-length text file (Flat File) into the JCAATs project.

- Confirm that the field has been cut out completely.
- Select "Next" to proceed.

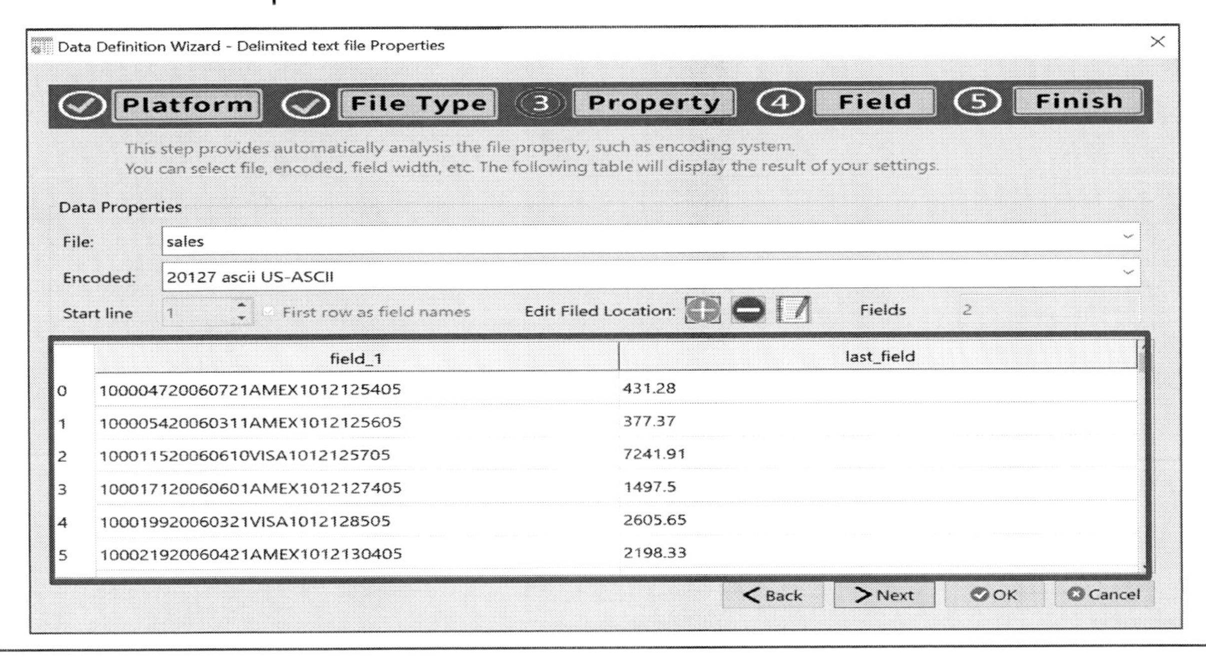

JCAATs-AI Audit Software
Copyright © 2023 JACKSOFT.

Exercise 4.7
Please practice importing a fixed-length text file (Flat File) into the JCAATs project.

- Field Definition: JCAATs can define the field name, display name, data type and data format of each field. After completing the settings, select "Next" to proceed.

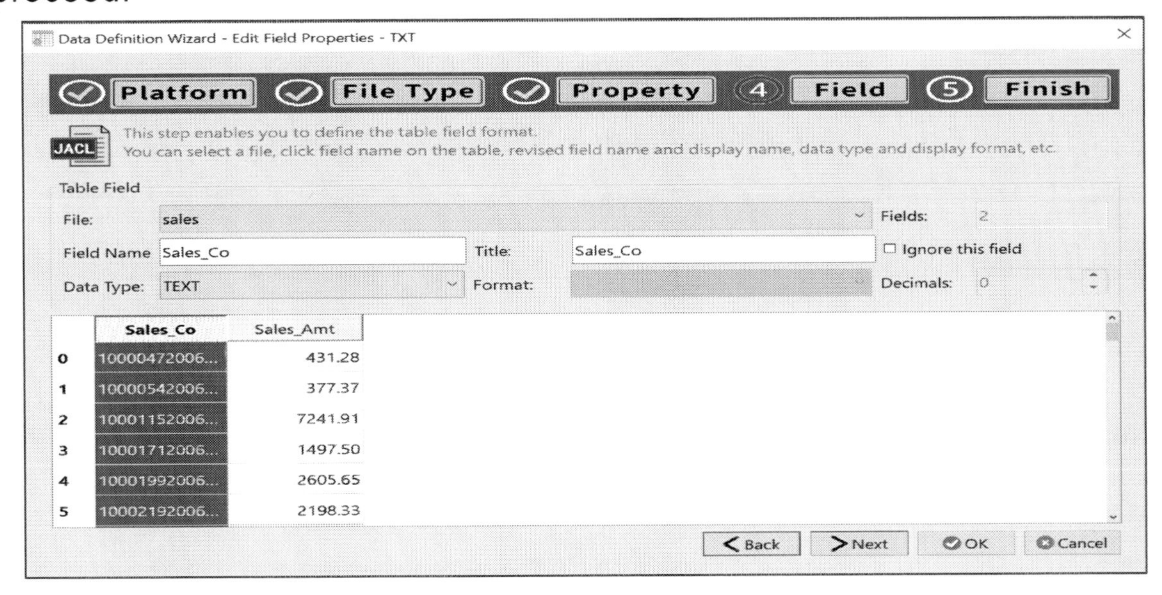

Exercise 4.7
Please practice importing a fixed-length text file (Flat File) into the JCAATs project.

- The default data file path is the project folder.
- After ensuring that the path and information are correct, select "Done".

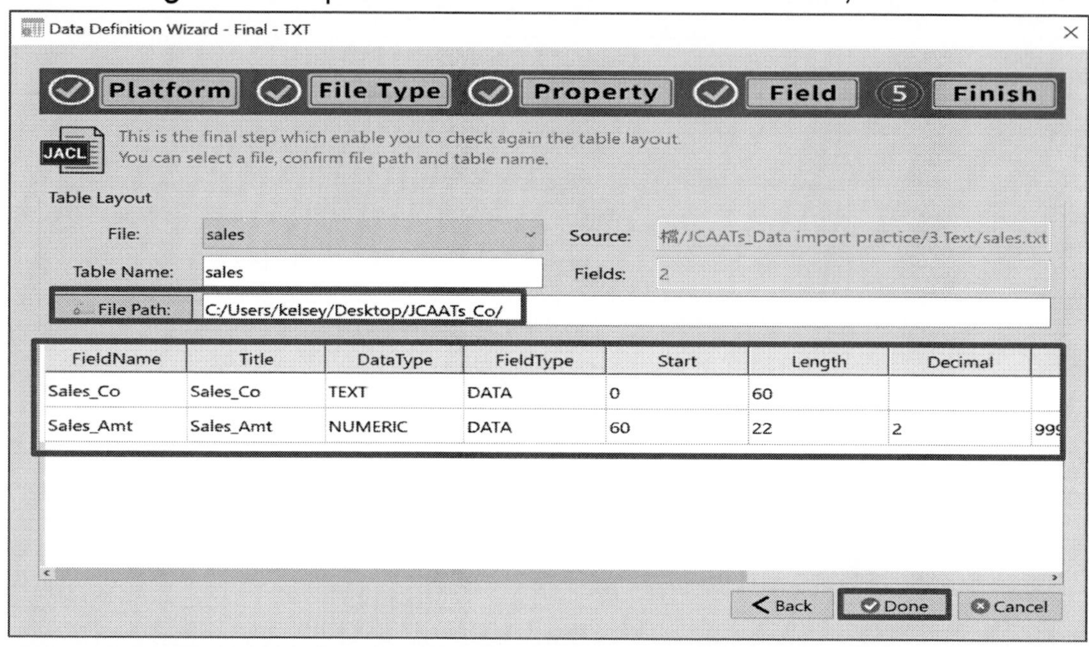

JCAATs-AI Audit Software

Exercise 4.7
Please practice importing a fixed-length text file (Flat File) into the JCAATs project.

- After the import progress is completed, we can see that the data table has been successfully imported.

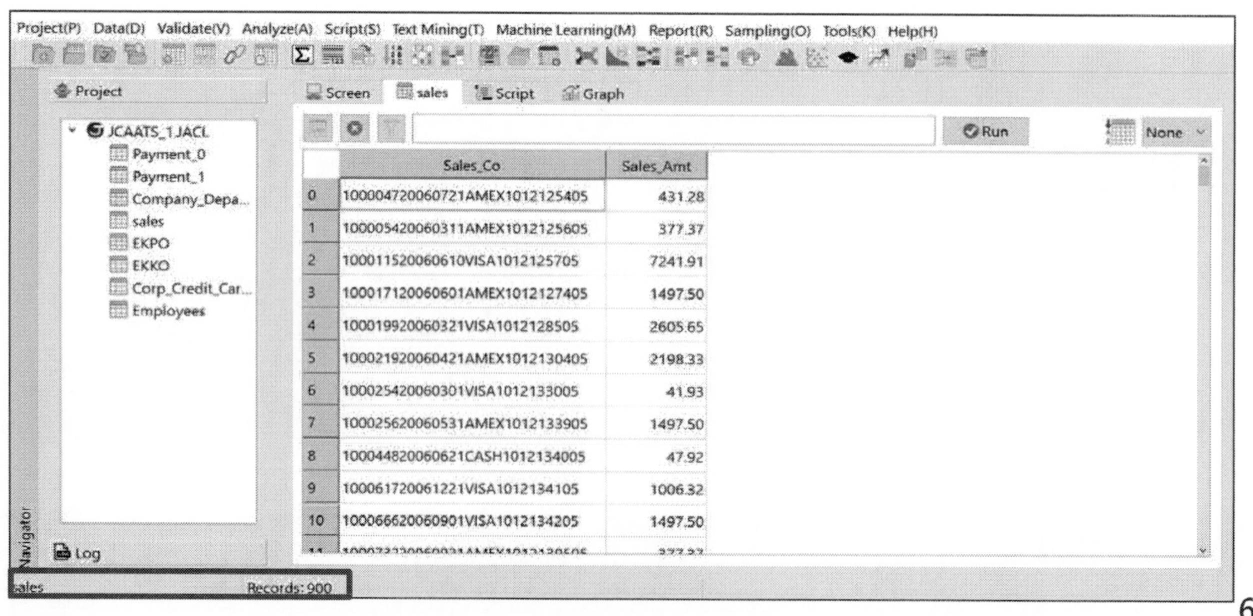

Chapter 4 - Exercise

- Exercise 4.8: Please practice importing OpenDocument (ODS) into the JCAATs project.
- Exercise 4.9: Please practice importing JSON into the JCAATs project.
- Exercise 4.10: Please practice importing PDF into the JCAATs project.

JCAATs - AI Audit Software

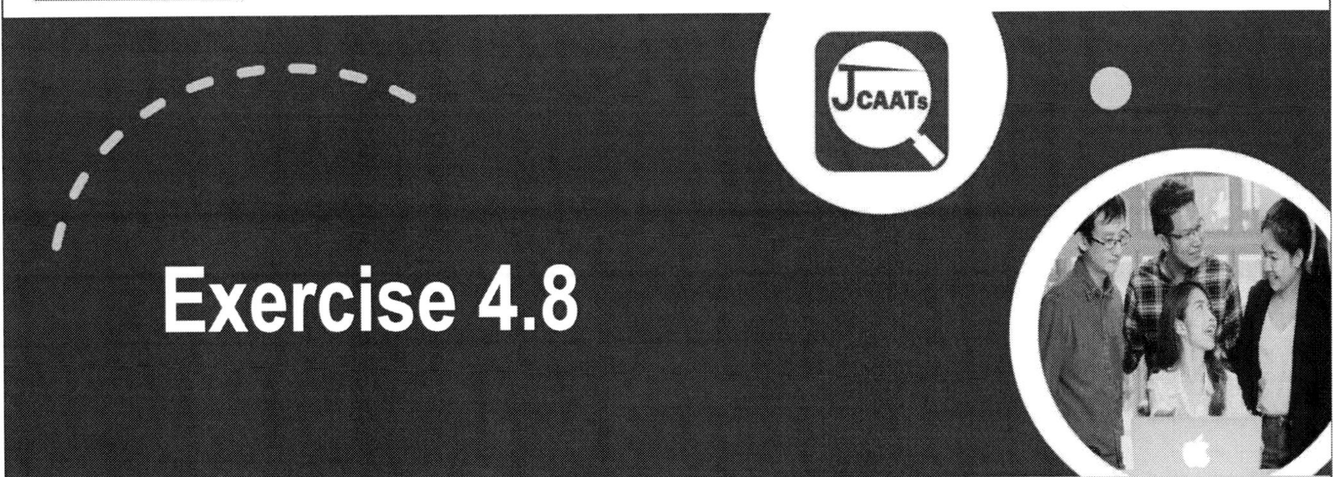

Please practice importing OpenDocument (ODS) into the JCAATs project.

11102 External employment Employing external resources Case count and headcount.ods

Exercise 4.8

Please practice importing OpenDocument (ODS) into the JCAATs project.

- Select "Data > New Table".
- After selecting the data source platform as "File", click "Next".

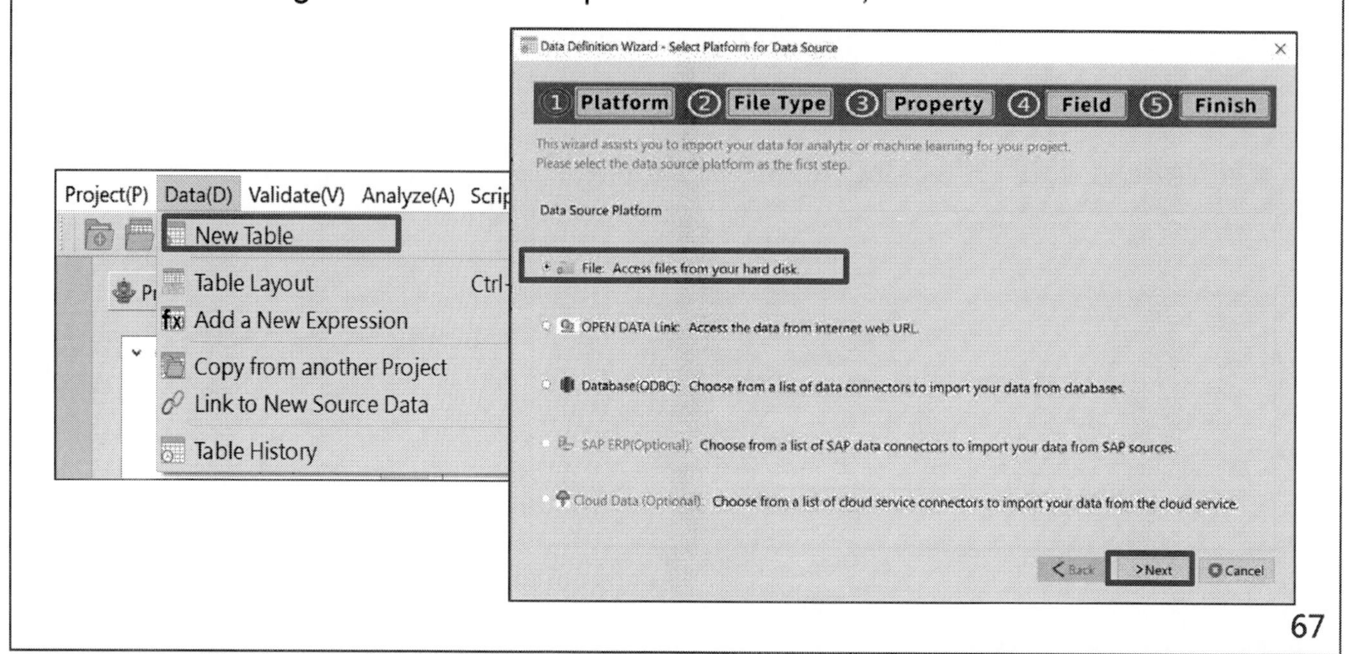

JCAATs-AI Audit Software
Copyright © 2023 JACKSOFT.

Exercise 4.8

Please practice importing OpenDocument (ODS) into the JCAATs project.

- Select 11102 External employment Employing external resources Case count and headcount.ods
- Click " Open"

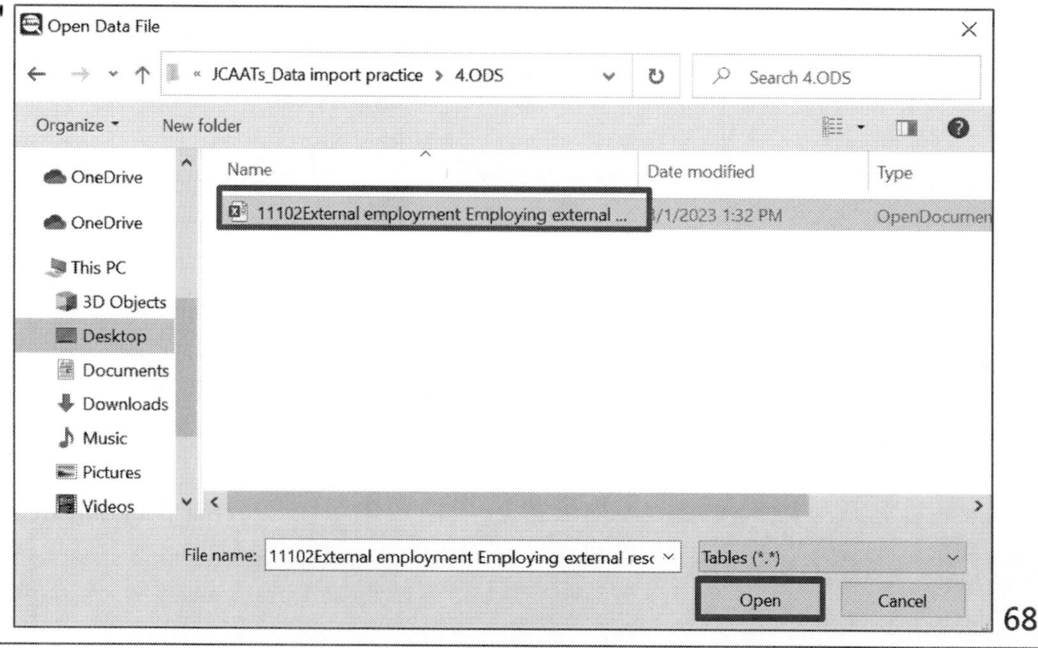

Exercise 4.8
Please practice importing OpenDocument (ODS) into the JCAATs project.

- JCAATs will detect the file format automatically.
- If there are no errors with the detected format, select "Next" to proceed.

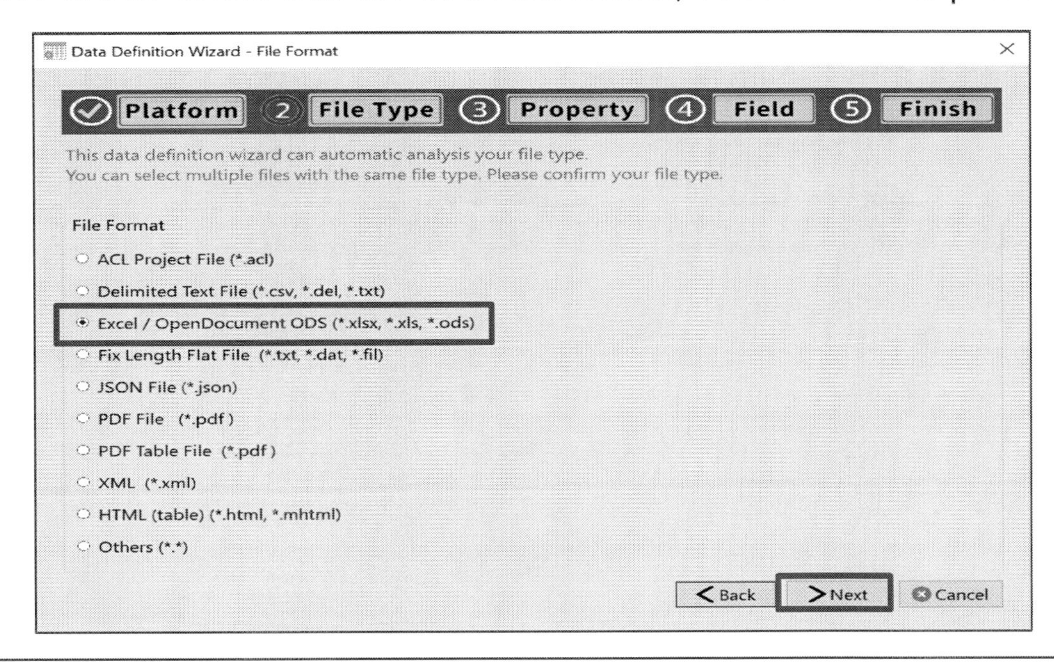

JCAATs-AI Audit Software
Copyright © 2023 JACKSOFT.

Exercise 4.8
Please practice importing OpenDocument (ODS) into the JCAATs project.

- Select 11102 and click "Next" to proceed.

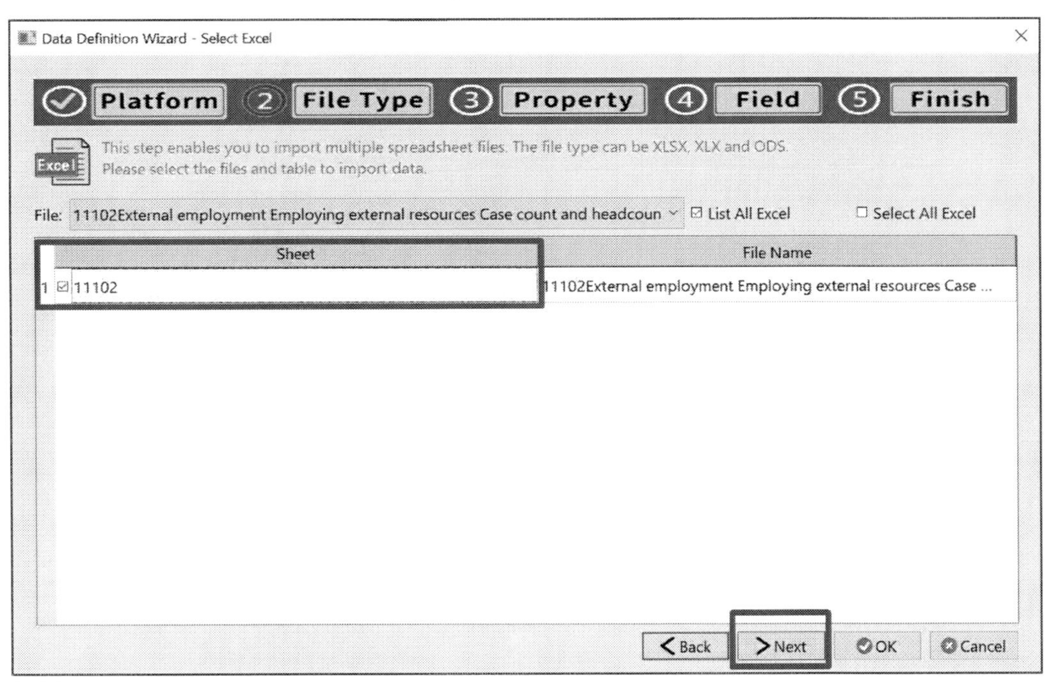

Exercise 4.8
Please practice importing OpenDocument (ODS) into the JCAATs project.

- Select "Next".

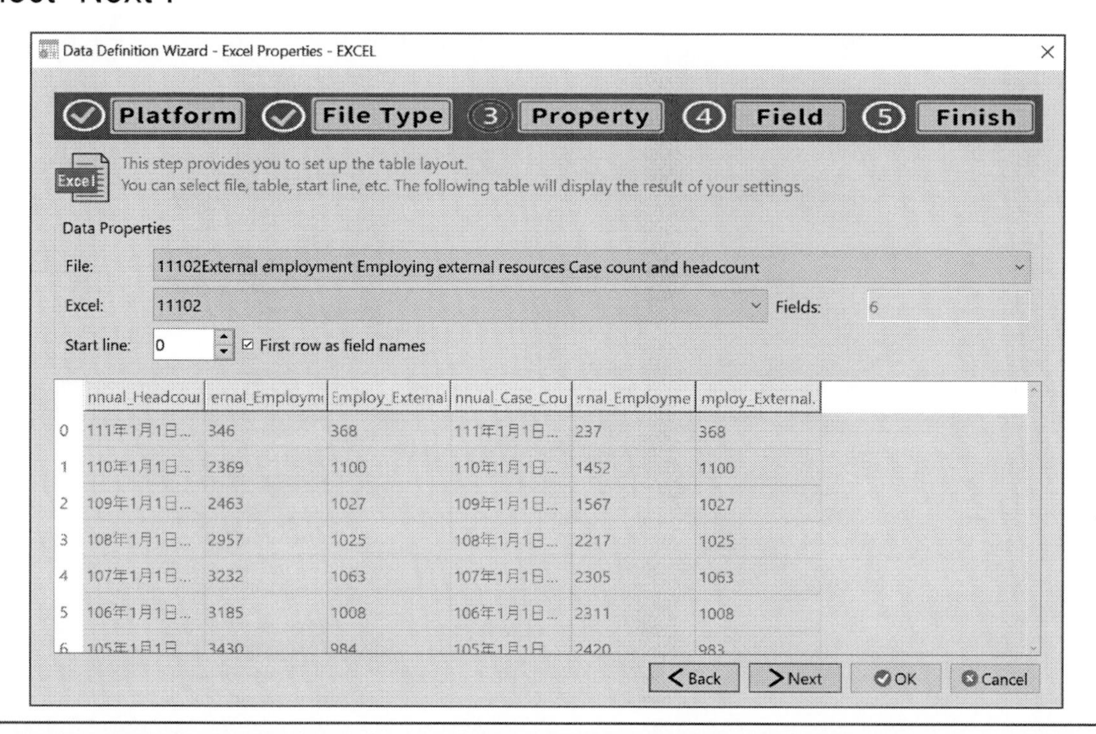

Exercise 4.8
Please practice importing OpenDocument (ODS) into the JCAATs project.

- Field Definition: JCAATs can define the field name, display name, data type and data format of each field. After completing the settings, select "Next" to proceed.

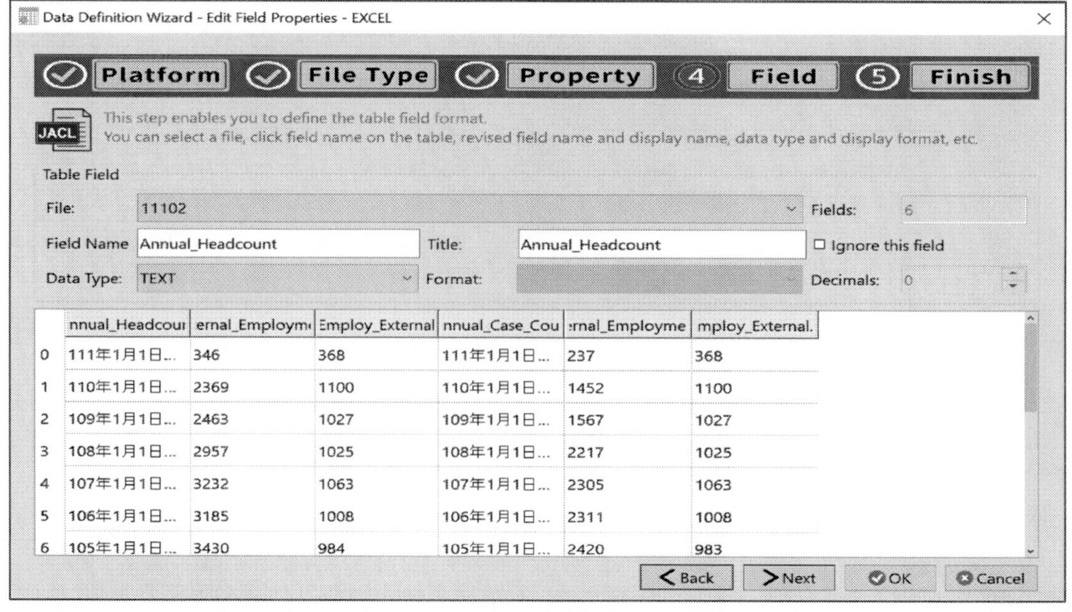

Exercise 4.8
Please practice importing OpenDocument (ODS) into the JCAATs project.

- The default data file path is the project folder
- After ensuring that the path and information are correct, select "Done".

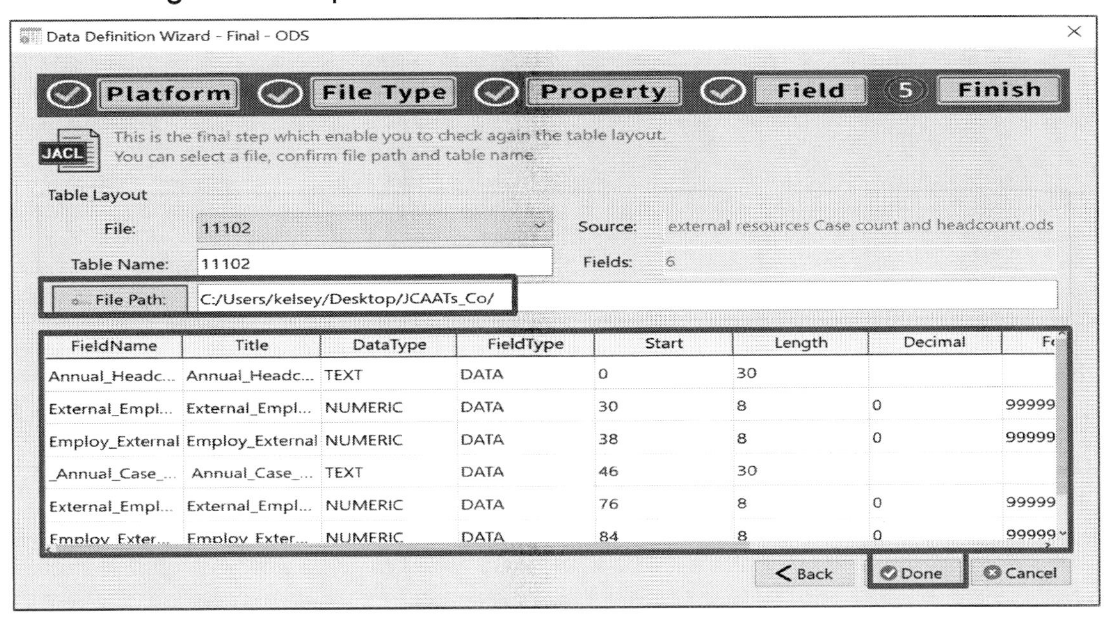

JCAATs-AI Audit Software
Copyright © 2023 JACKSOFT.

Exercise 4.8
Please practice importing OpenDocument (ODS) into the JCAATs project.

- After the import progress is completed, we can see that the data table has been successfully imported.

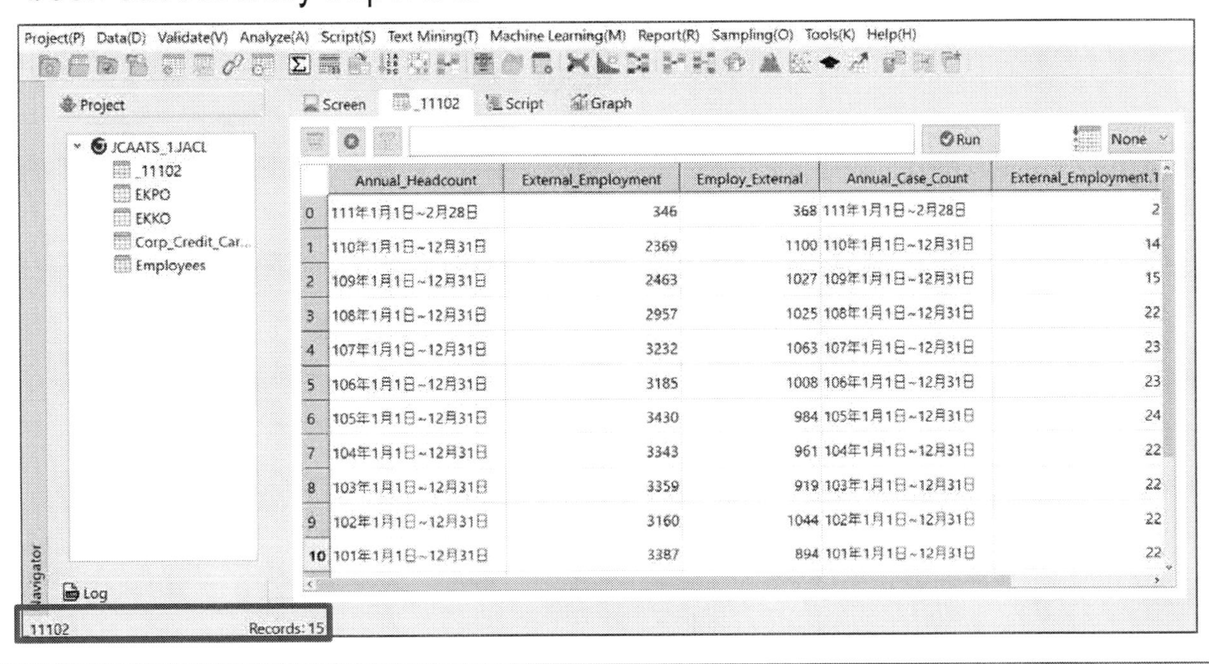

Exercise 4.9

Please practice importing JSON into the JCAATs project.
superHeroes.JSON

Exercise 4.9
Please practice importing JSON into the JCAATs project.

- Select "Data" > "New Table.
- After selecting the data source platform as "File, " click "Next".

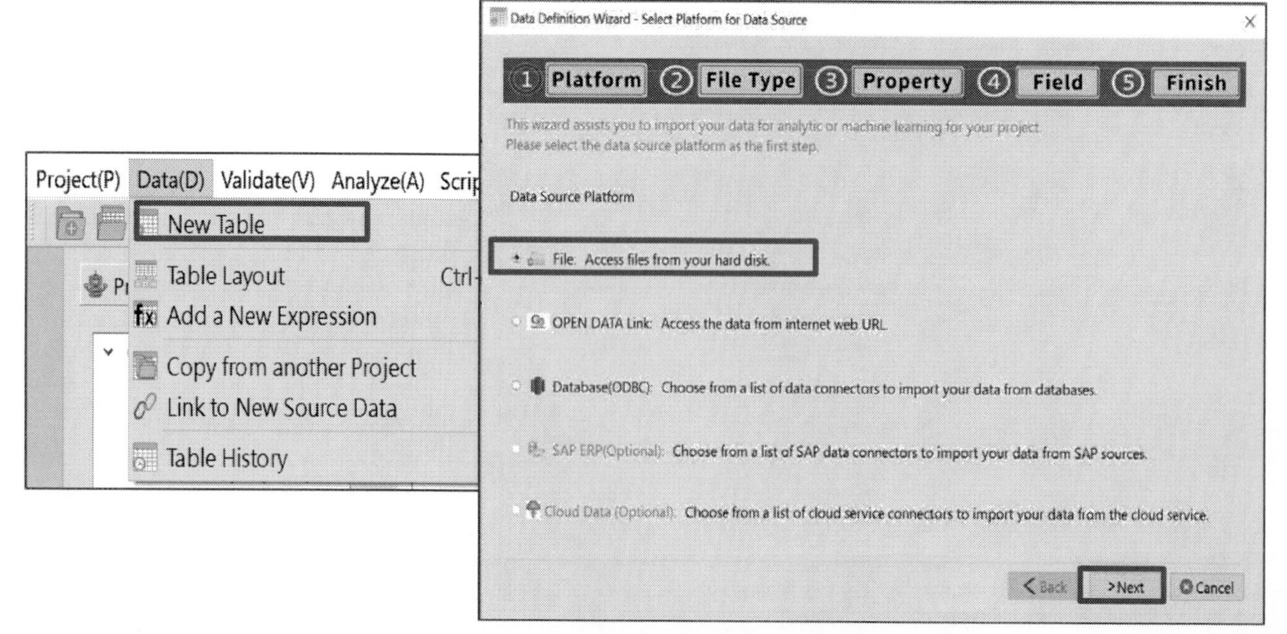

Exercise 4.9
Please practice importing JSON into the JCAATs project.

- Select superHeroes.JSON to import.
- JCAATs will detect the file format automatically. If there are no errors with the detected format, select "Next" to proceed.

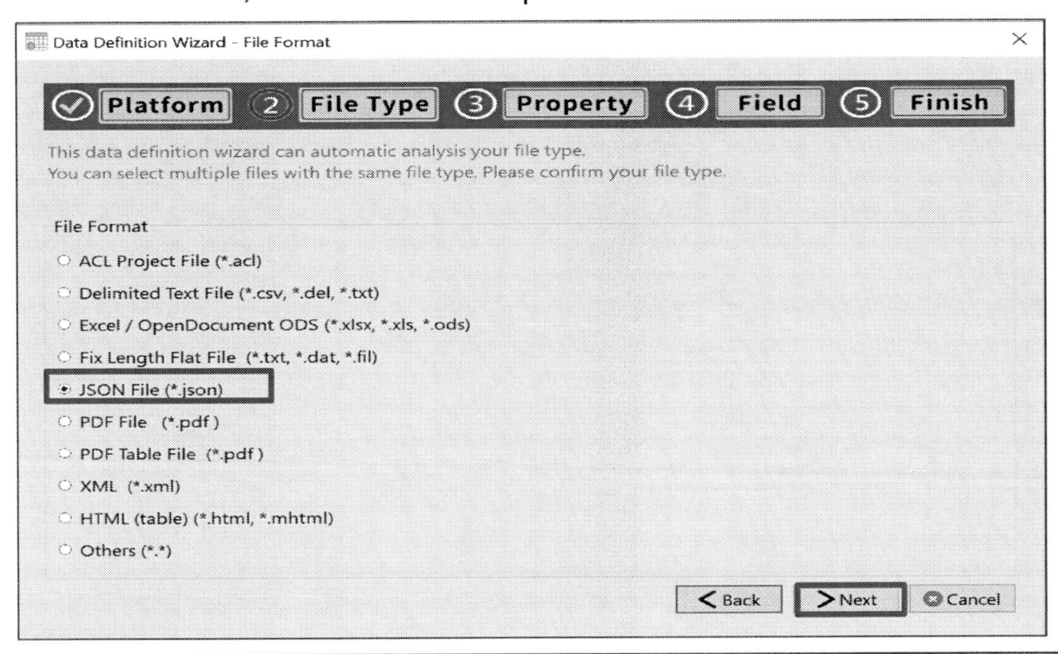

JCAATs-AI Audit Software
Copyright © 2023 JACKSOFT.

Exercise 4.9
Please practice importing JSON into the JCAATs project.

- JCAATs will detect file character encoding automatically.
- After completing the settings, select "Next" to proceed.

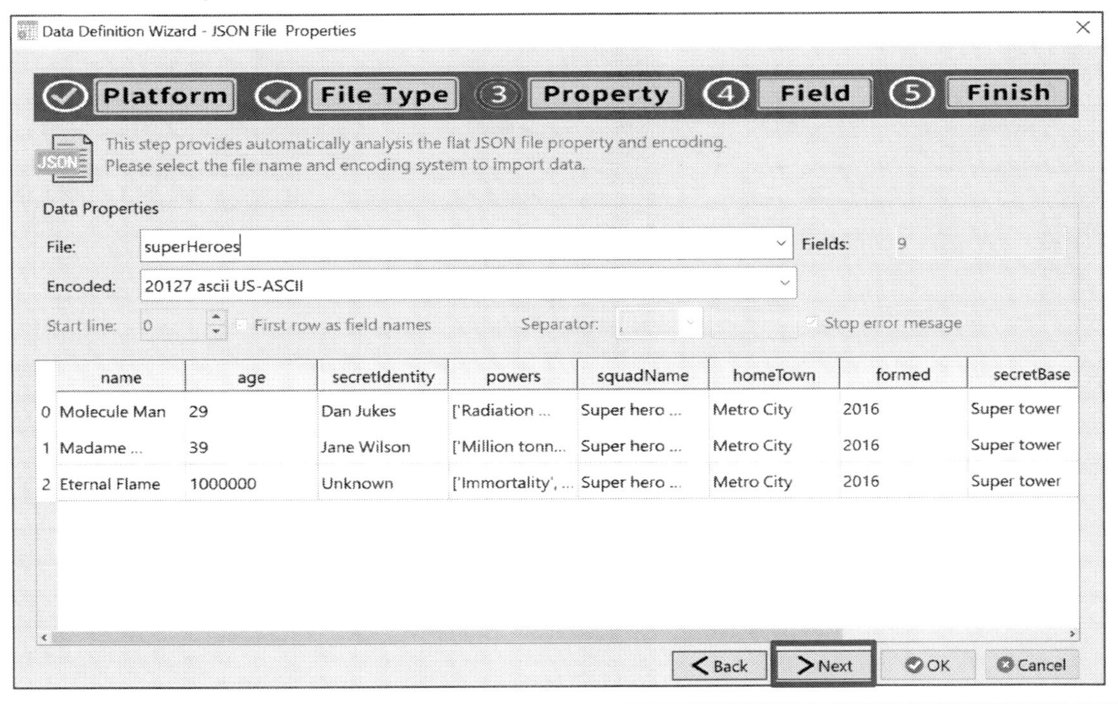

Exercise 4.9

Please practice importing JSON into the JCAATs project.

- Field Definition: JCAATs can define the field name, display name, data type and data format of each field. After completing the settings, select "Next" to proceed.

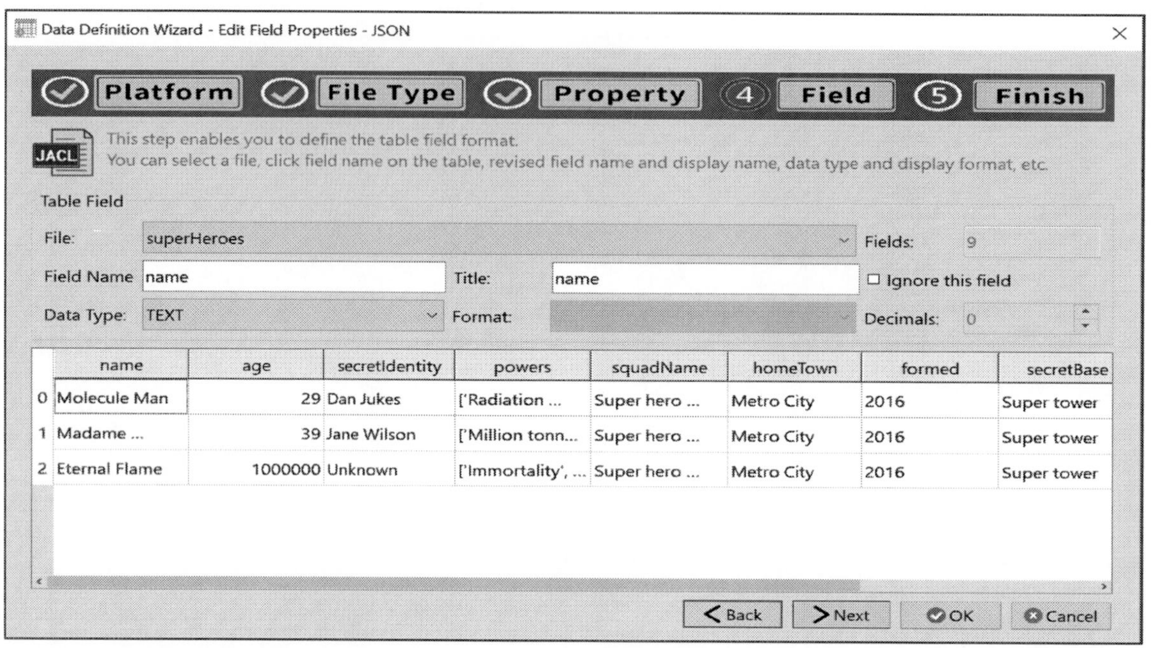

JCAATs-AI Audit Software
Copyright © 2023 JACKSOFT.

Exercise 4.9

Please practice importing JSON into the JCAATs project.

- The default data file path is the project folder
- After ensuring that the path and information are correct, select "Done".

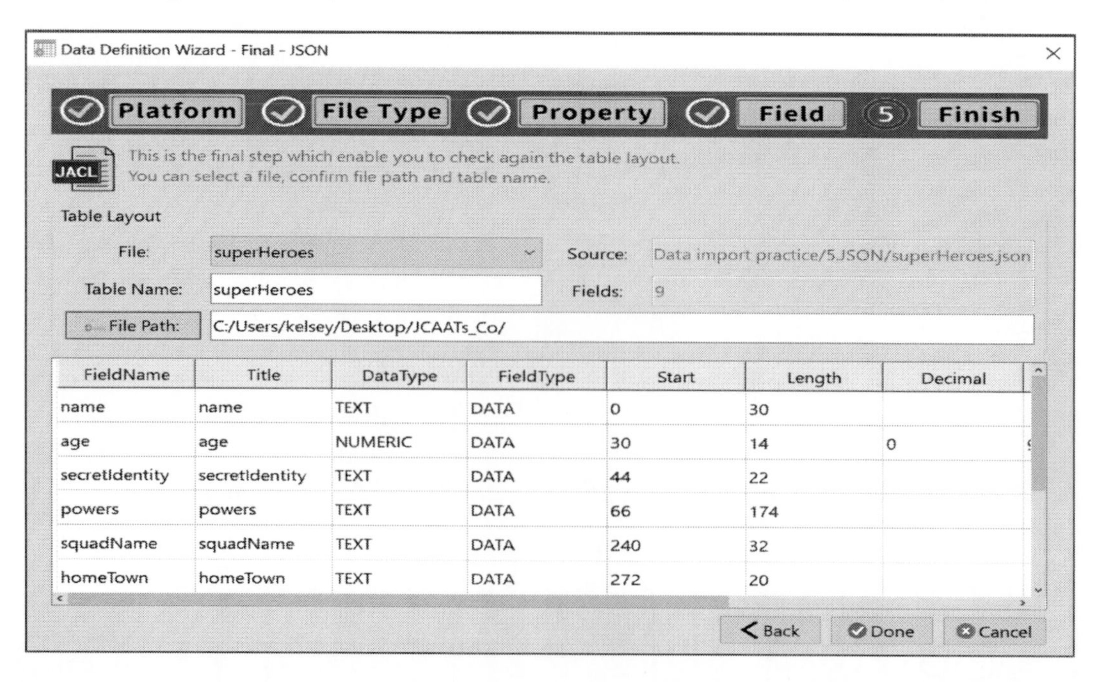

Exercise 4.9
Please practice importing JSON into the JCAATs project.

- After the import progress is completed, we can see that the data table has been successfully imported.

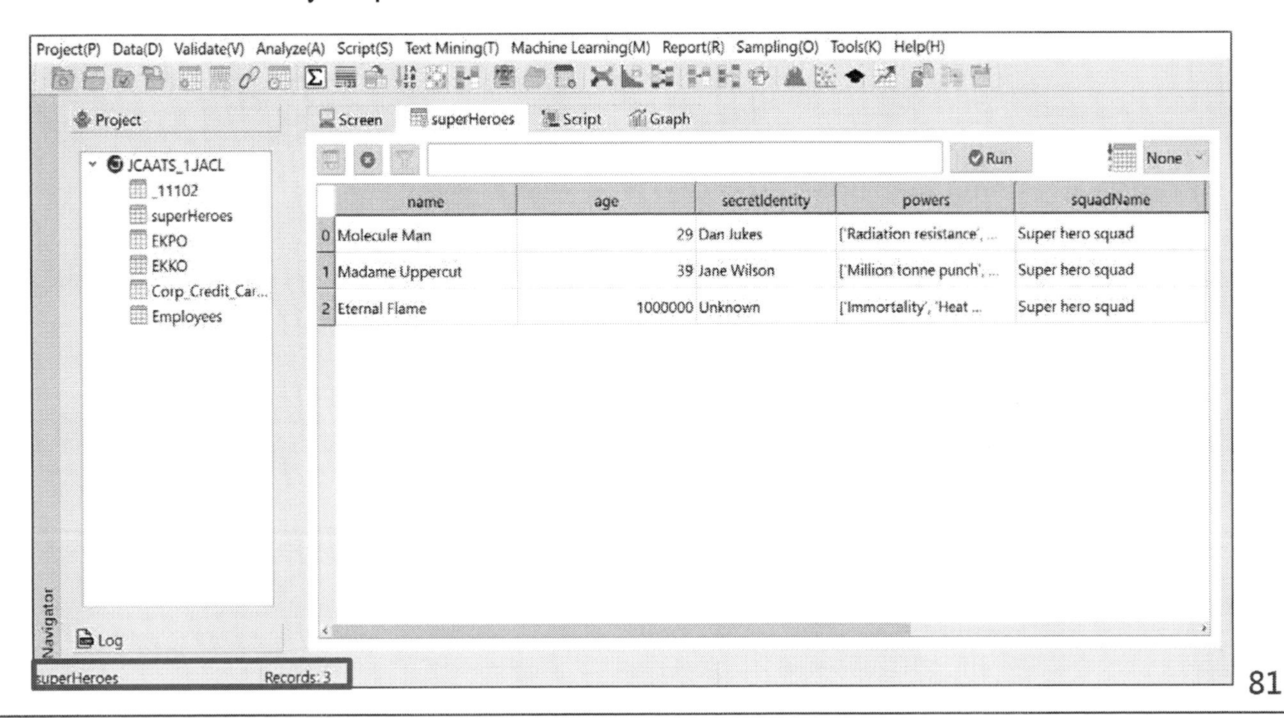

Please practice importing PDF into the JCAATs project.

Consulting services contract.pdf
Inventory_table.pdf
Importing multiple PDF files of the ten penalty cases.

Exercise 4.10.1
Please practice importing PDF into the JCAATs project

- Select "Data" > "New Table.
- After selecting the data source platform as "File, " click "Next".

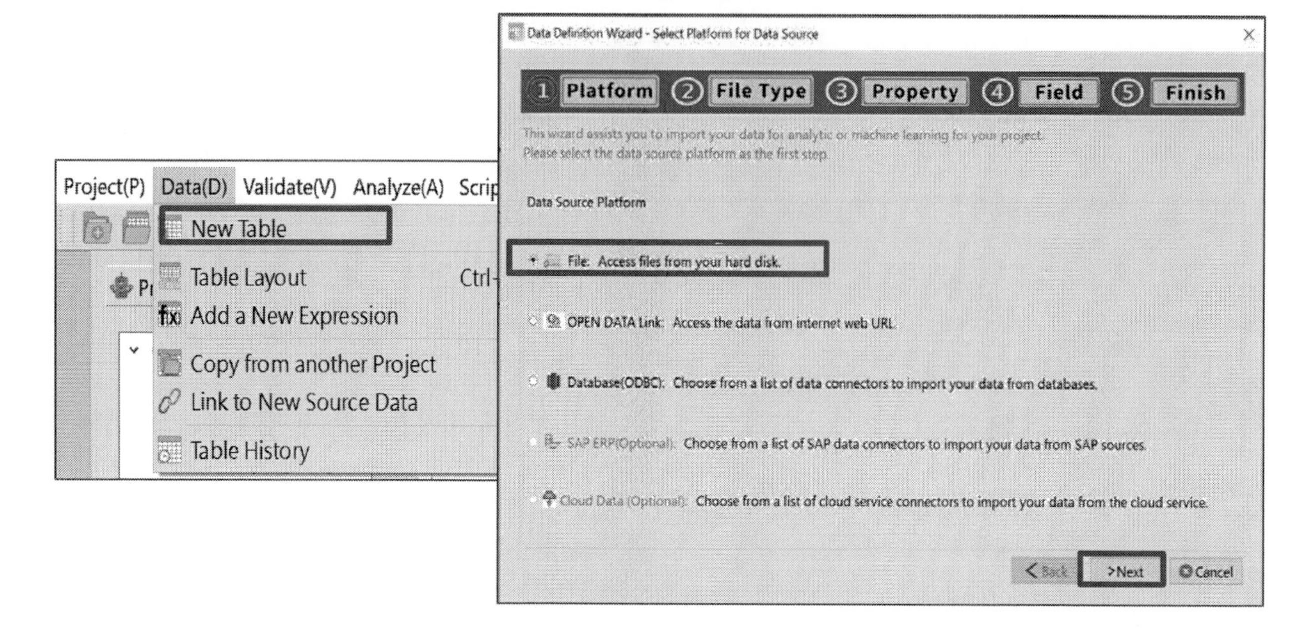

JCAATs-AI Audit Software
Copyright © 2023 JACKSOFT.

Exercise 4.10.1
Please practice importing PDF into the JCAATs project

- Select Consulting services contract.pdf
- Click " Open "

Exercise 4.10.1
Please practice importing PDF into the JCAATs project

- JCAATs will detect the file format automatically.
- If there are no errors with the detected format, select "Next" to proceed.

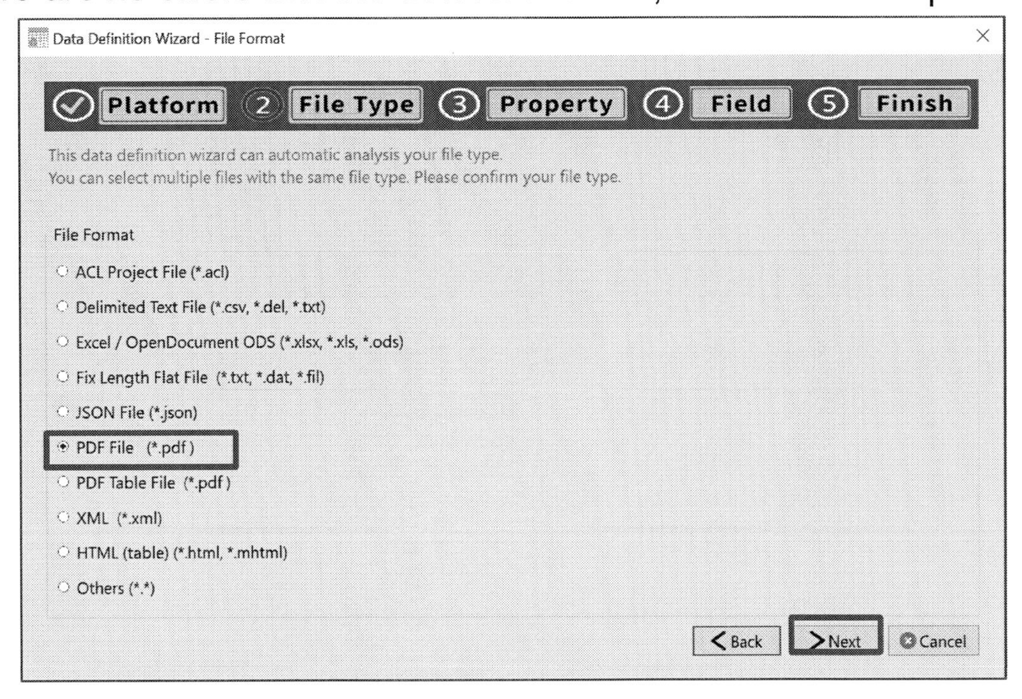

JCAATs-AI Audit Software
Copyright © 2023 JACKSOFT.

Exercise 4.10.1
Please practice importing PDF into the JCAATs project

- JCAATs will detect file character encoding automatically.
- After completing the settings, select "Next" to proceed.

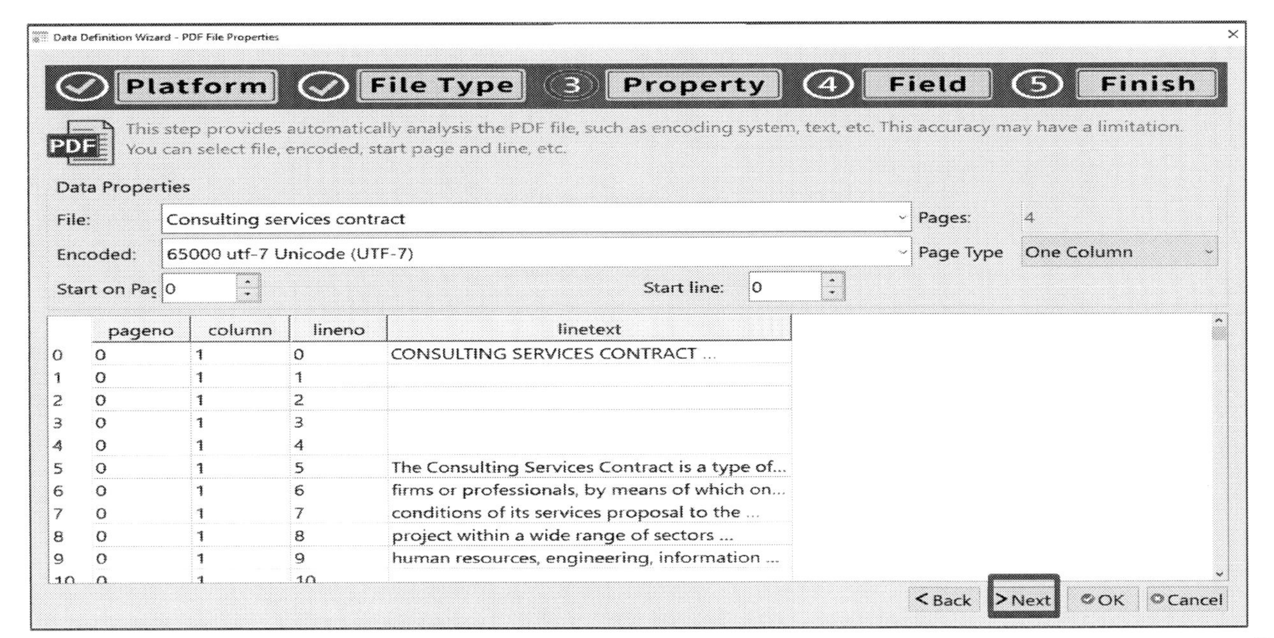

Exercise 4.10.1
Please practice importing PDF into the JCAATs project

- Field Definition: JCAATs can define the field name, display name, data type and data format of each field. After completing the settings, select "Next" to proceed.

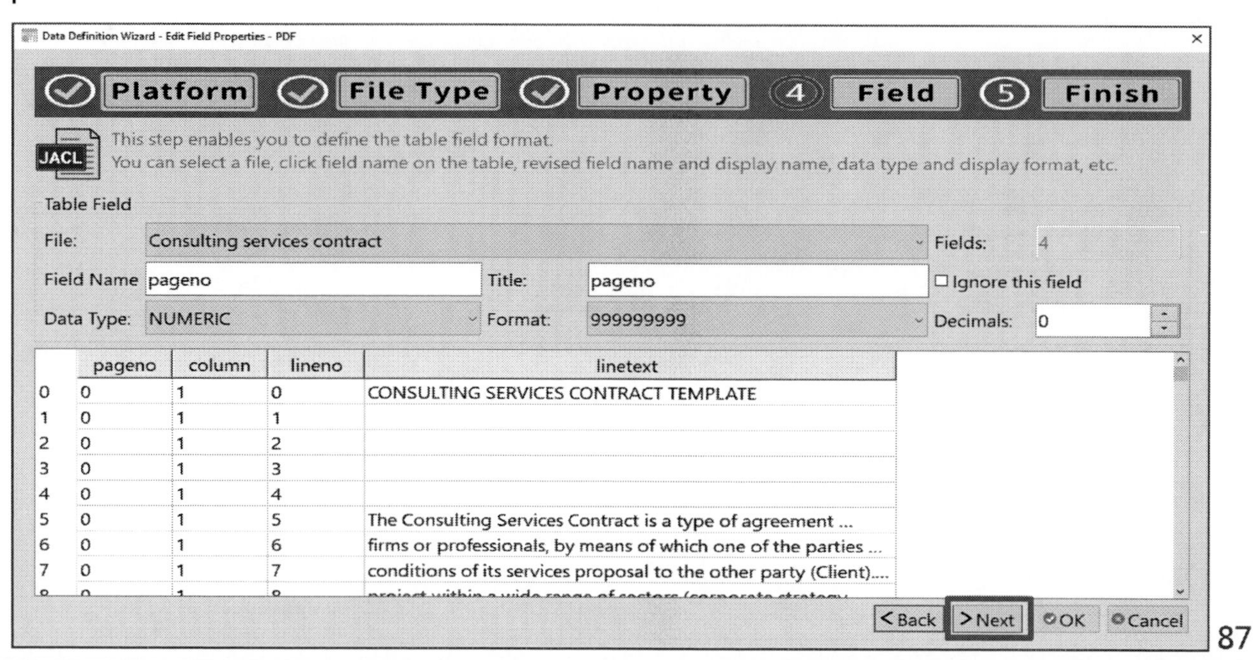

JCAATs-AI Audit Software
Copyright © 2023 JACKSOFT.

Exercise 4.10.1
Please practice importing PDF into the JCAATs project

- The default data file path is the project folder
- After ensuring that the path and information are correct, select "Done".

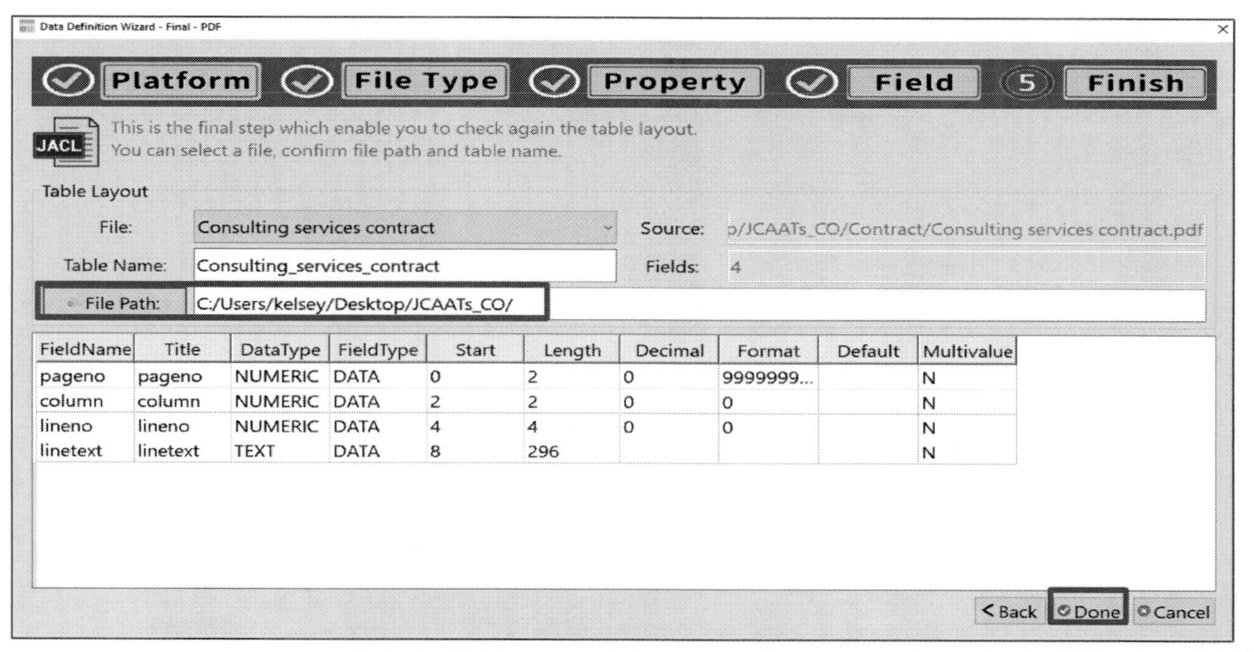

FieldName	Title	DataType	FieldType	Start	Length	Decimal	Format	Default	Multivalue
pageno	pageno	NUMERIC	DATA	0	2	0	9999999...		N
column	column	NUMERIC	DATA	2	2	0	0		N
lineno	lineno	NUMERIC	DATA	4	4	0	0		N
linetext	linetext	TEXT	DATA	8	296				N

Exercise 4.10.1
Please practice importing PDF into the JCAATs project

- After the import progress is completed, we can see that the data table has been successfully imported.

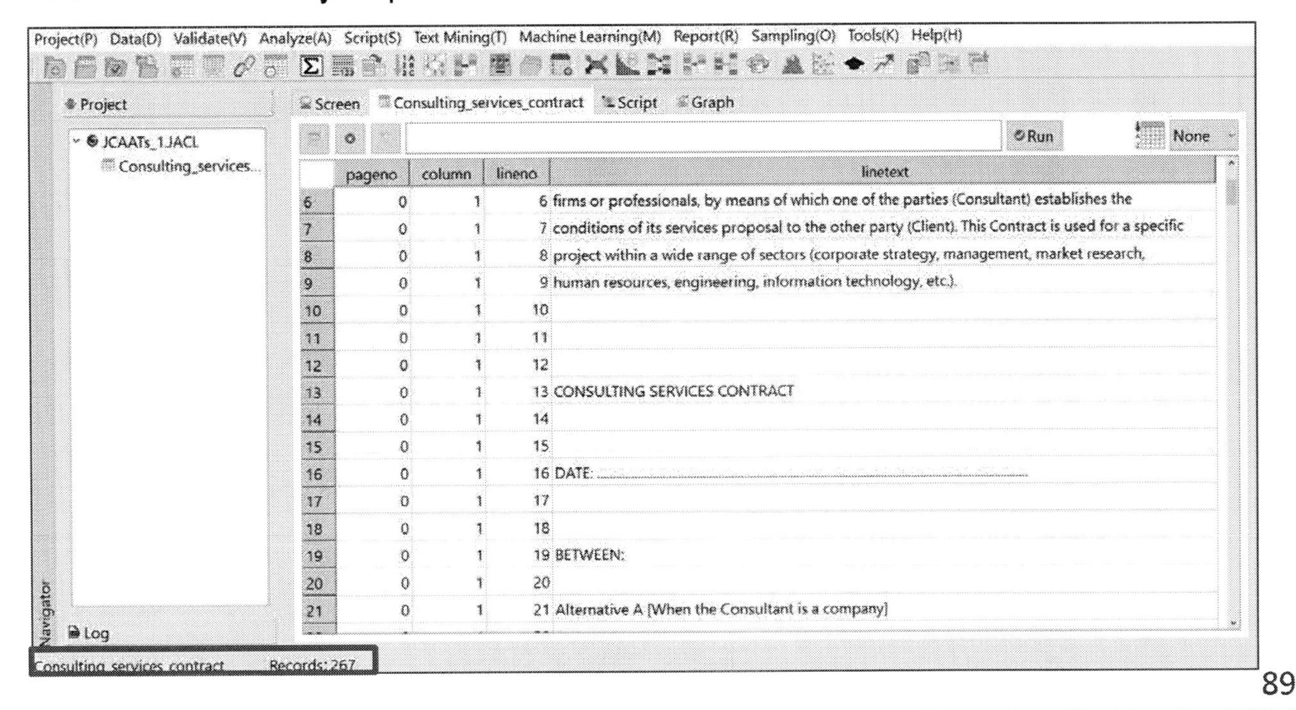

Exercise 4.10.2
Please practice importing PDF into the JCAATs project

- Select "Data" > "New Table.
- After selecting the data source platform as "File, " click "Next".

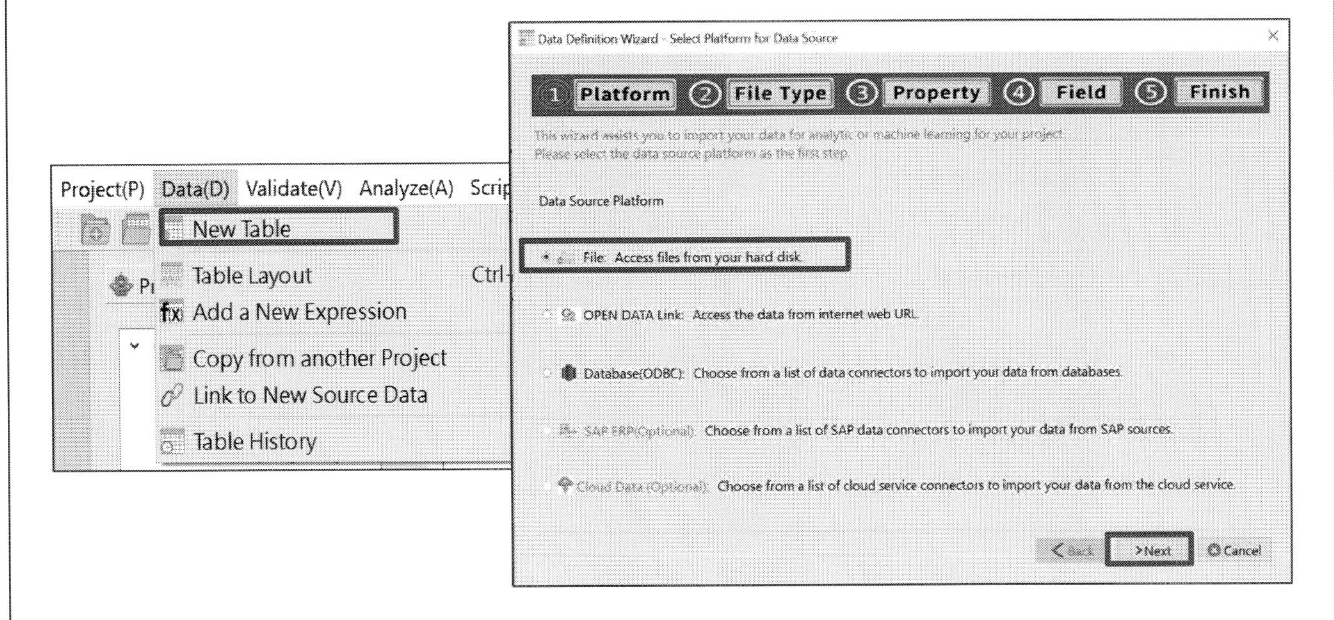

Exercise 4.10.2
Please practice importing PDF into the JCAATs project

- Select Inventory_table.pdf, and click "open" to proceed.

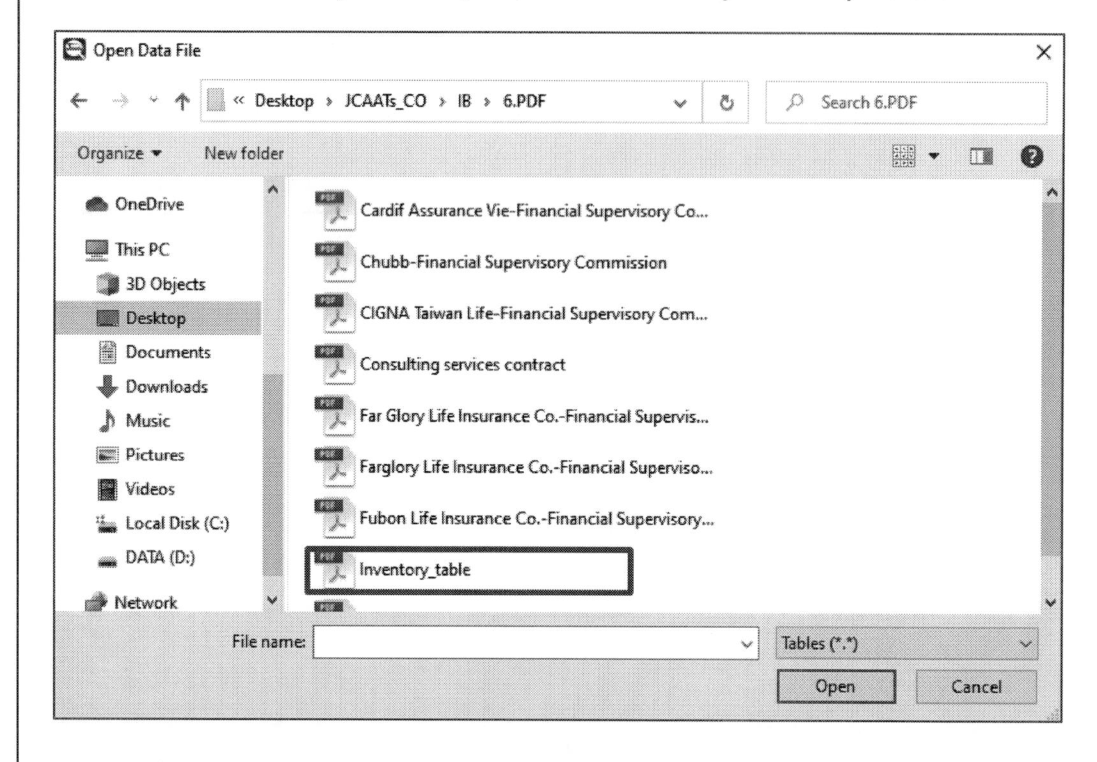

JCAATs-AI Audit Software
Copyright © 2023 JACKSOFT.

Exercise 4.10.2
Please practice importing PDF into the JCAATs project

- Select"PDF Table File", and click "next" to proceed.

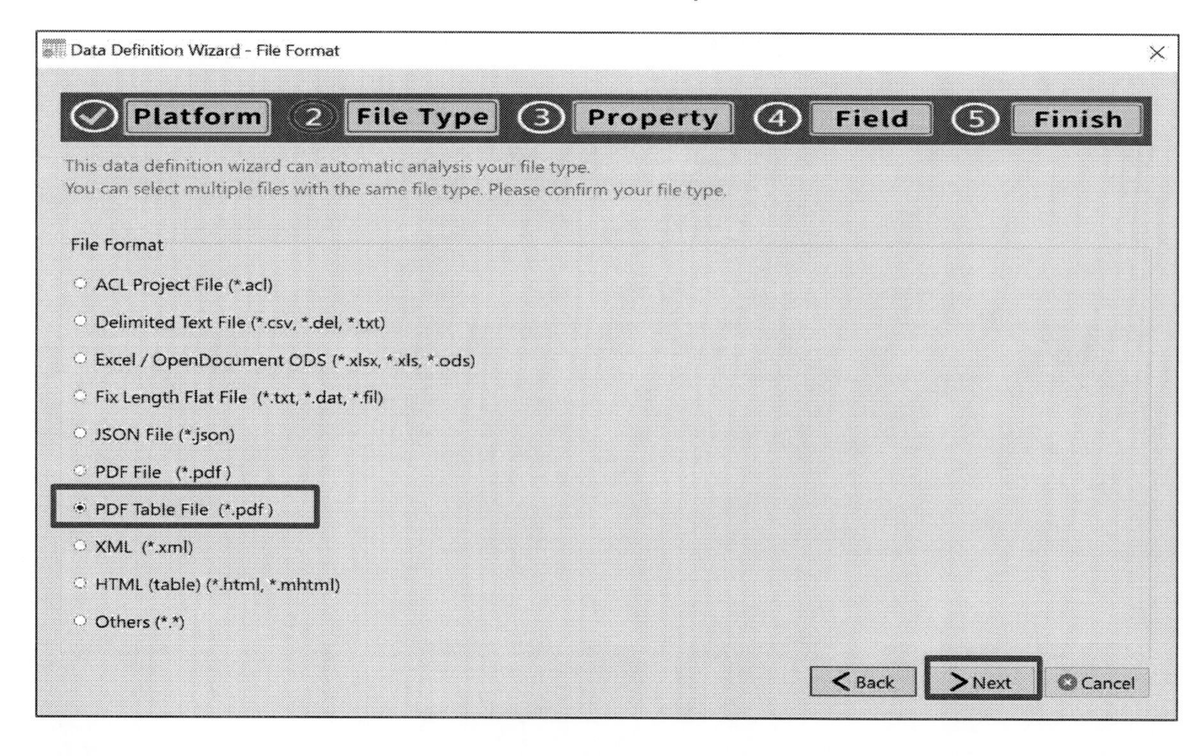

Exercise 4.10.2
Please practice importing PDF into the JCAATs project

- JCAATs will detect file character encoding automatically.
- Select "Select All Table" to import multiply tables at once.
- After completing the settings, select 'Next' to proceed.

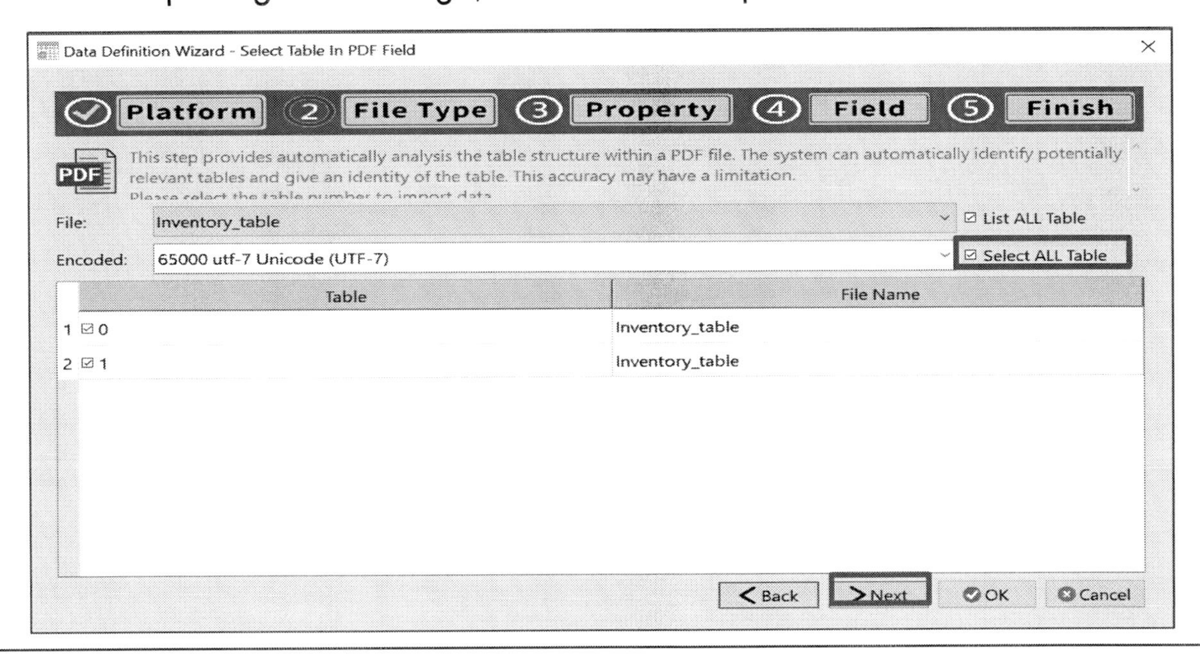

Exercise 4.10.2
Please practice importing PDF into the JCAATs project

- Data Features: JCAATs will identify the file encoding method automatically.
- Click "Next" after completing the settings.

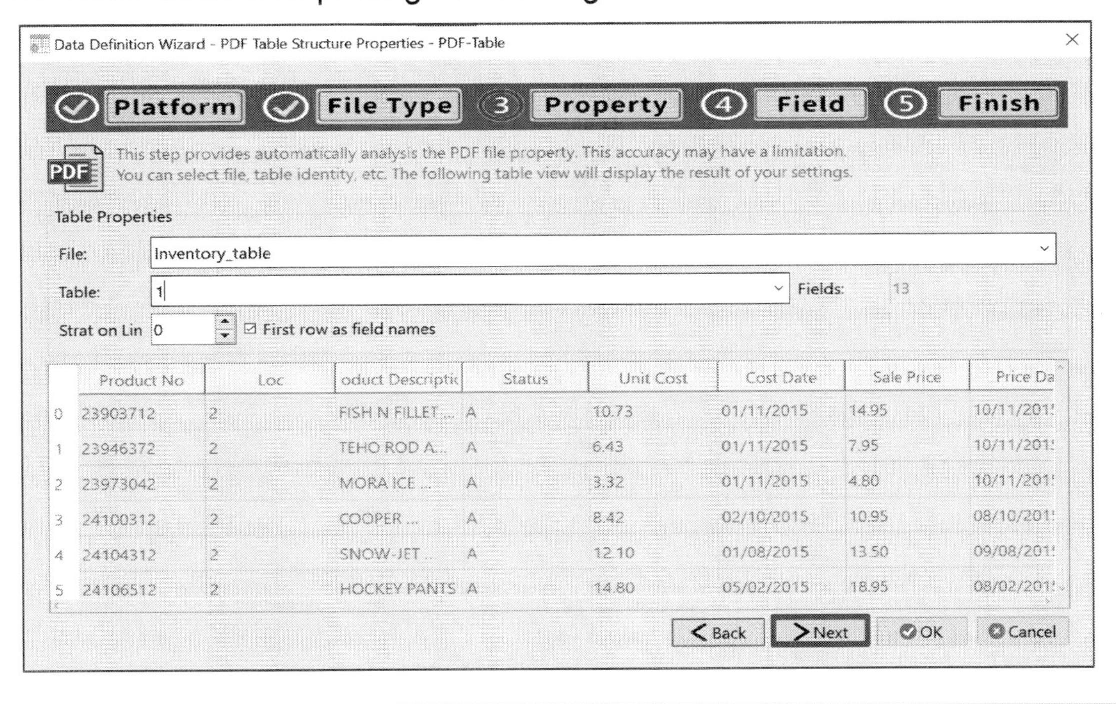

Exercise 4.10.2
Please practice importing PDF into the JCAATs project

- Field Definition: JCAATs can define the field name, display name, data type and data format of each field. After completing the settings, select "Next" to proceed.

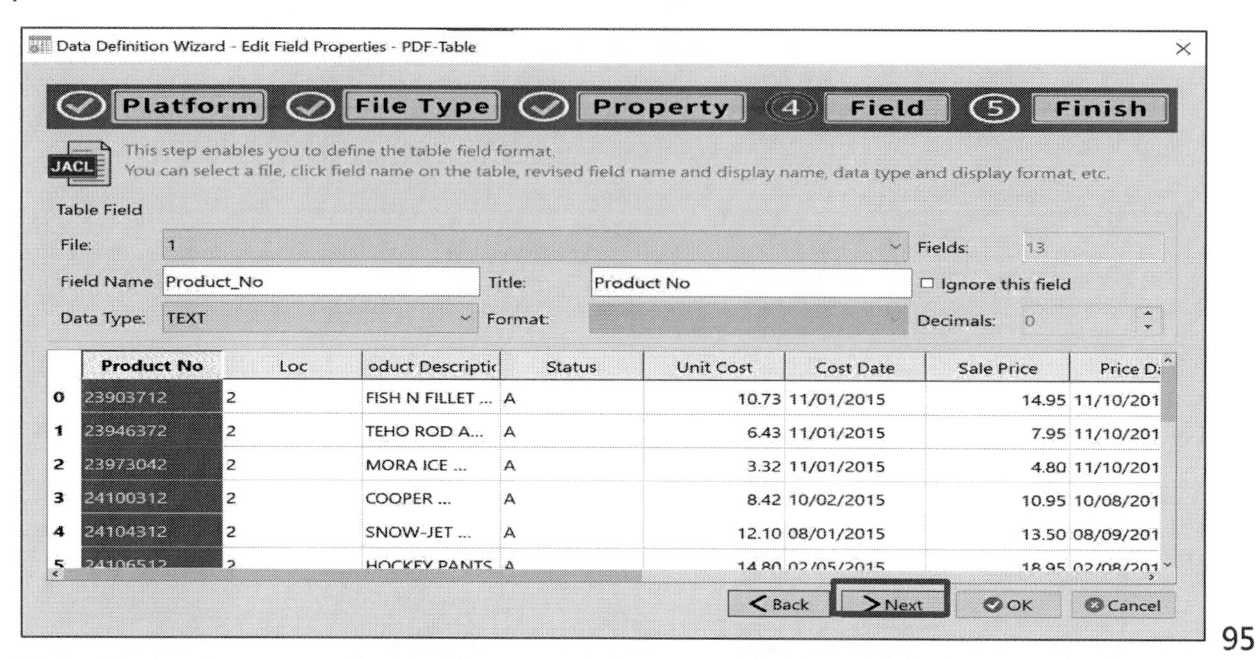

JCAATs-AI Audit Software

Copyright © 2023 JACKSOFT.

Exercise 4.10.2
Please practice importing PDF into the JCAATs project

- The default data file path is the project folder
- After ensuring that the path and information are correct, select "Done".

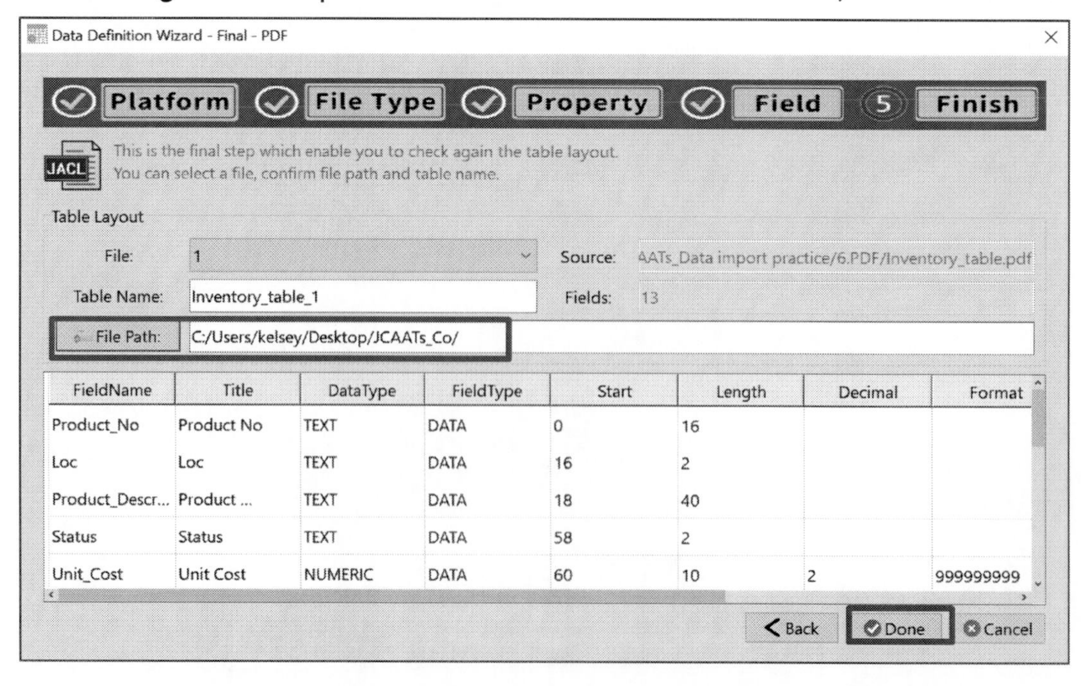

Exercise 4.10.2
Please practice importing PDF into the JCAATs project

- After the import progress is completed, we can see that the data table has been successfully imported.

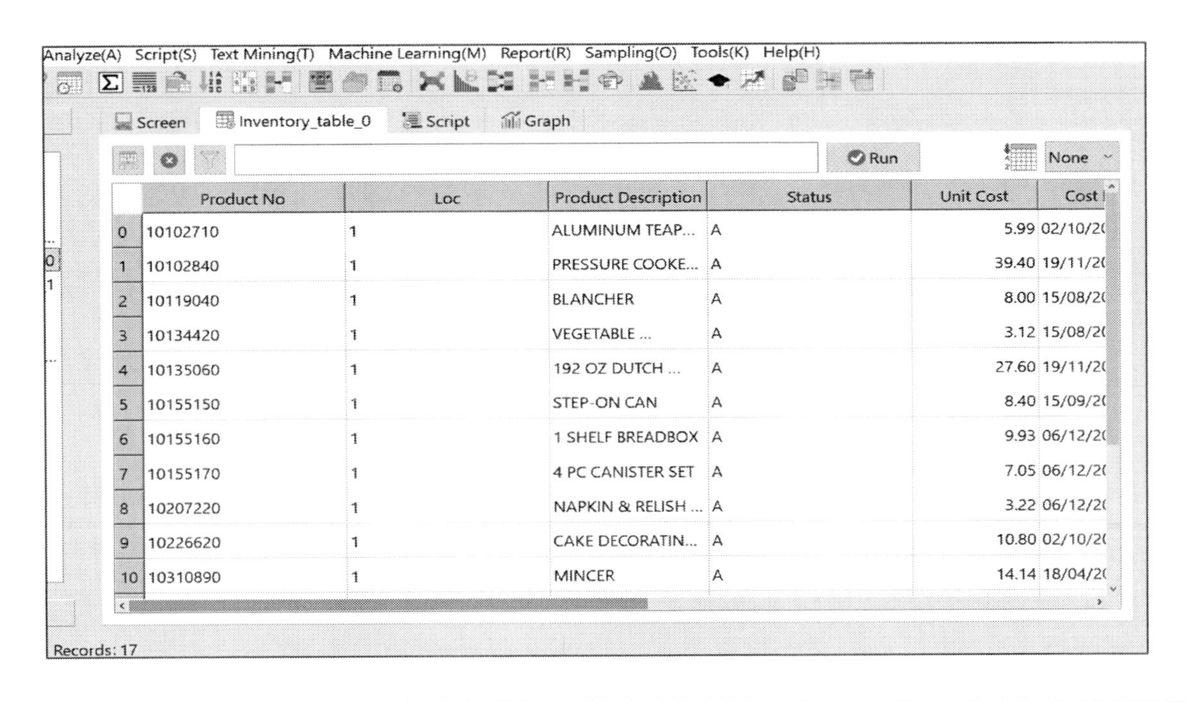

	Product No	Loc	Product Description	Status	Unit Cost	Cost
0	10102710	1	ALUMINUM TEAP...	A	5.99	02/10/2
1	10102840	1	PRESSURE COOKE...	A	39.40	19/11/2
2	10119040	1	BLANCHER	A	8.00	15/08/2
3	10134420	1	VEGETABLE ...	A	3.12	15/08/2
4	10135060	1	192 OZ DUTCH ...	A	27.60	19/11/2
5	10155150	1	STEP-ON CAN	A	8.40	15/09/2
6	10155160	1	1 SHELF BREADBOX	A	9.93	06/12/2
7	10155170	1	4 PC CANISTER SET	A	7.05	06/12/2
8	10207220	1	NAPKIN & RELISH ...	A	3.22	06/12/2
9	10226620	1	CAKE DECORATIN...	A	10.80	02/10/2
10	10310890	1	MINCER	A	14.14	18/04/2

Exercise 4.10.3
Please practice importing PDF into the JCAATs project

- Select "Data" > "New Table.
- After selecting the data source platform as "File, " click "Next".

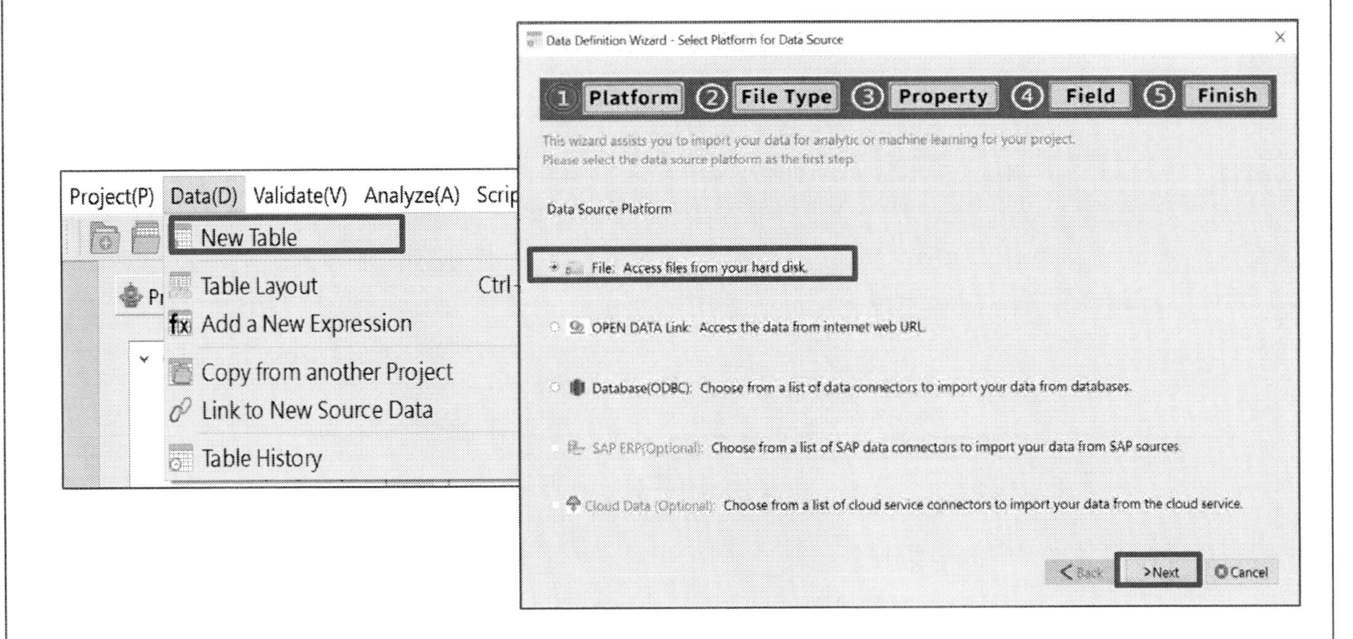

Exercise 4.10.3
Please practice importing PDF into the JCAATs project

- Select 10 PDF files related to penalty cases under the path JCAATs_CO/DATA.
- Click " Open " to proceed.

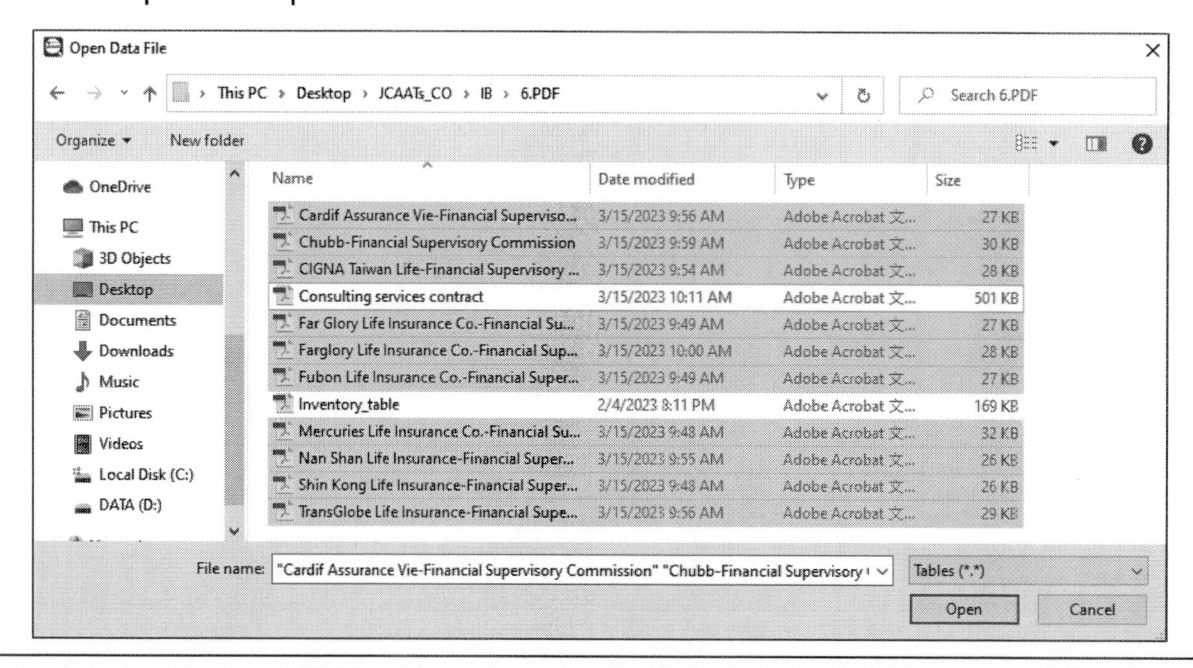

Exercise 4.10.3
Please practice importing PDF into the JCAATs project

- JCAATs will detect the file format automatically. Ensure that the file format is set to PDF File.
- If there are no errors with the detected format, select "Next" to proceed.

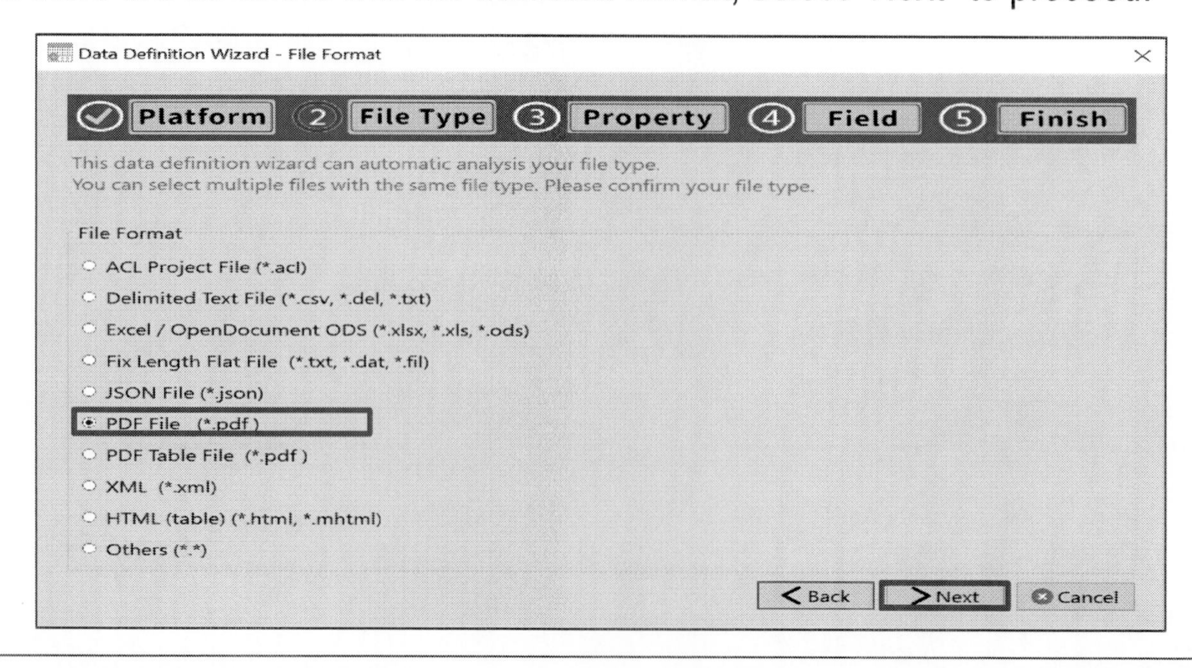

Exercise 4.10.3
Please practice importing PDF into the JCAATs project

- JCAATs will detect file character encoding automatically.
- After completing the settings, select "Next" to proceed.

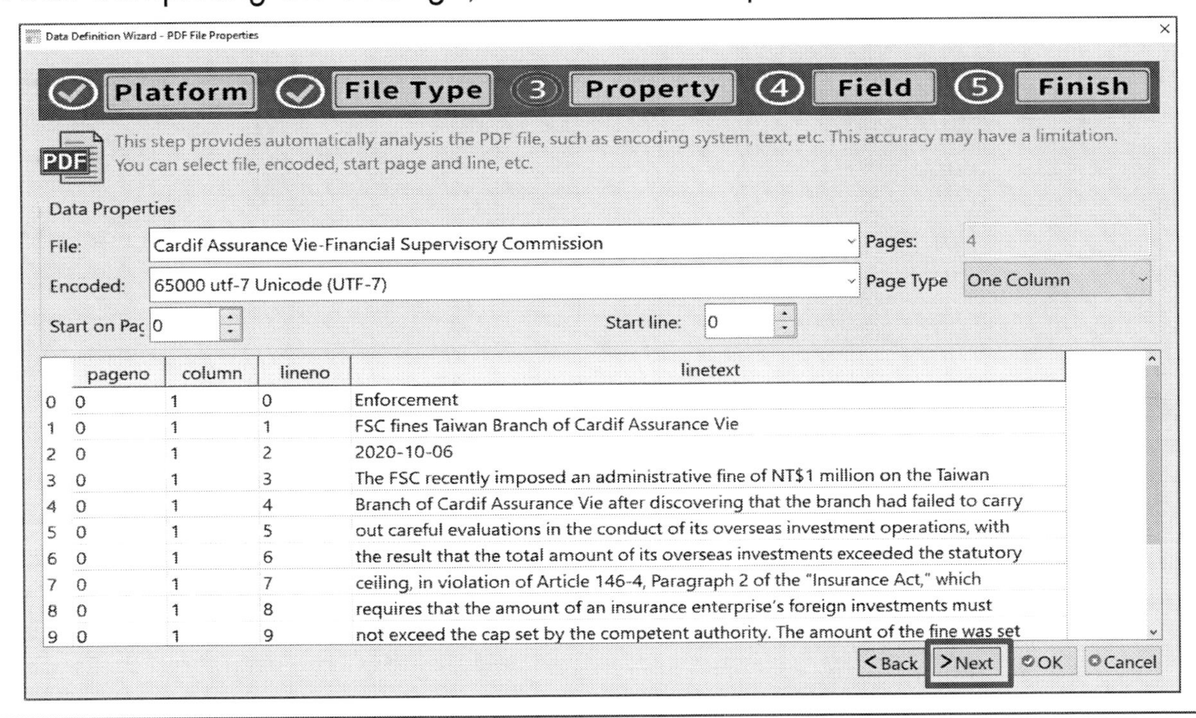

JCAATs-AI Audit Software

Exercise 4.10.3
Please practice importing PDF into the JCAATs project

- Field Definition: JCAATs can define the field name, display name, data type and data format of each field. After completing the settings, select "Next" to proceed.

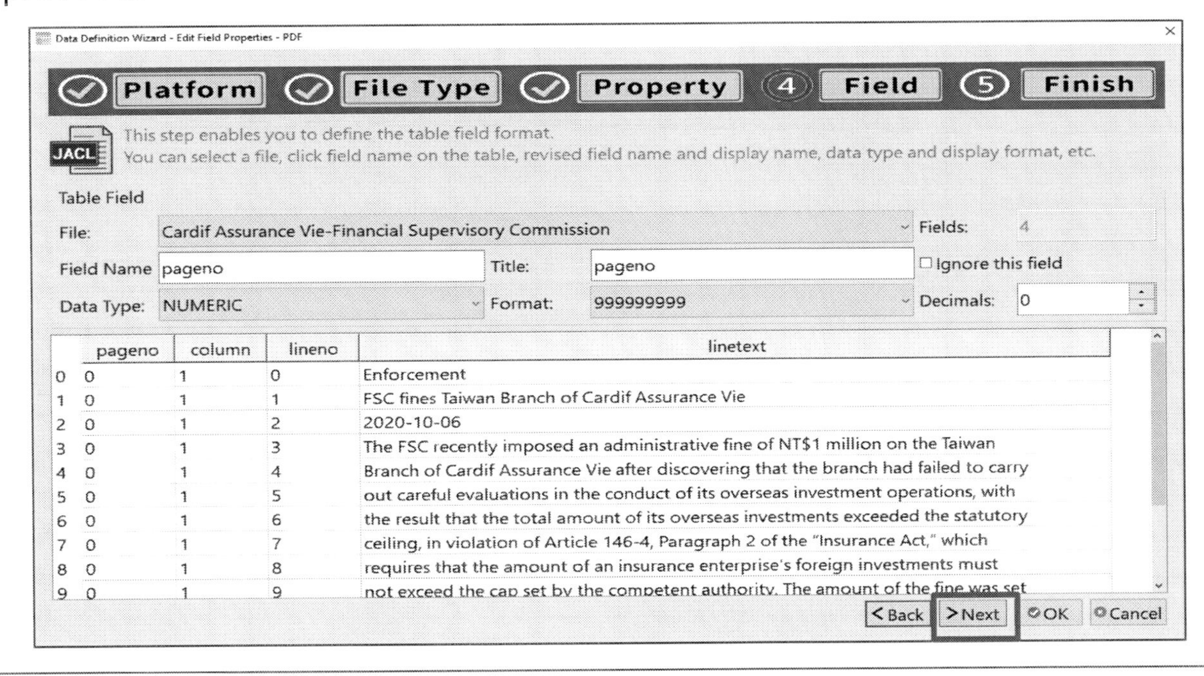

Exercise 4.10.3
Please practice importing PDF into the JCAATs project

- The default data file path is the project folder
- After ensuring that the path and information are correct, select "Done".

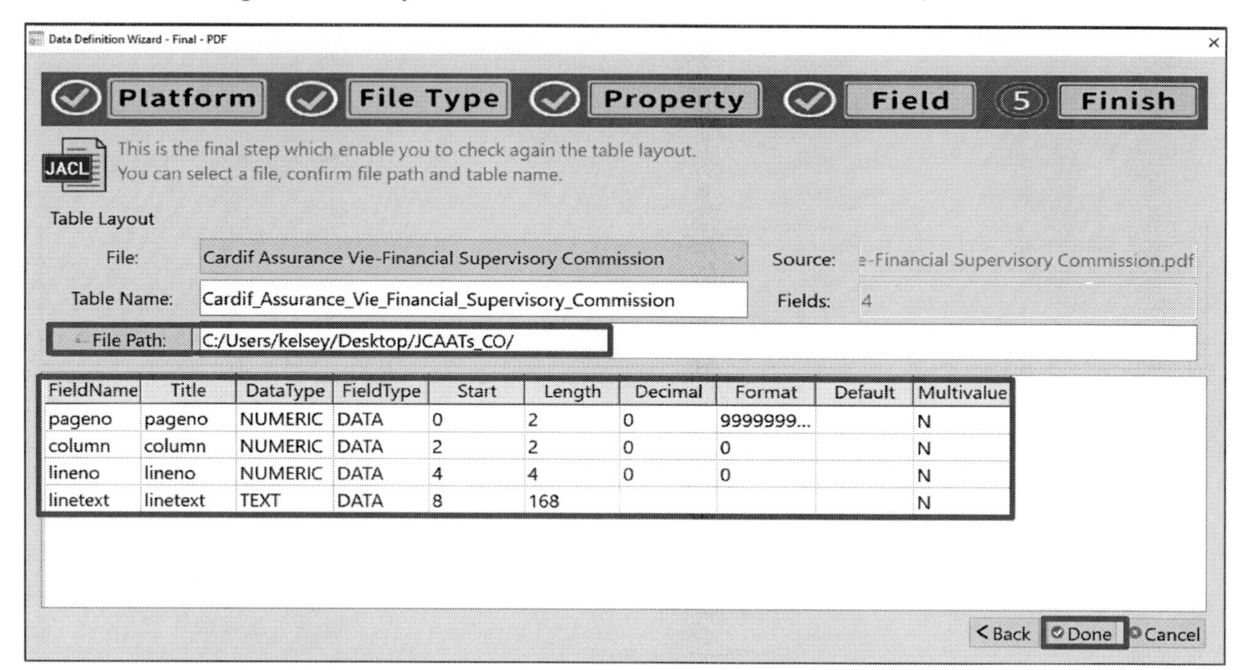

JCAATs-AI Audit Software
Copyright © 2023 JACKSOFT.

Exercise 4.10.3
Please practice importing PDF into the JCAATs project

- After the import progress is completed, we can see that the data table has been successfully imported.

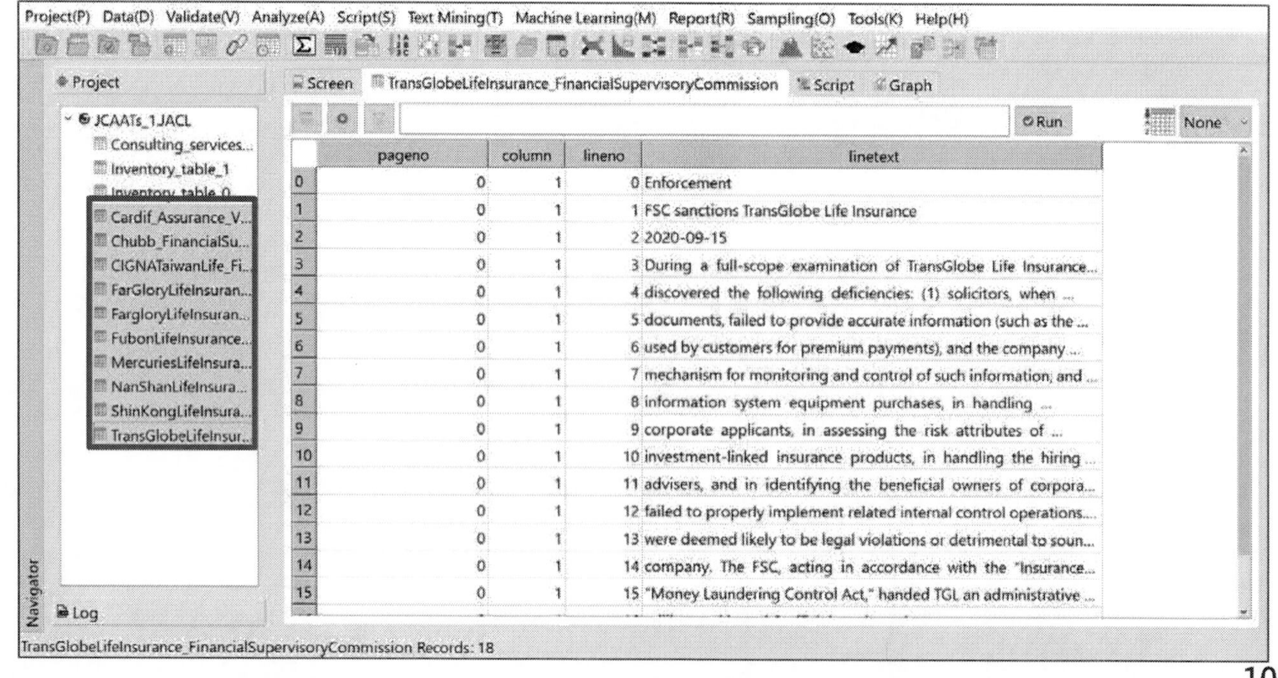

Chapter 4 - Exercise

- Exercise 4.11: Please practice importing XML into a JCAATs project.
- Exercise 4.12: Importing OPEN DATA - US Treasury Department's SDN sanctions list.
- Exercise 4.13: Importing OPEN DATA – Taiwan Government Procurement Network's public announcement of rejection list.

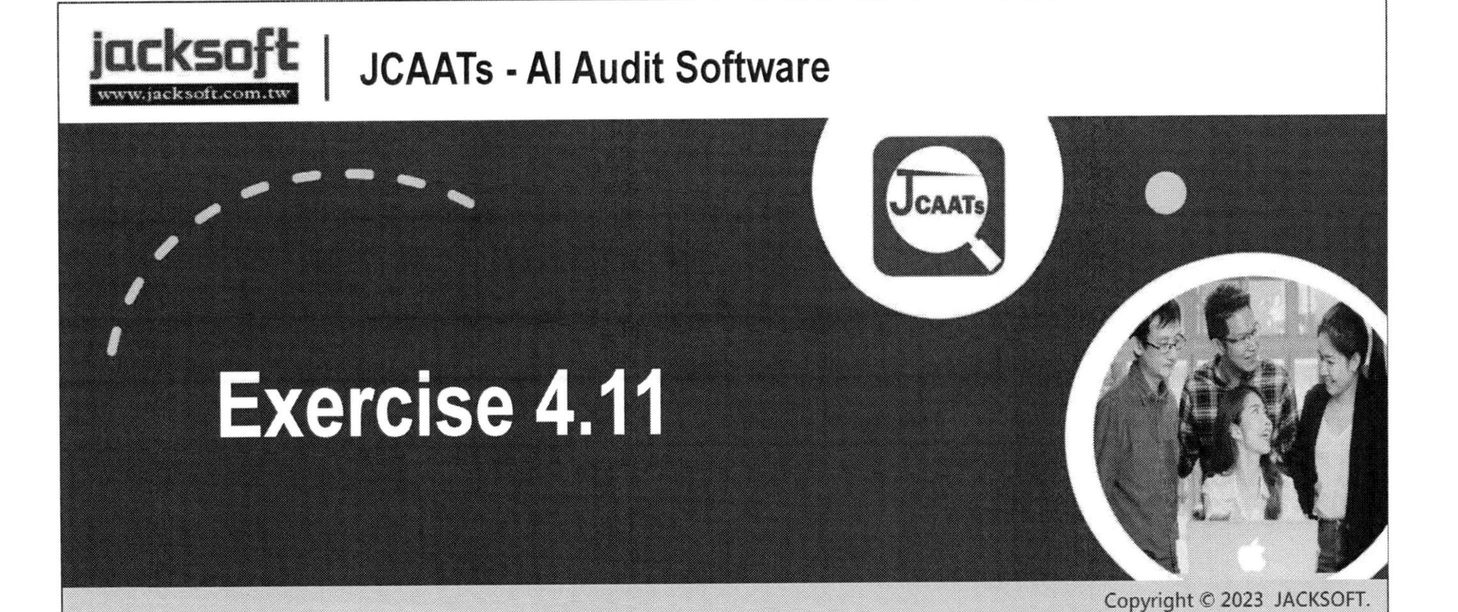

Please practice importing XML into a JCAATs project.

sdn.xml

Exercise 4.11

Please practice importing XML into a JCAATs project.

- Select "Data" > "New Table.
- After selecting the data source platform as "File, " click "Next".

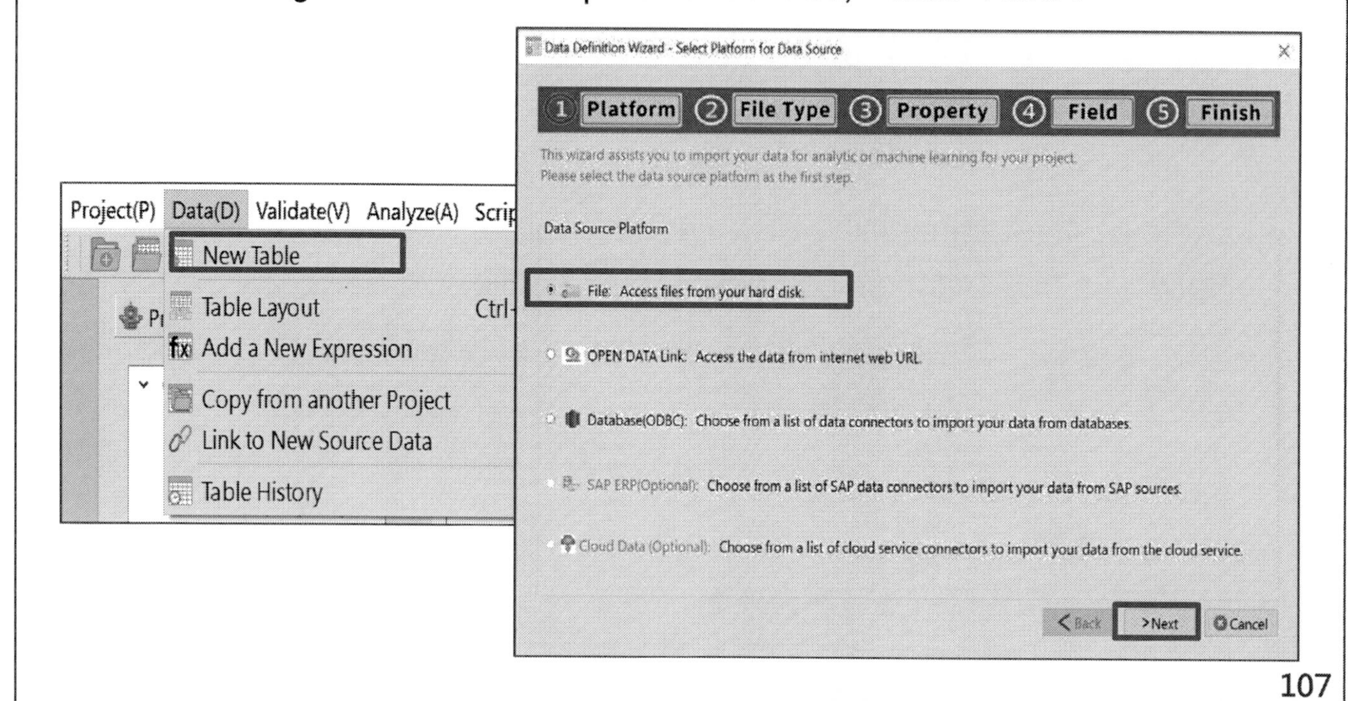

JCAATs-AI Audit Software

Exercise 4.11

Please practice importing XML into a JCAATs project.

- Select sdn.xml file to import.
- JCAATs will automatically detect the file format to ensure accuracy. If there are no errors, select "Next ".

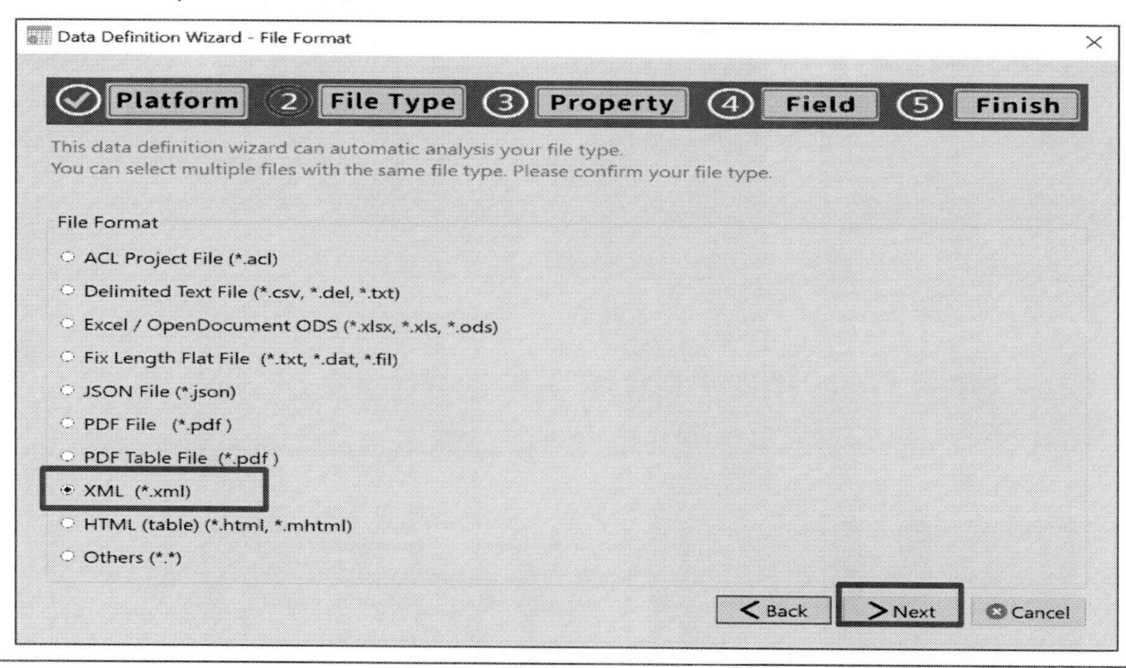

Exercise 4.11
Please practice importing XML into a JCAATs project.

- There are different data structures in XML data format. After selecting "sdnEntry," click "Next" to proceed.

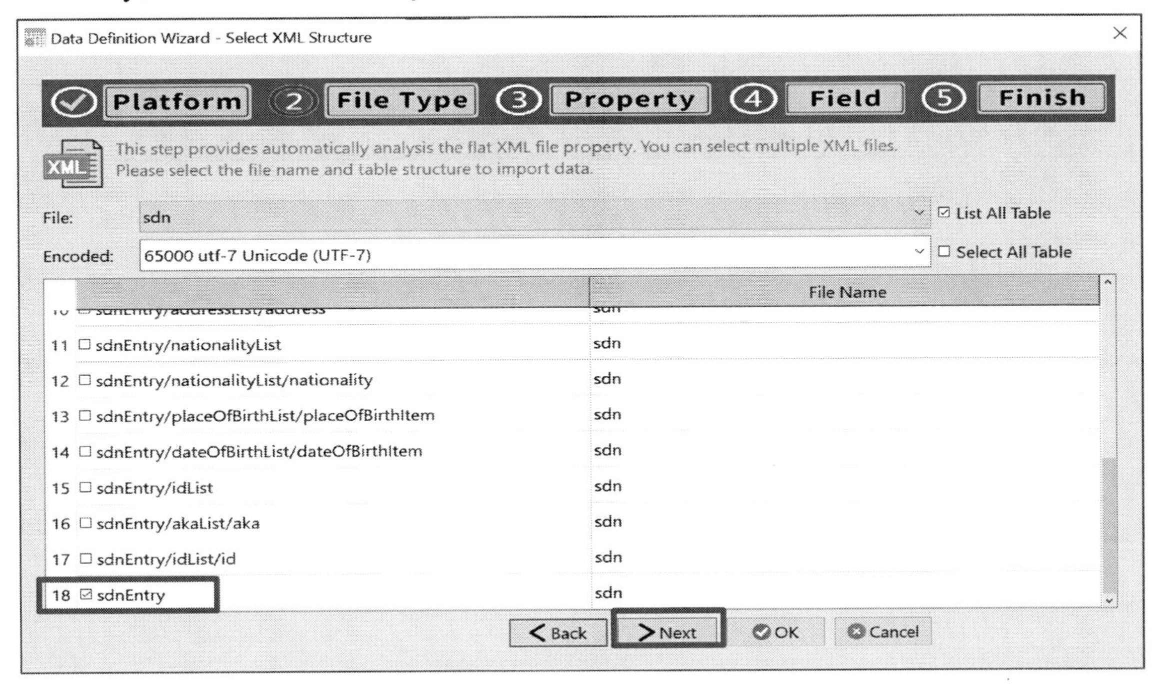

JCAATs-AI Audit Software Copyright © 2023 JACKSOFT.

Exercise 4.11
Please practice importing XML into a JCAATs project.

- Select "Next"

Exercise 4.11
Please practice importing XML into a JCAATs project.

- Field Definition: JCAATs can define the field name, display name, data type and data format of each field. After completing the settings, select "Next" to proceed.

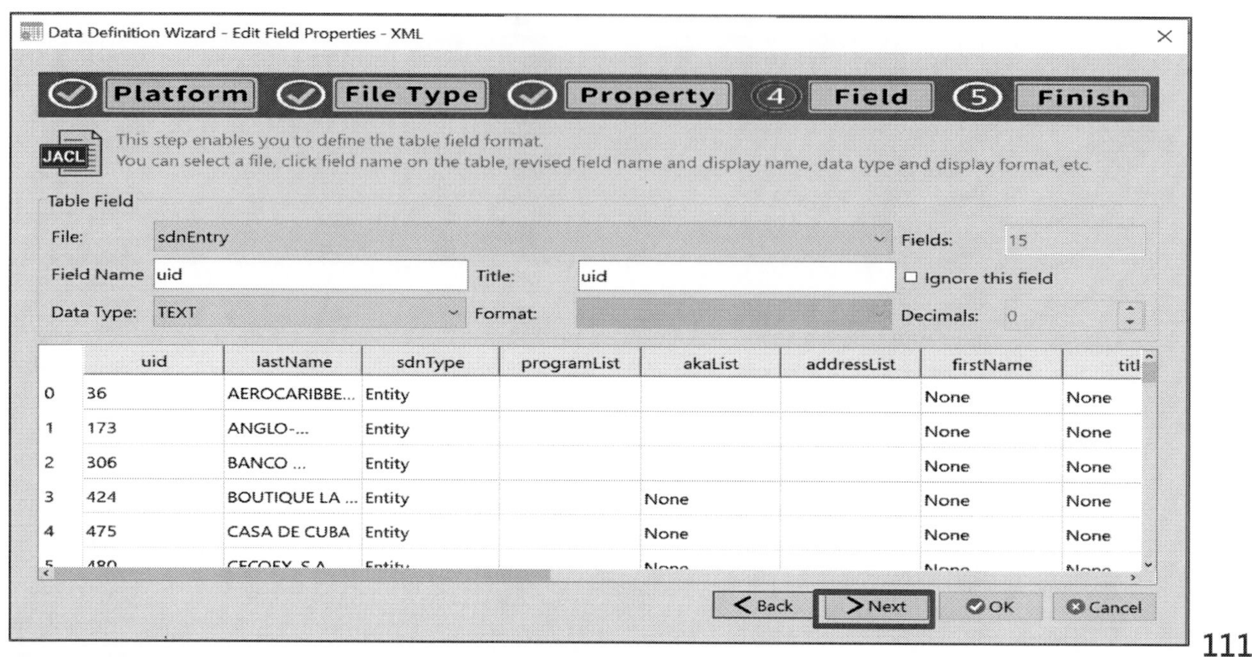

Exercise 4.11
Please practice importing XML into a JCAATs project.

- The default data file path is the project folder
- After ensuring that the path and information are correct, select "Done".

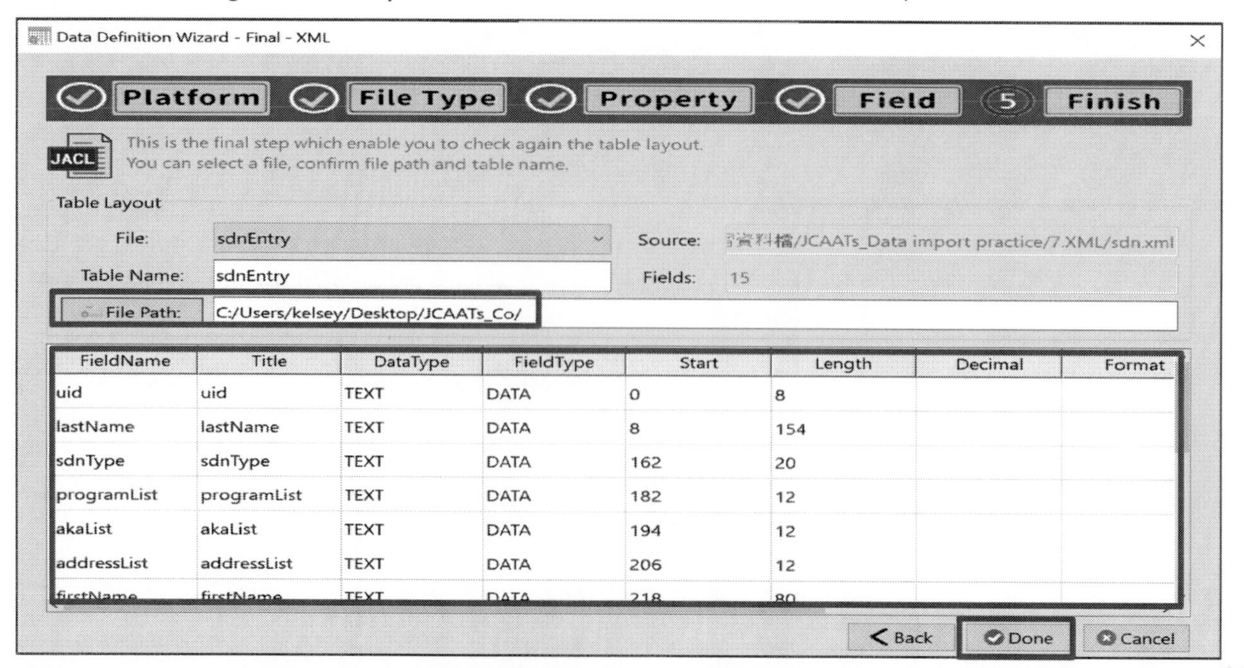

Exercise 4.11
Please practice importing XML into a JCAATs project.

- After the import progress is completed, we can see that the data table has been successfully imported.

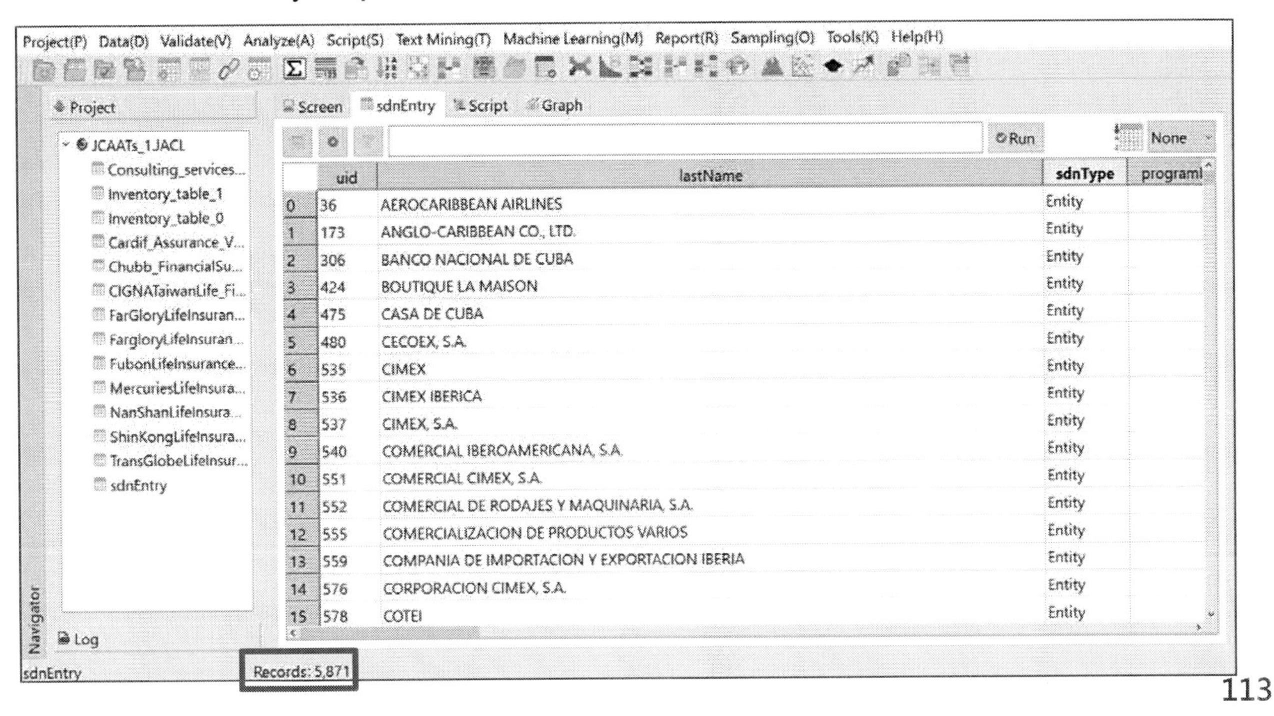

Importing OPEN DATA
–US Treasury Department's SDN sanctions list.

Exercise 4.12
Importing OPEN DATA - US Treasury Department's SDN sanctions list.

Importing OPEN DATA - OFAC SDN

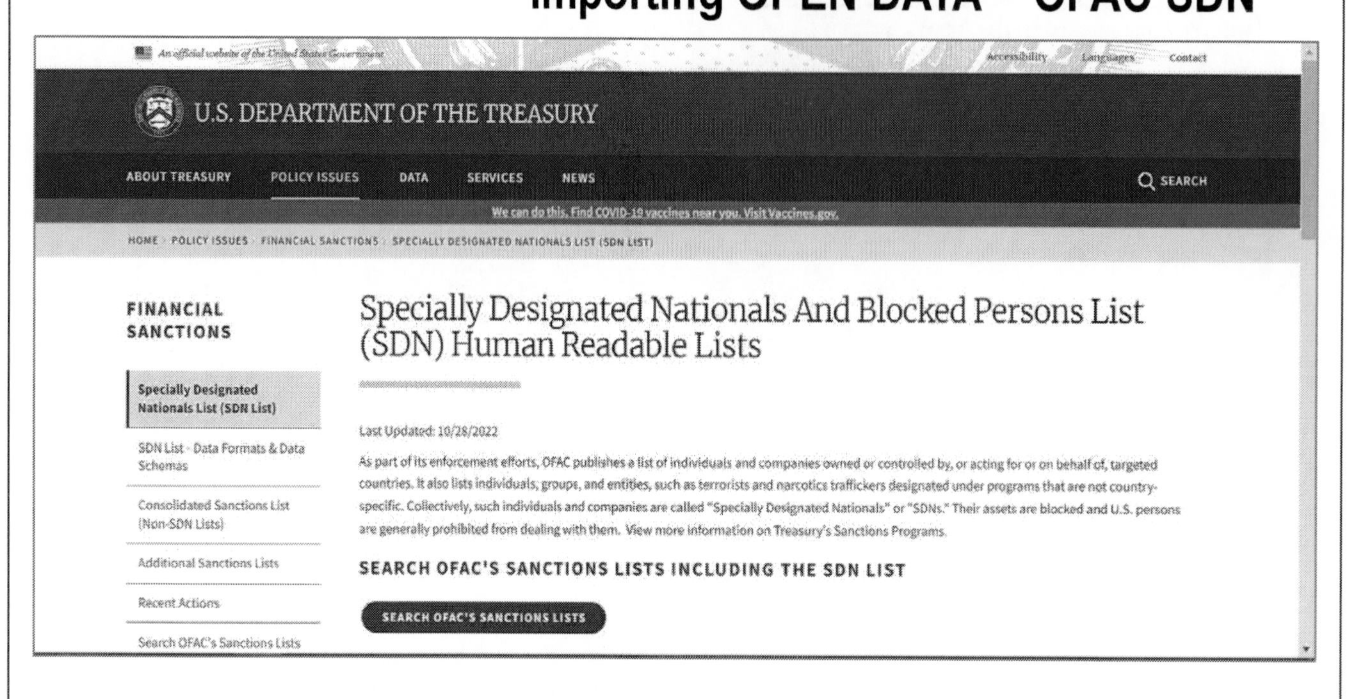

Link : https://home.treasury.gov/policy-issues/financial-sanctions/specially-designated-nationals-list-data-formats-data-schemas

Copy Link

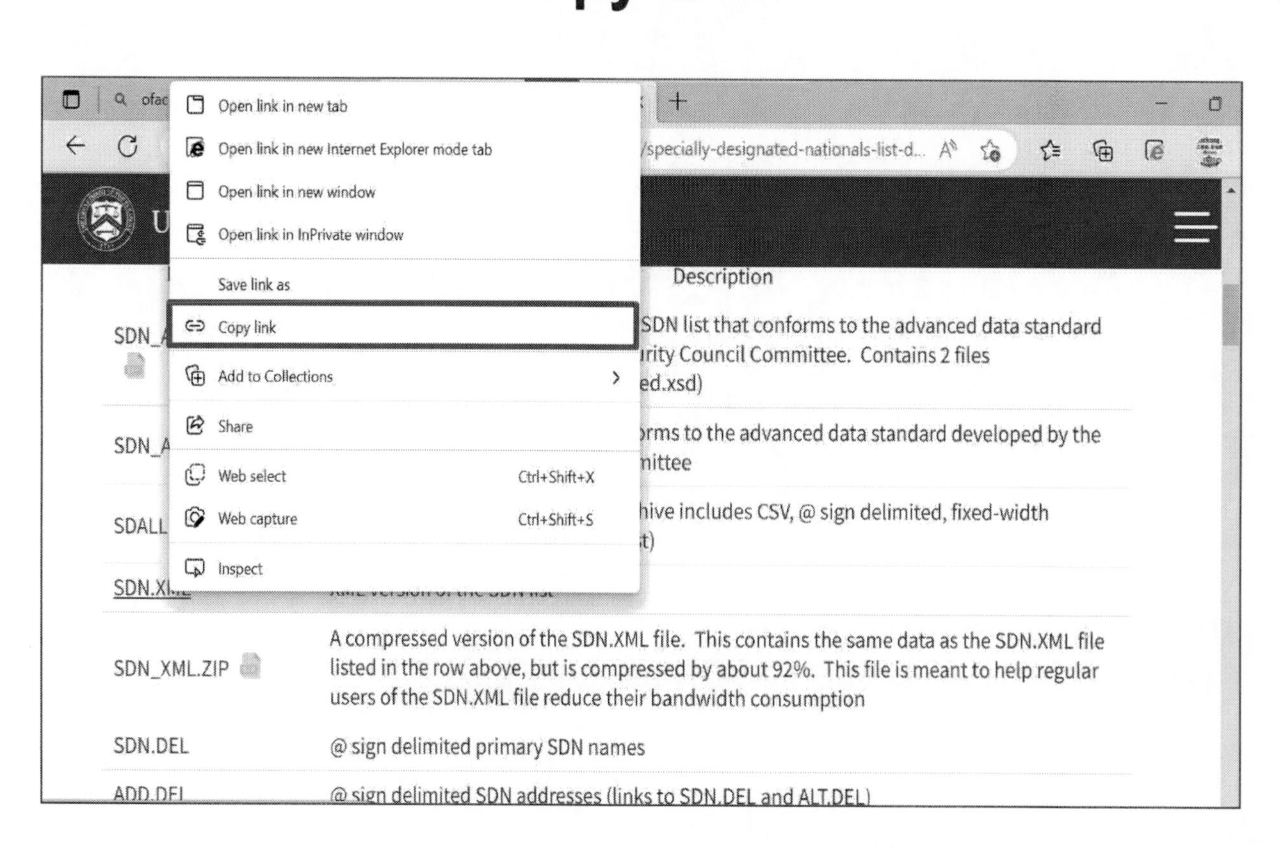

Data Definition Wizard —OPEN DATA Link

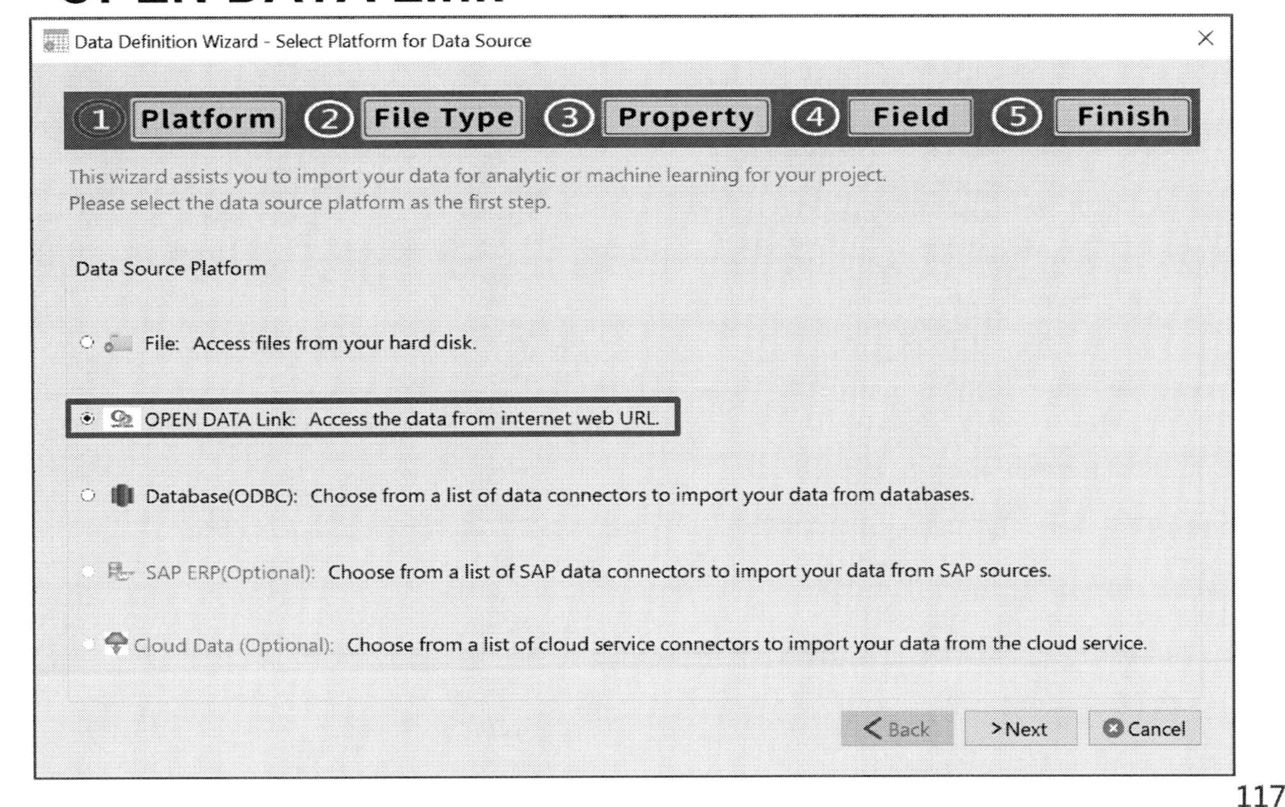

Paste the URL of the announcement data and select the file type

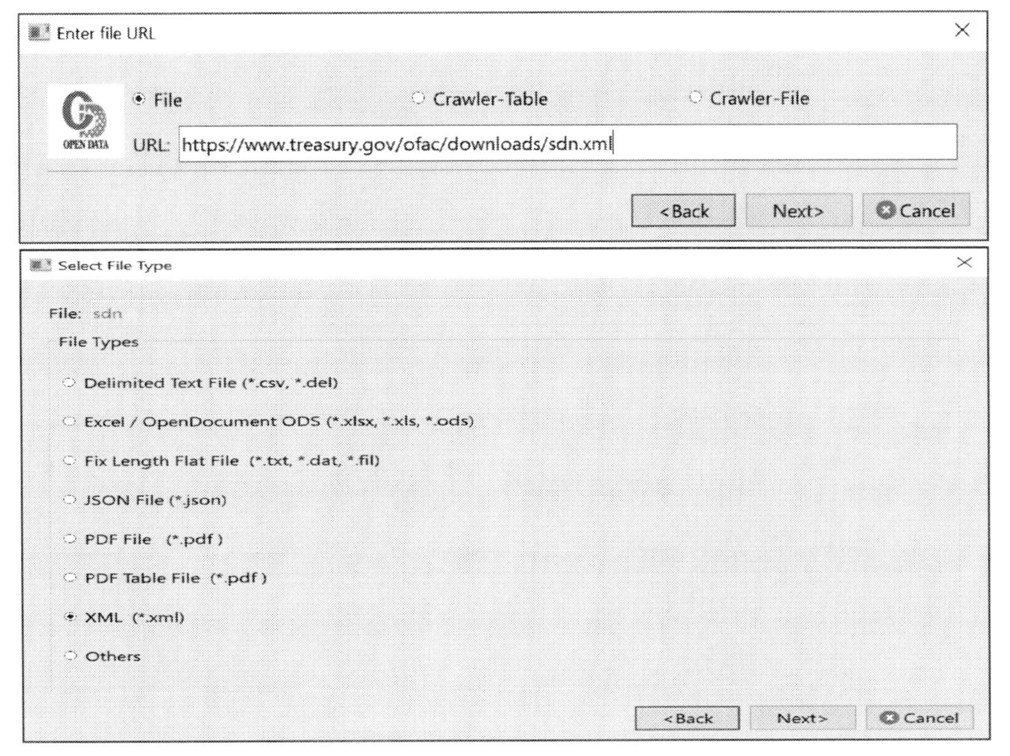

Link：https://www.treasury.gov/ofac/downloads/sdn.xml

Select Import Table—SDN Entry

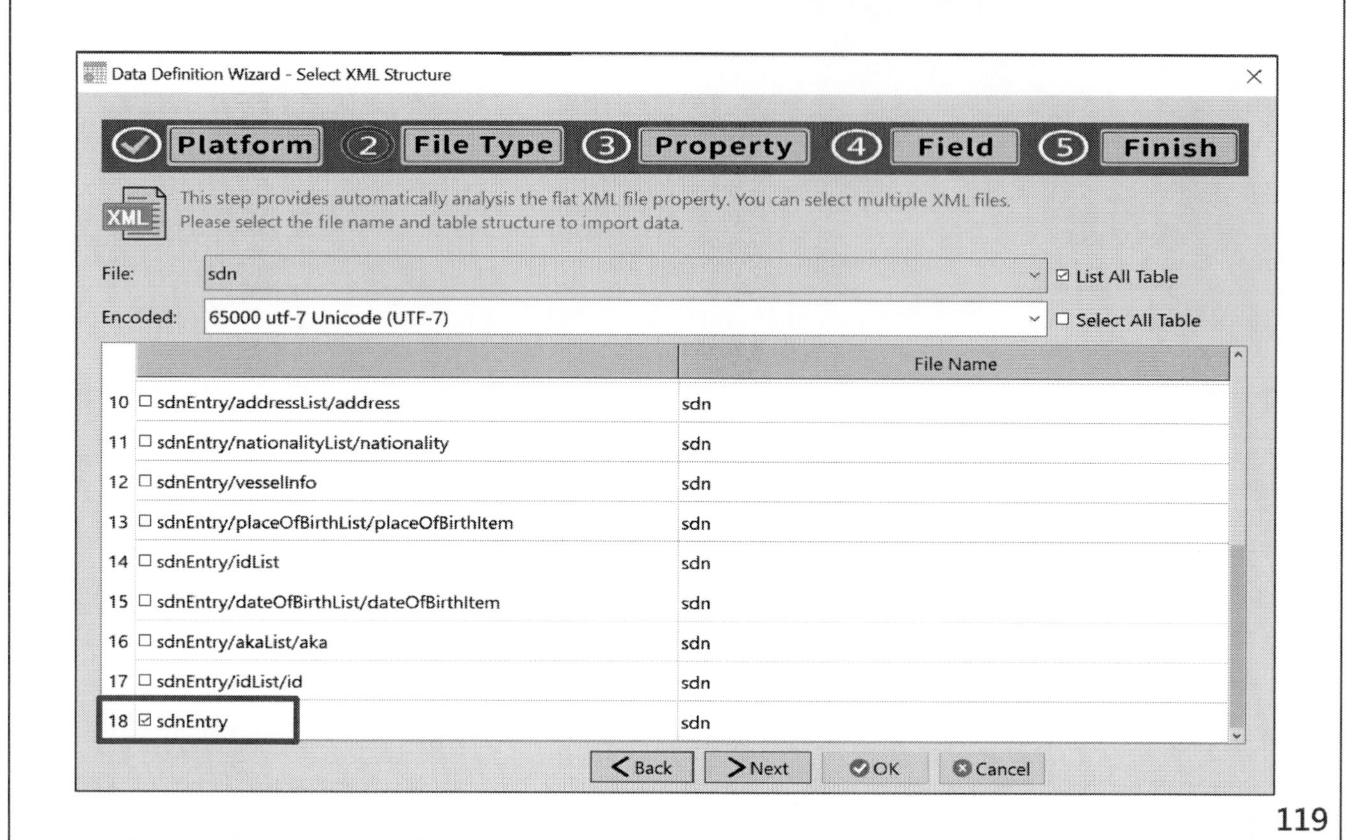

Follow the Data Definition Wizard instructions to complete the process in sequence.

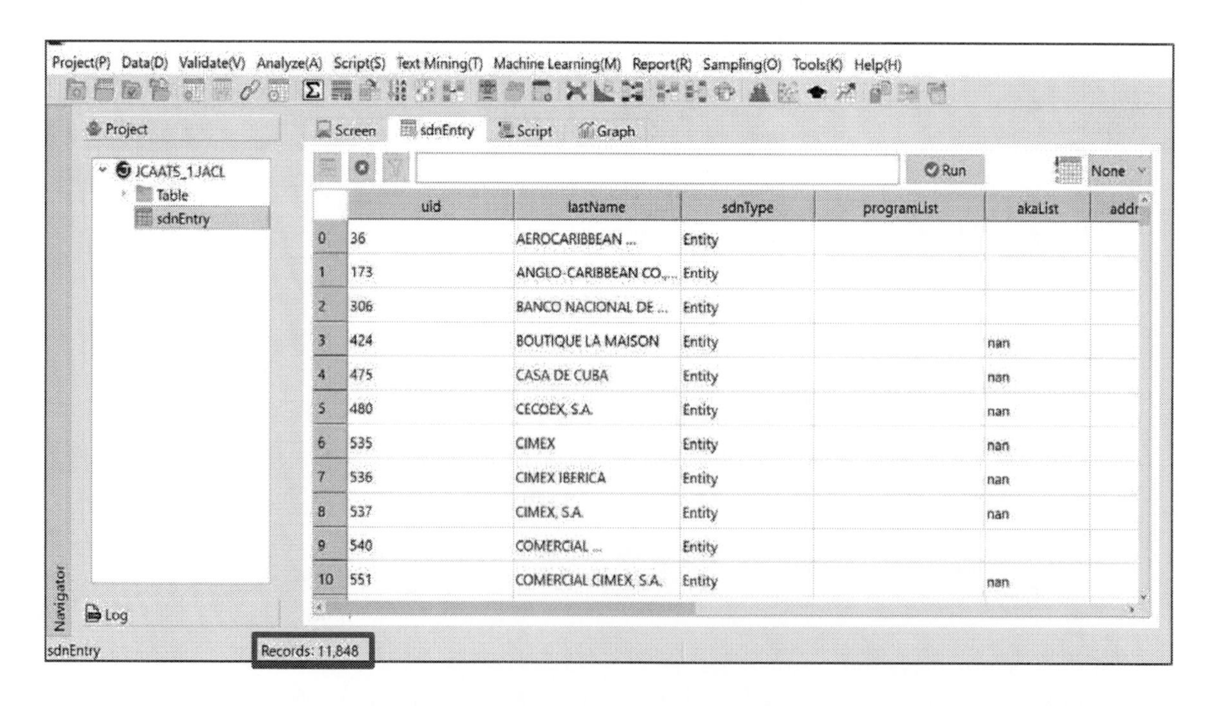

The records above will be different based on time changing.

JCAATs - AI Audit Software

Importing OPEN DATA
– UNITED NATIONS SECURITY COUNCIL Sanctions List

Exercise 4.13
Importing OPEN DATA-UNITED NATIONS SECURITY COUNCIL Announcement of Sanctions List

Importing OPEN DATA — Announcement of Sanctions List

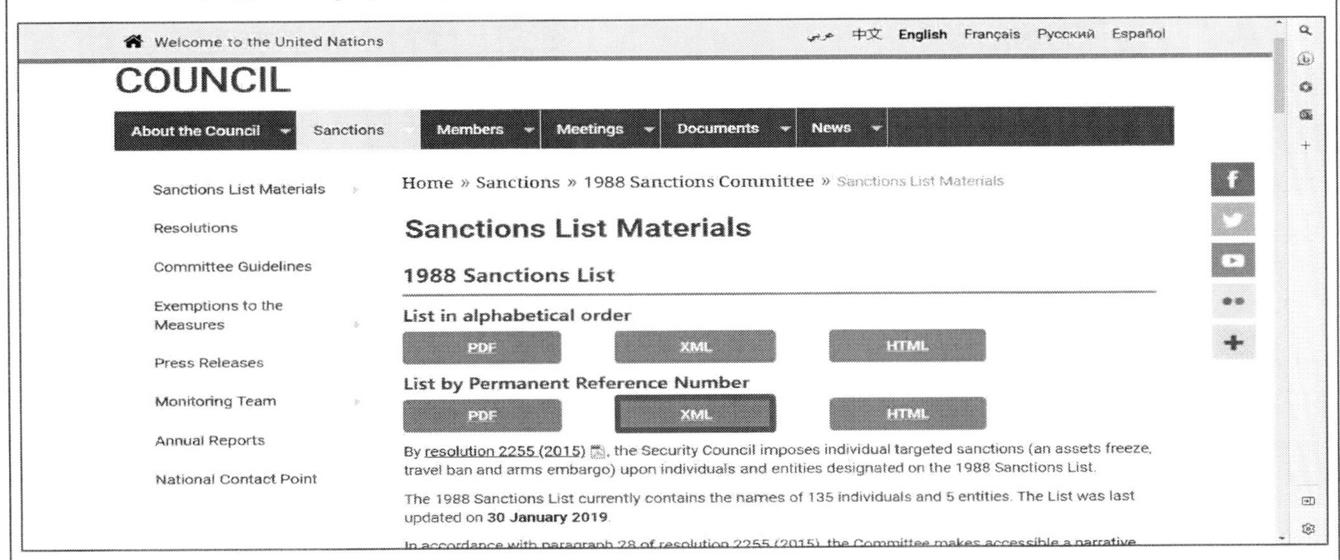

https://www.un.org/securitycouncil/sanctions/1988/materials

Import Results

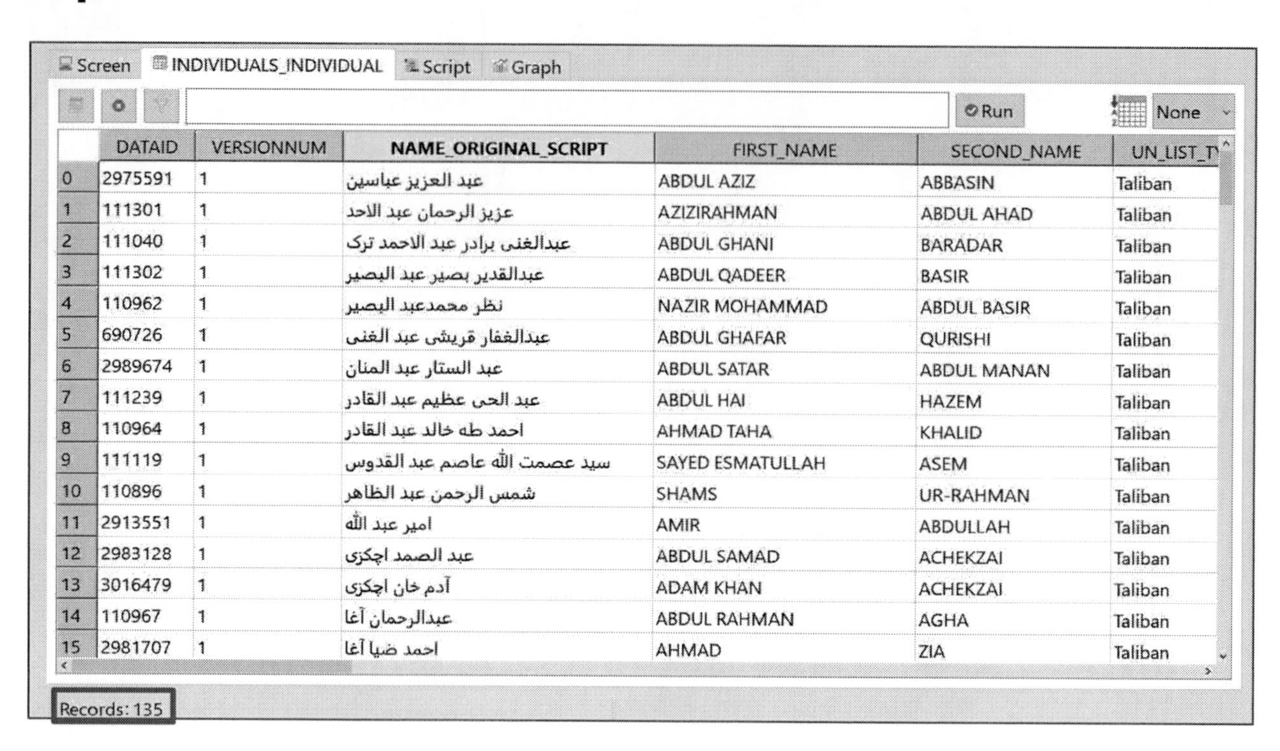

	DATAID	VERSIONNUM	NAME_ORIGINAL_SCRIPT	FIRST_NAME	SECOND_NAME	UN_LIST_TY
0	2975591	1	عبد العزيز عباسين	ABDUL AZIZ	ABBASIN	Taliban
1	111301	1	عزيز الرحمان عبد الاحد	AZIZIRAHMAN	ABDUL AHAD	Taliban
2	111040	1	عبدالغنی برادر عبد الاحمد ترک	ABDUL GHANI	BARADAR	Taliban
3	111302	1	عبدالقدير بصير عبد البصير	ABDUL QADEER	BASIR	Taliban
4	110962	1	نظر محمدعبد البصير	NAZIR MOHAMMAD	ABDUL BASIR	Taliban
5	690726	1	عبدالغفار قریشی عبد الغنی	ABDUL GHAFAR	QURISHI	Taliban
6	2989674	1	عبد الستار عبد المنان	ABDUL SATAR	ABDUL MANAN	Taliban
7	111239	1	عبد الحی عظیم عبد القادر	ABDUL HAI	HAZEM	Taliban
8	110964	1	احمد طه خالد عبد القادر	AHMAD TAHA	KHALID	Taliban
9	111119	1	سيد عصمت الله عاصم عبد القدوس	SAYED ESMATULLAH	ASEM	Taliban
10	110896	1	شمس الرحمن عبد الظاهر	SHAMS	UR-RAHMAN	Taliban
11	2913551	1	امير عبد الله	AMIR	ABDULLAH	Taliban
12	2983128	1	عبد الصمد اچکزی	ABDUL SAMAD	ACHEKZAI	Taliban
13	3016479	1	آدم خان اچکزی	ADAM KHAN	ACHEKZAI	Taliban
14	110967	1	عبدالرحمان آغا	ABDUL RAHMAN	AGHA	Taliban
15	2981707	1	احمد ضيا آغا	AHMAD	ZIA	Taliban

Records: 135

JCAATs Learning Note:

JCAATs - AI Audit Software

Chapter 5 - Exercise

Chapter 5 - Exercise

- Exercise 5.1: Please open the Corp_Credit_Cards table in the project file.
- Exercise 5.2: Use Quick Sort: Ascending to sort the CARDNUM column from smallest to largest.
- Exercise 5.3: Use Quick Filter to filter the EXPDT column by date 20050801.
- Exercise 5.4: Use Filter to set the condition as NEWBAL = 0.
- Exercise 5.5: Use Filter to set the condition as NEWBAL > CREDLIM.

JCAATs - AI Audit Software

Open the "Corp_Credit_Cards" table in the project file.

Exercise 5.1
Please open the "Corp_Credit_Cards" table in the project file.

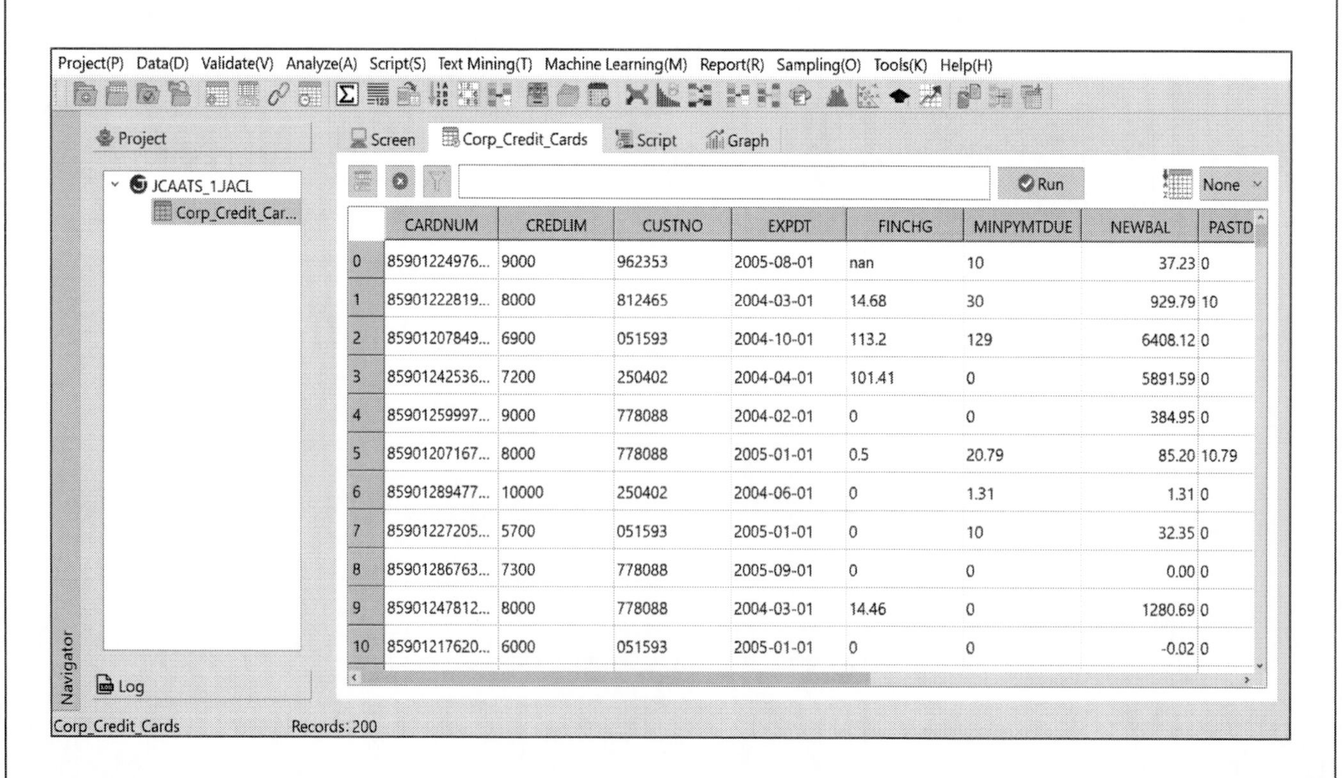

	CARDNUM	CREDLIM	CUSTNO	EXPDT	FINCHG	MINPYMTDUE	NEWBAL	PASTD
0	85901224976...	9000	962353	2005-08-01	nan	10	37.23	0
1	85901222819...	8000	812465	2004-03-01	14.68	30	929.79	10
2	85901207849...	6900	051593	2004-10-01	113.2	129	6408.12	0
3	85901242536...	7200	250402	2004-04-01	101.41	0	5891.59	0
4	85901259997...	9000	778088	2004-02-01	0	0	384.95	0
5	85901207167...	8000	778088	2005-01-01	0.5	20.79	85.20	10.79
6	85901289477...	10000	250402	2004-06-01	0	1.31	1.31	0
7	85901227205...	5700	051593	2005-01-01	0	10	32.35	0
8	85901286763...	7300	778088	2005-09-01	0	0	0.00	0
9	85901247812...	8000	778088	2004-03-01	14.46	0	1280.69	0
10	85901217620...	6000	051593	2005-01-01	0	0	-0.02	0

Use Quick Sort: Asending
Sort the "CARDNUM" column in ascending order.

Exercise 5.2
Use Quick Sort: Asending Sort the "CARDNUM" column in ascending order.

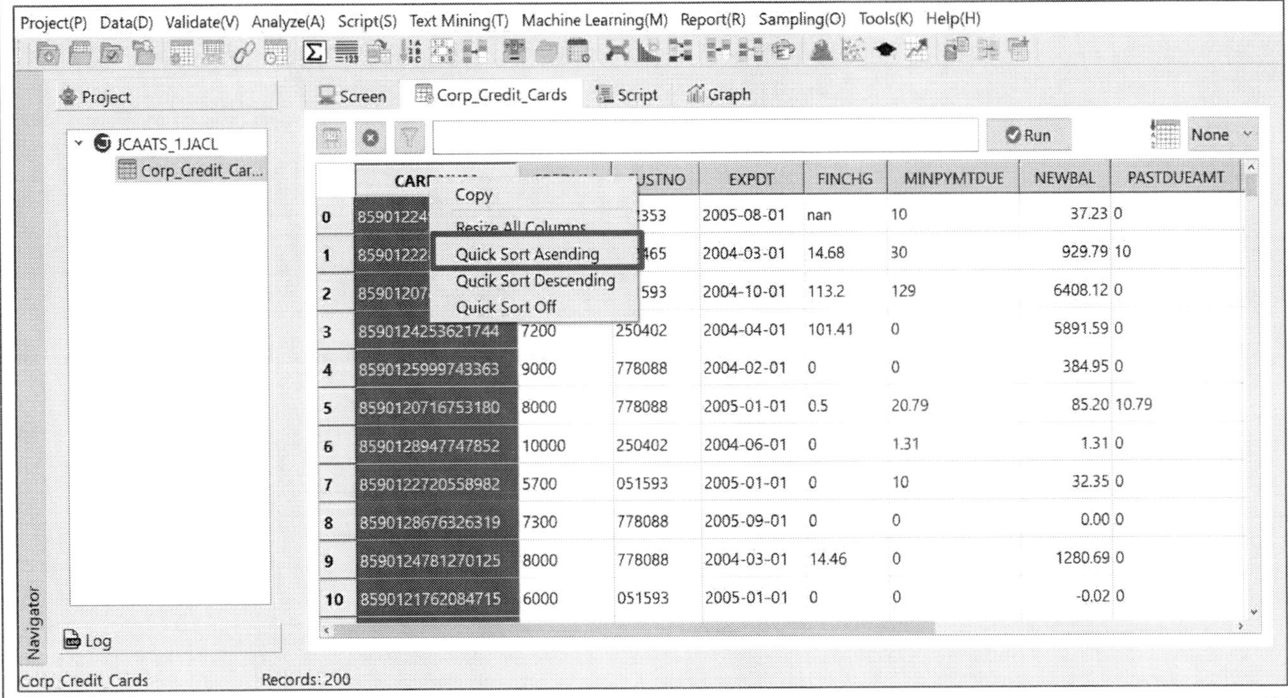

Exercise 5.2

Use Quick Sort: Asending Sort the "CARDNUM" column in ascending order.

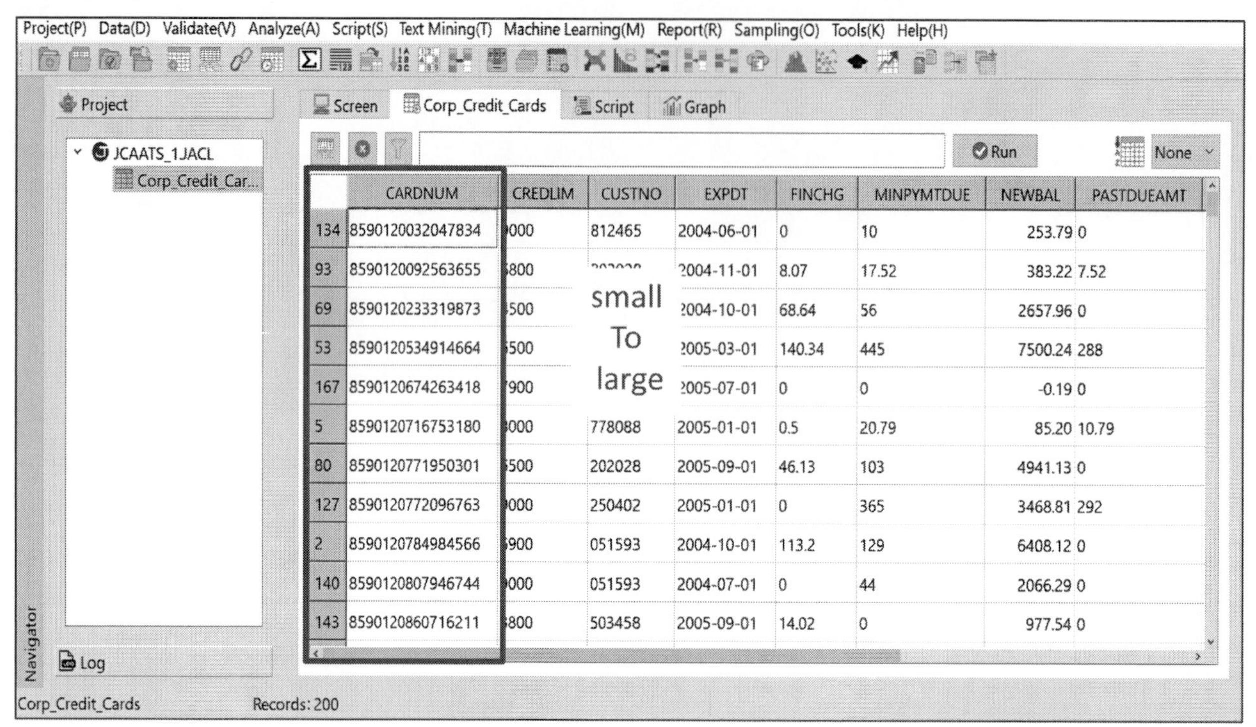

	CARDNUM	CREDLIM	CUSTNO	EXPDT	FINCHG	MINPYMTDUE	NEWBAL	PASTDUEAMT
134	8590120032047834	000	812465	2004-06-01	0	10	253.79	0
93	8590120092563655	800		2004-11-01	8.07	17.52	383.22	7.52
69	8590120233319873	500		2004-10-01	68.64	56	2657.96	0
53	8590120534914664	500		2005-03-01	140.34	445	7500.24	288
167	8590120674263418	900		2005-07-01	0	0	-0.19	0
5	8590120716753180	000	778088	2005-01-01	0.5	20.79	85.20	10.79
80	8590120771950301	500	202028	2005-09-01	46.13	103	4941.13	0
127	8590120772096763	000	250402	2005-01-01	0	365	3468.81	292
2	8590120784984566	900	051593	2004-10-01	113.2	129	6408.12	0
140	8590120807946744	000	051593	2004-07-01	0	44	2066.29	0
143	8590120860716211	800	503458	2005-09-01	14.02	0	977.54	0

JCAATs - AI Audit Software

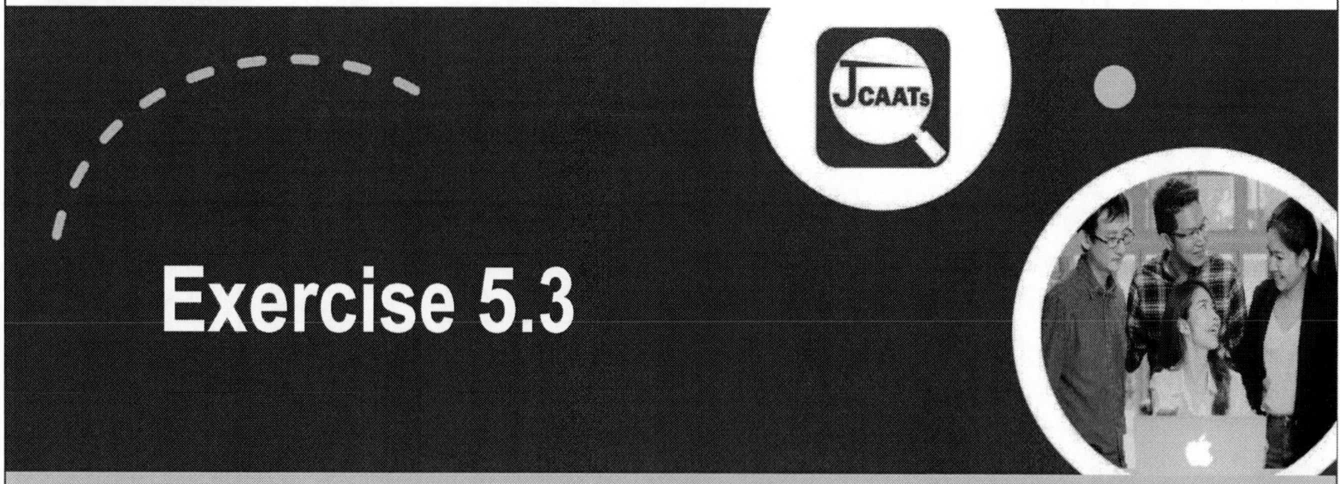

Use Quick Filer to filter the "EXPDT" column by the date "20050801"

Exercise 5.3
Right-click and use Quick Filter to filter the "EXPDT" column by the date "20050801"

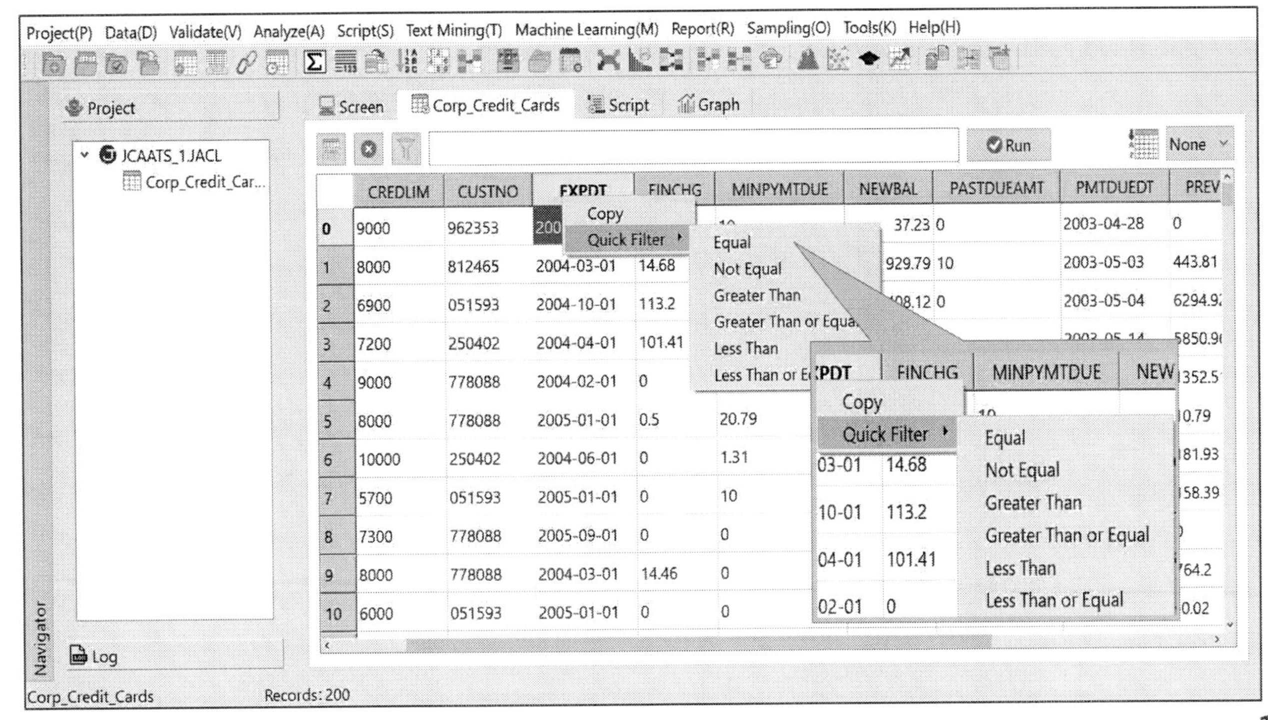

Exercise 5.3
Right-click and use Quick Filter to filter the "EXPDT" column by the date "20050801"

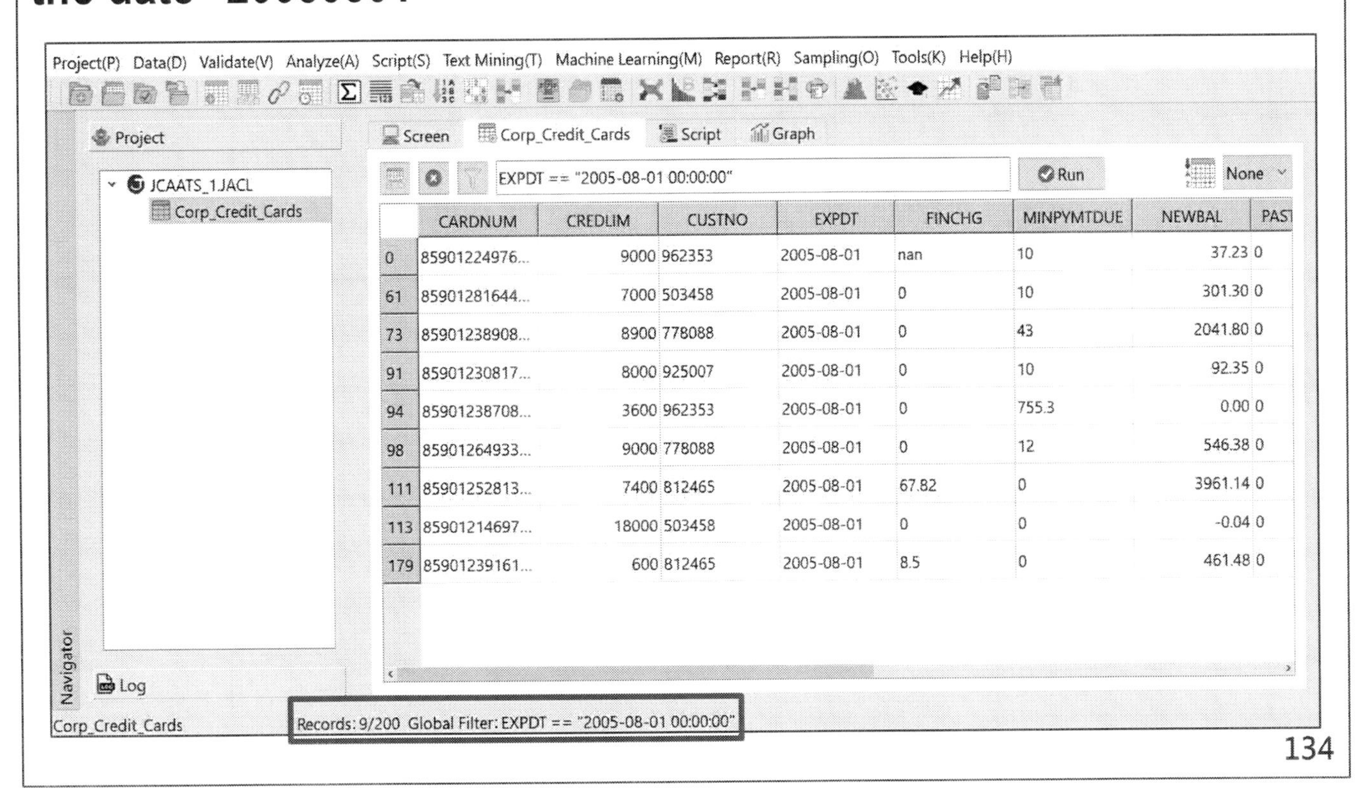

Use the Filter to set the condition as NEWBAL = 0

Exercise 5.4
Set the condition as NEWBAL = 0 and perform a syntax check.

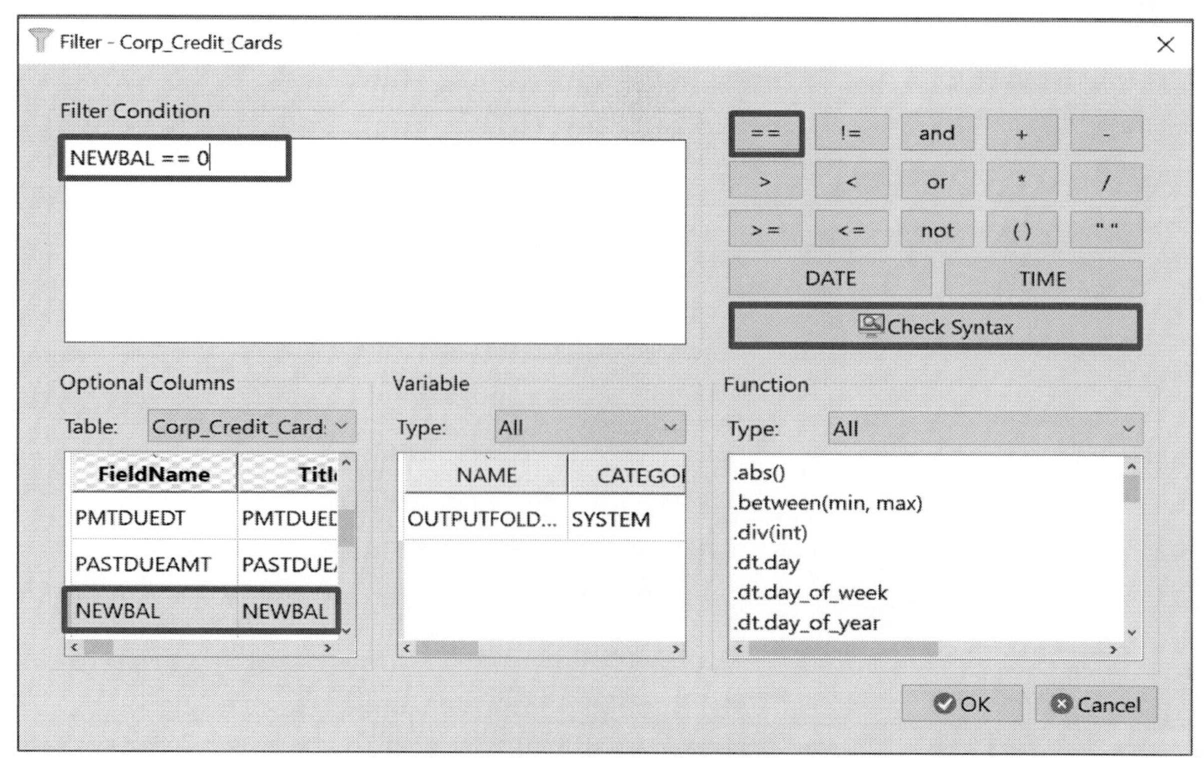

Exercise 5.4

After completing the filtering, there are a total of 16 records with a balance of 0.

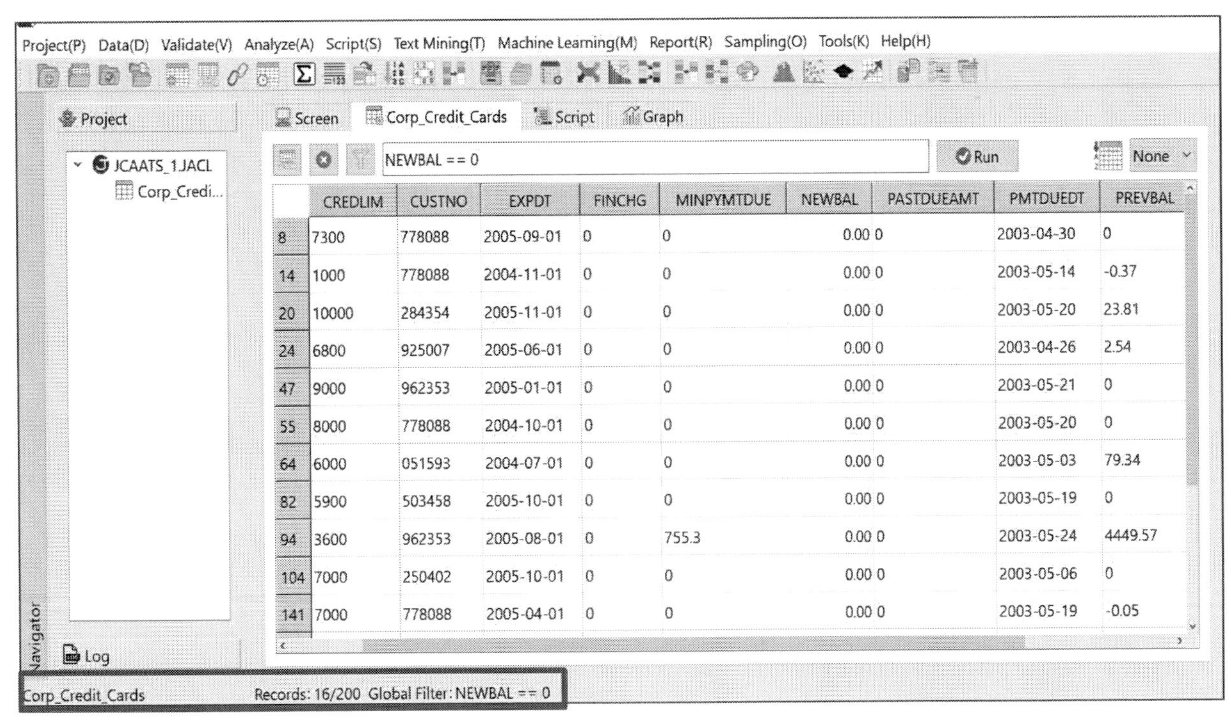

	CREDLIM	CUSTNO	EXPDT	FINCHG	MINPYMTDUE	NEWBAL	PASTDUEAMT	PMTDUEDT	PREVBAL
8	7300	778088	2005-09-01	0	0	0.00	0	2003-04-30	0
14	1000	778088	2004-11-01	0	0	0.00	0	2003-05-14	-0.37
20	10000	284354	2005-11-01	0	0	0.00	0	2003-05-20	23.81
24	6800	925007	2005-06-01	0	0	0.00	0	2003-04-26	2.54
47	9000	962353	2005-01-01	0	0	0.00	0	2003-05-21	0
55	8000	778088	2004-10-01	0	0	0.00	0	2003-05-20	0
64	6000	051593	2004-07-01	0	0	0.00	0	2003-05-03	79.34
82	5900	503458	2005-10-01	0	0	0.00	0	2003-05-19	0
94	3600	962353	2005-08-01	0	755.3	0.00	0	2003-05-24	4449.57
104	7000	250402	2005-10-01	0	0	0.00	0	2003-05-06	0
141	7000	778088	2005-04-01	0	0	0.00	0	2003-05-19	-0.05

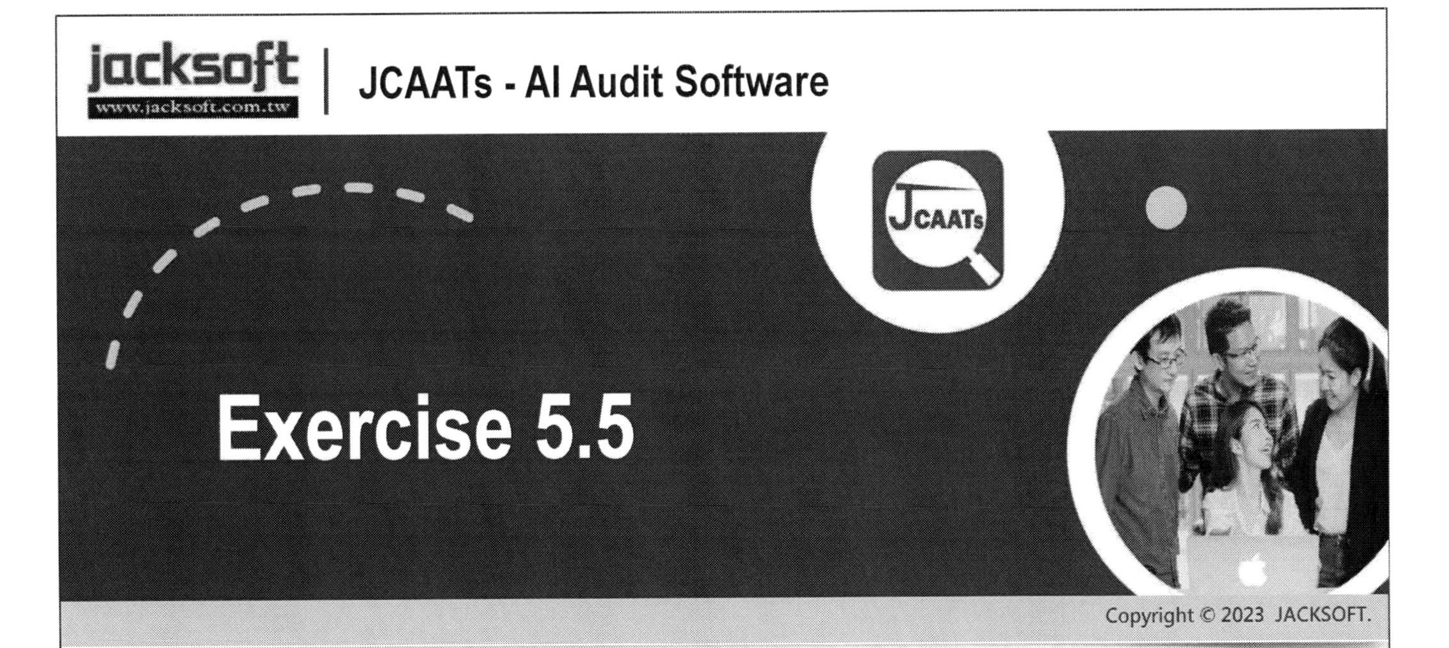

Use the Filter to set the condition as NEWBAL > CREDLIM.

Exercise 5.5

Set the condition as NEWBAL > CREDLIM and perform a syntax check.

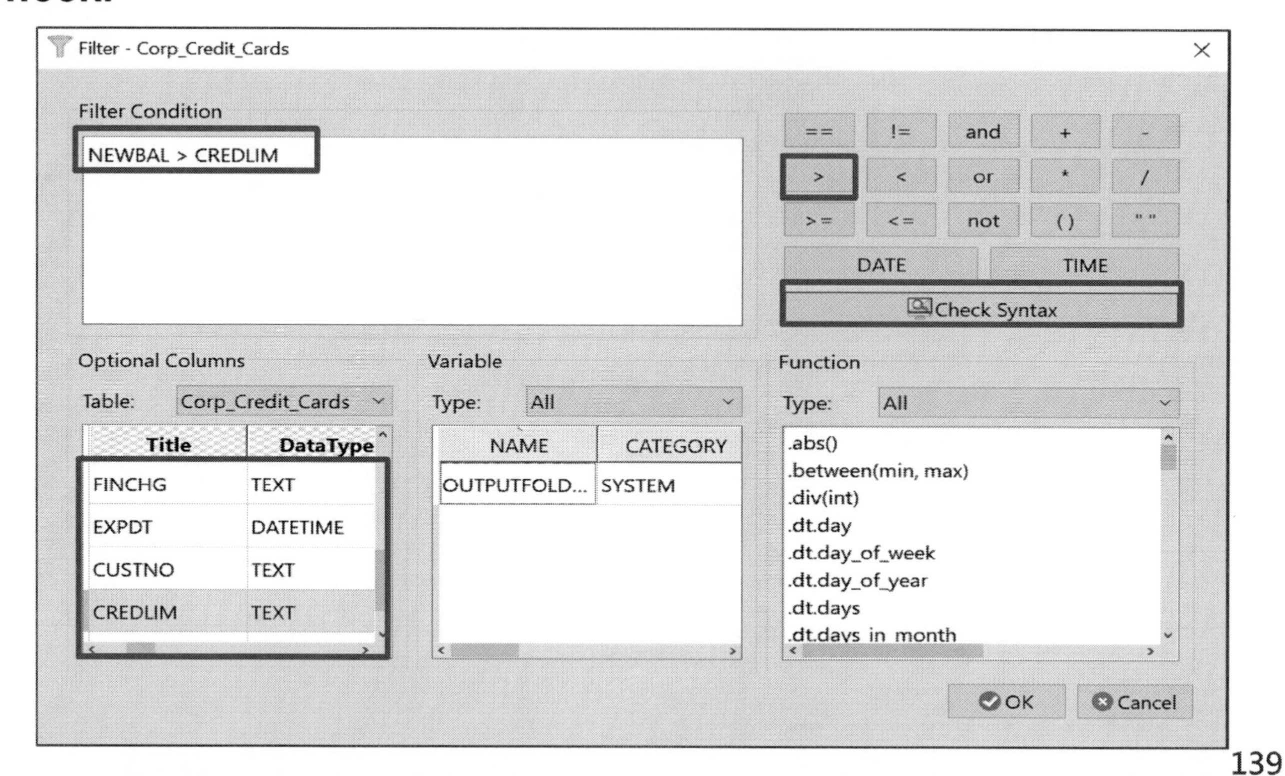

Exercise 5.5

After completing the filtering, there are a total of 10 records with a balance greater than the credit limit.

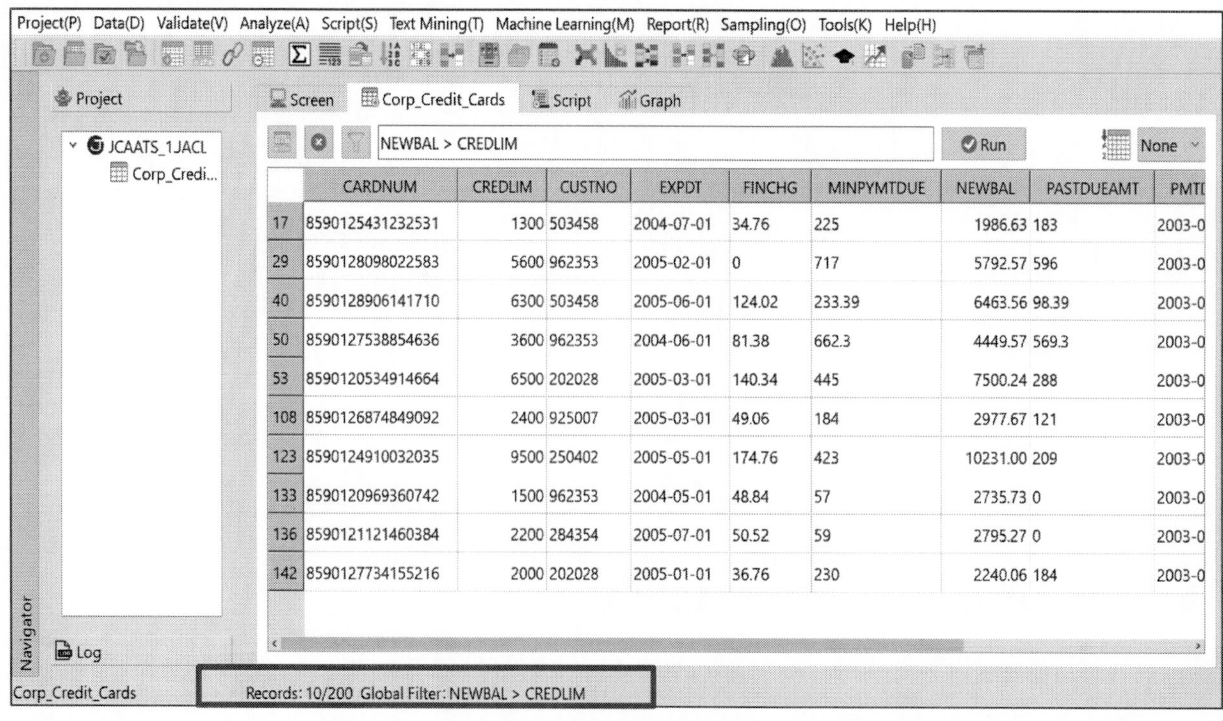

	CARDNUM	CREDLIM	CUSTNO	EXPDT	FINCHG	MINPYMTDUE	NEWBAL	PASTDUEAMT	PMT
17	8590125431232531	1300	503458	2004-07-01	34.76	225	1986.63	183	2003-0
29	8590128098022583	5600	962353	2005-02-01	0	717	5792.57	596	2003-0
40	8590128906141710	6300	503458	2005-06-01	124.02	233.39	6463.56	98.39	2003-0
50	8590127538854636	3600	962353	2004-06-01	81.38	662.3	4449.57	569.3	2003-0
53	8590120534914664	6500	202028	2005-03-01	140.34	445	7500.24	288	2003-0
108	8590126874849092	2400	925007	2005-03-01	49.06	184	2977.67	121	2003-0
123	8590124910032035	9500	250402	2005-05-01	174.76	423	10231.00	209	2003-0
133	8590120969360742	1500	962353	2004-05-01	48.84	57	2735.73	0	2003-0
136	8590121121460384	2200	284354	2005-07-01	50.52	59	2795.27	0	2003-0
142	8590127734155216	2000	202028	2005-01-01	36.76	230	2240.06	184	2003-0

Chapter 5 - Exercise

- Exercise 5.6: Please open the table structure (Table Layout) of the 'Corp_Credit_Cards' data table in the project.
- Exercise 5.7: Add a formula field (Computed field) to perform date difference calculations.
- Exercise 5.8: Please check if the formula field calculation result is displayed correctly in the viewing area.

JCAATs - AI Audit Software

Please open the Table Layout of the "Corp_Credit_Cards" table in the project.

Exercise 5.6

Select "Data" from the menu→ Table Layout

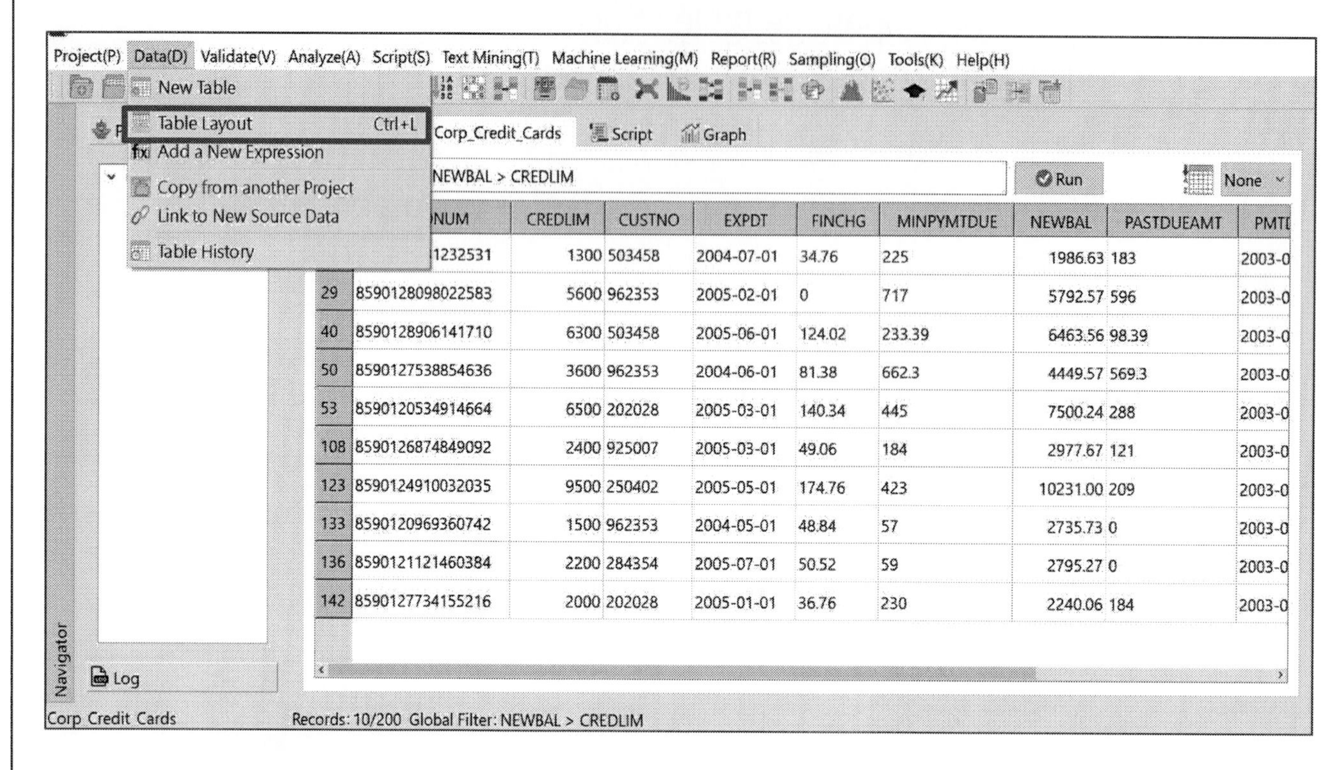

	CARDNUM	CREDLIM	CUSTNO	EXPDT	FINCHG	MINPYMTDUE	NEWBAL	PASTDUEAMT	PMT
	1232531	1300	503458	2004-07-01	34.76	225	1986.63	183	2003-0
29	8590128098022583	5600	962353	2005-02-01	0	717	5792.57	596	2003-0
40	8590128906141710	6300	503458	2005-06-01	124.02	233.39	6463.56	98.39	2003-0
50	8590127538854636	3600	962353	2004-06-01	81.38	662.3	4449.57	569.3	2003-0
53	8590120534914664	6500	202028	2005-03-01	140.34	445	7500.24	288	2003-0
108	8590126874849092	2400	925007	2005-03-01	49.06	184	2977.67	121	2003-0
123	8590124910032035	9500	250402	2005-05-01	174.76	423	10231.00	209	2003-0
133	8590120969360742	1500	962353	2004-05-01	48.84	57	2735.73	0	2003-0
136	8590121121460384	2200	284354	2005-07-01	50.52	59	2795.27	0	2003-0
142	8590127734155216	2000	202028	2005-01-01	36.76	230	2240.06	184	2003-0

Exercise 5.6

Table Layout allows you to view detailed structural information of each column, and you can also edit, add and delete columns.

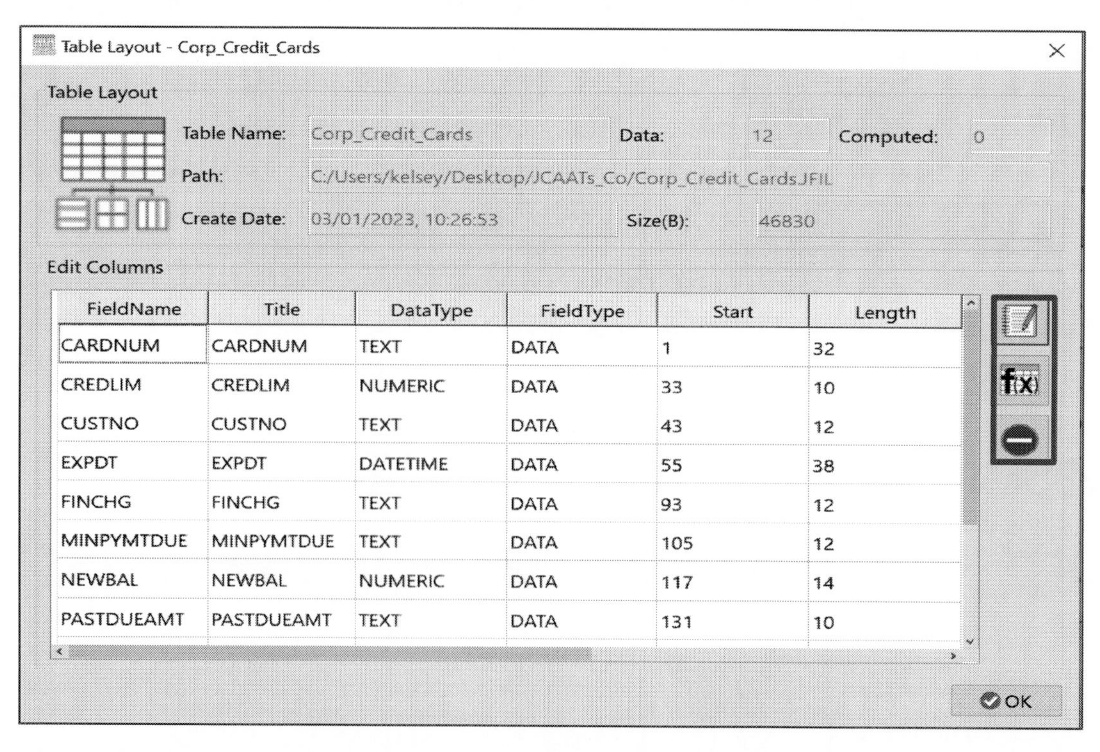

FieldName	Title	DataType	FieldType	Start	Length
CARDNUM	CARDNUM	TEXT	DATA	1	32
CREDLIM	CREDLIM	NUMERIC	DATA	33	10
CUSTNO	CUSTNO	TEXT	DATA	43	12
EXPDT	EXPDT	DATETIME	DATA	55	38
FINCHG	FINCHG	TEXT	DATA	93	12
MINPYMTDUE	MINPYMTDUE	TEXT	DATA	105	12
NEWBAL	NEWBAL	NUMERIC	DATA	117	14
PASTDUEAMT	PASTDUEAMT	TEXT	DATA	131	10

JCAATs - AI Audit Software

Add a formula column named "GAP_DAY" to perform the calculation of the difference in dates, which is PMTDUEDT - STMTDT.

Exercise 5.7

Click on F(X) to add a formula column named "GAP_DAY" , and the formula is PMTDUEDT - STMTDT.

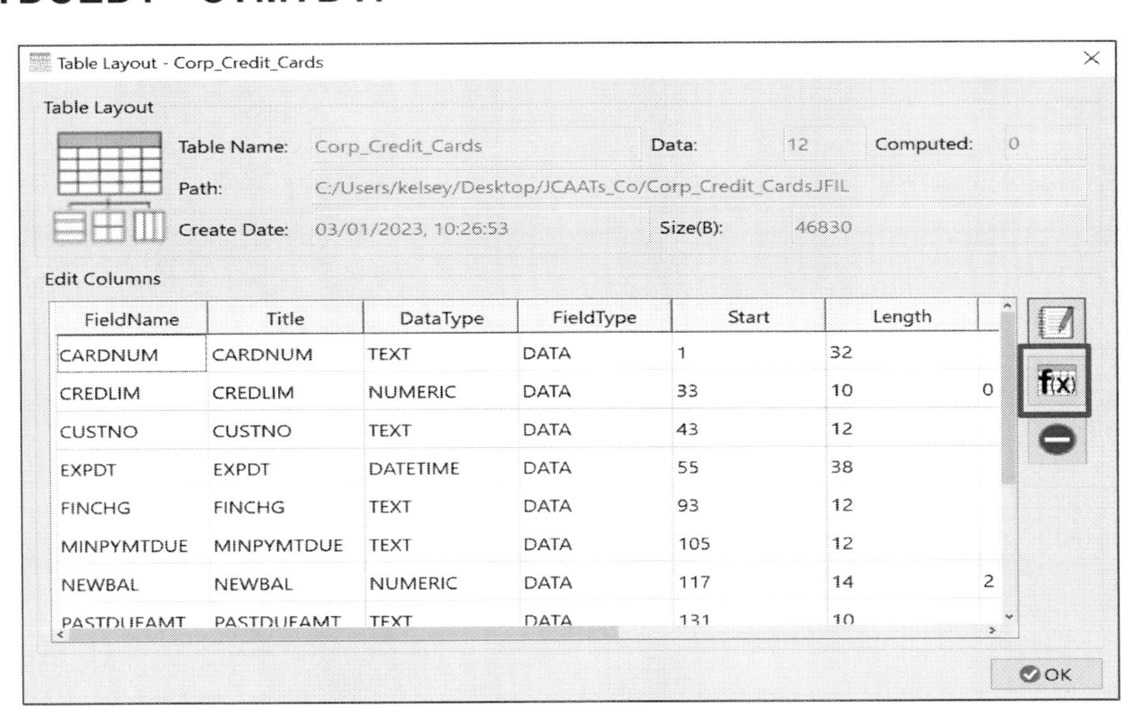

FieldName	Title	DataType	FieldType	Start	Length	
CARDNUM	CARDNUM	TEXT	DATA	1	32	
CREDLIM	CREDLIM	NUMERIC	DATA	33	10	0
CUSTNO	CUSTNO	TEXT	DATA	43	12	
EXPDT	EXPDT	DATETIME	DATA	55	38	
FINCHG	FINCHG	TEXT	DATA	93	12	
MINPYMTDUE	MINPYMTDUE	TEXT	DATA	105	12	
NEWBAL	NEWBAL	NUMERIC	DATA	117	14	2
PASTDUEAMT	PASTDUEAMT	TEXT	DATA	131	10	

Exercise 5.7

Click on F(X) to set the formula content for the initial value.

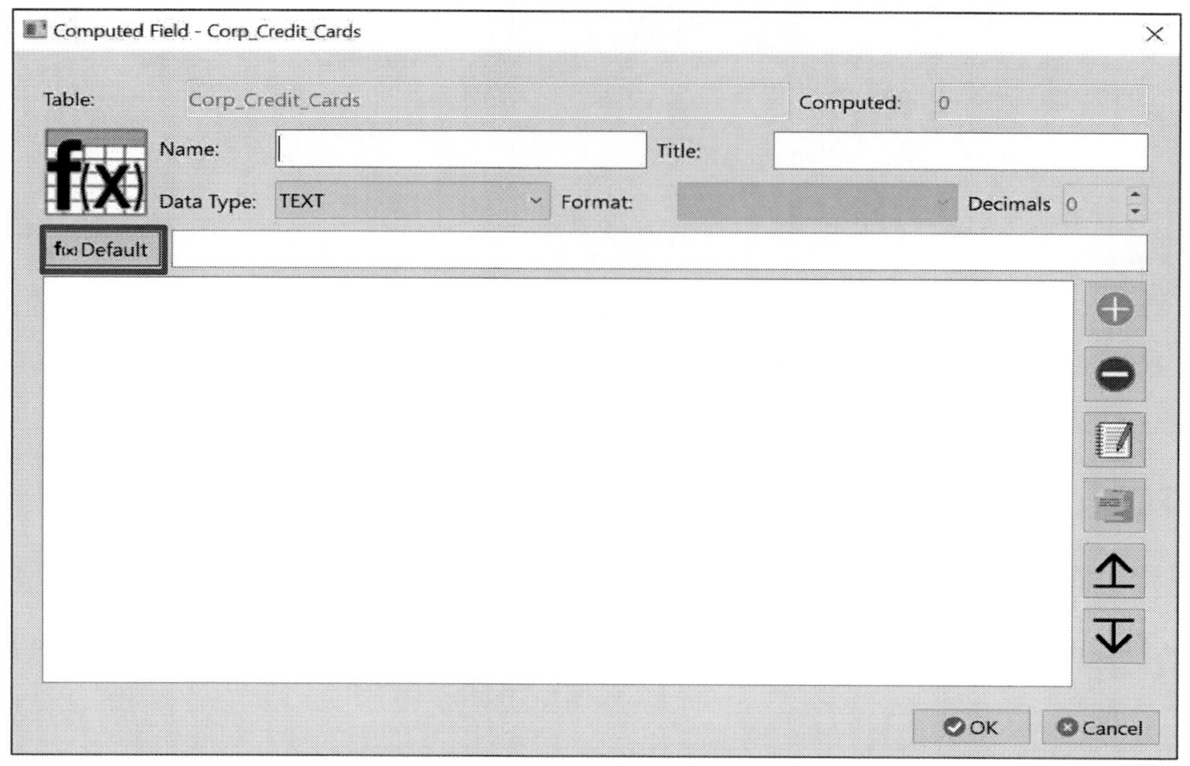

Exercise 5.7

Set the formula content as PMTDUEDT-STMTDT and use the conversion function .dt.days to convert the calculation result to days.

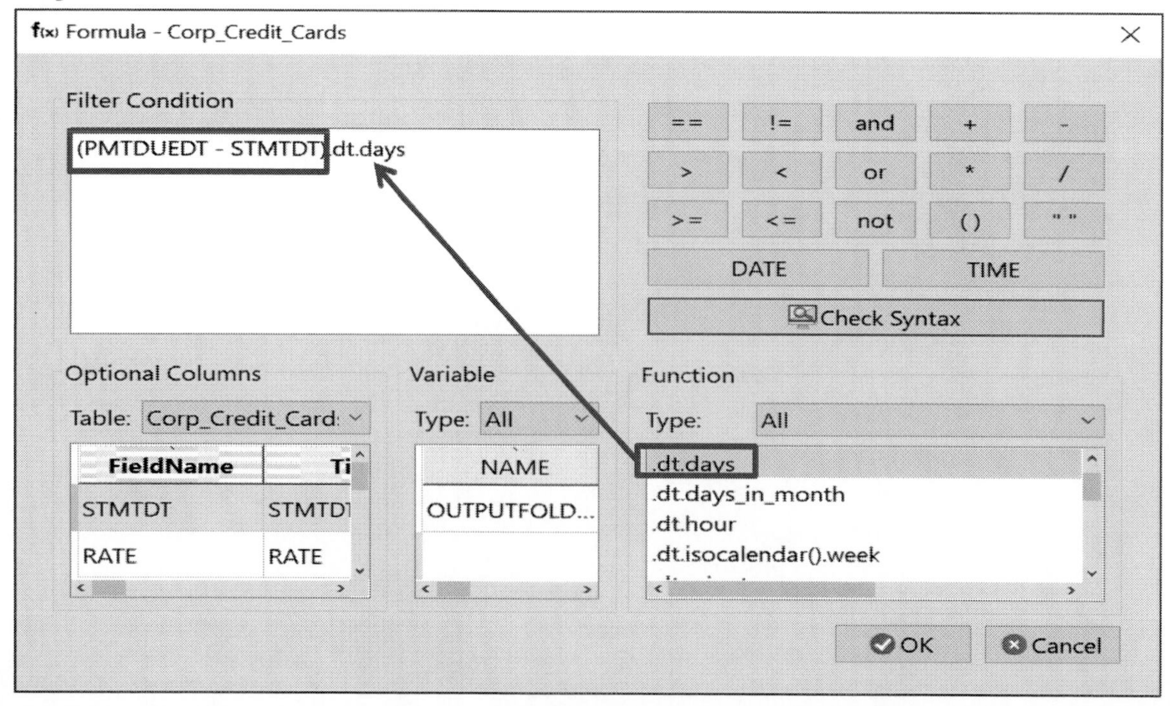

Exercise 5.7
After completing the relevant settings for the newly added formula column, click "OK".

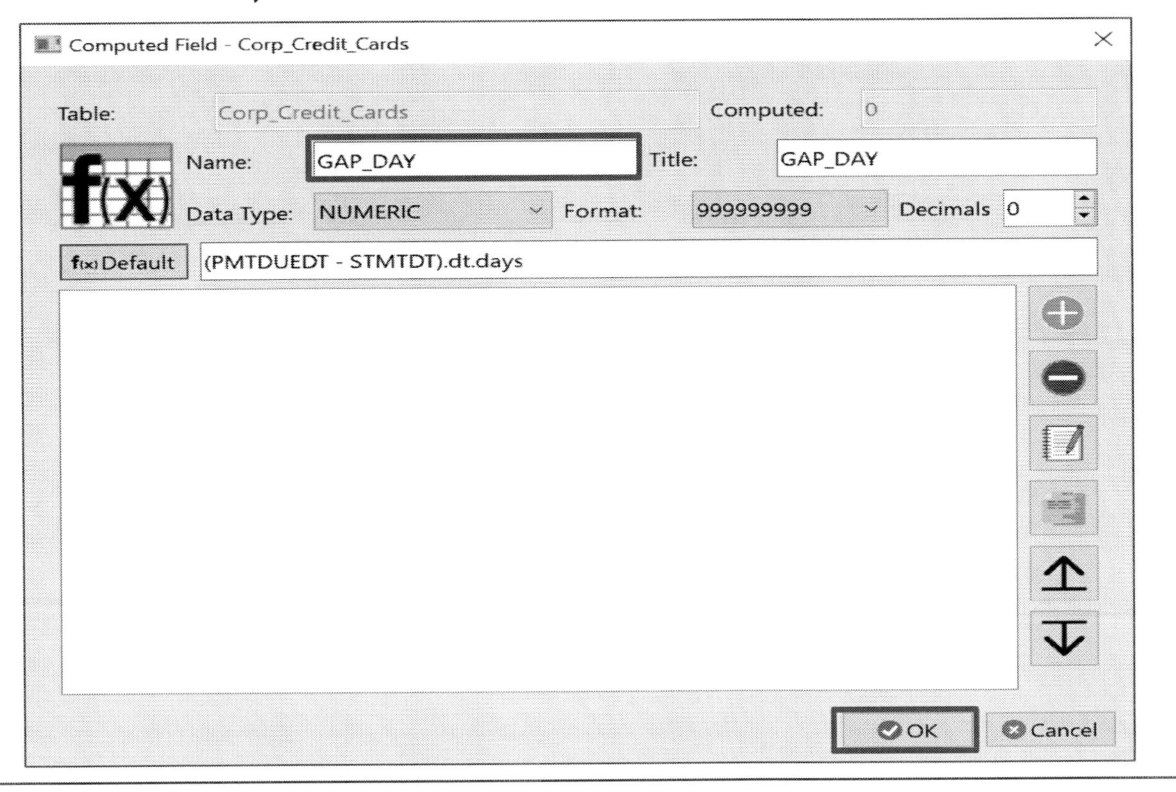

Exercise 5.7
The newly added formula column will now appear in the table structure, and its field type is "COMPUTED".

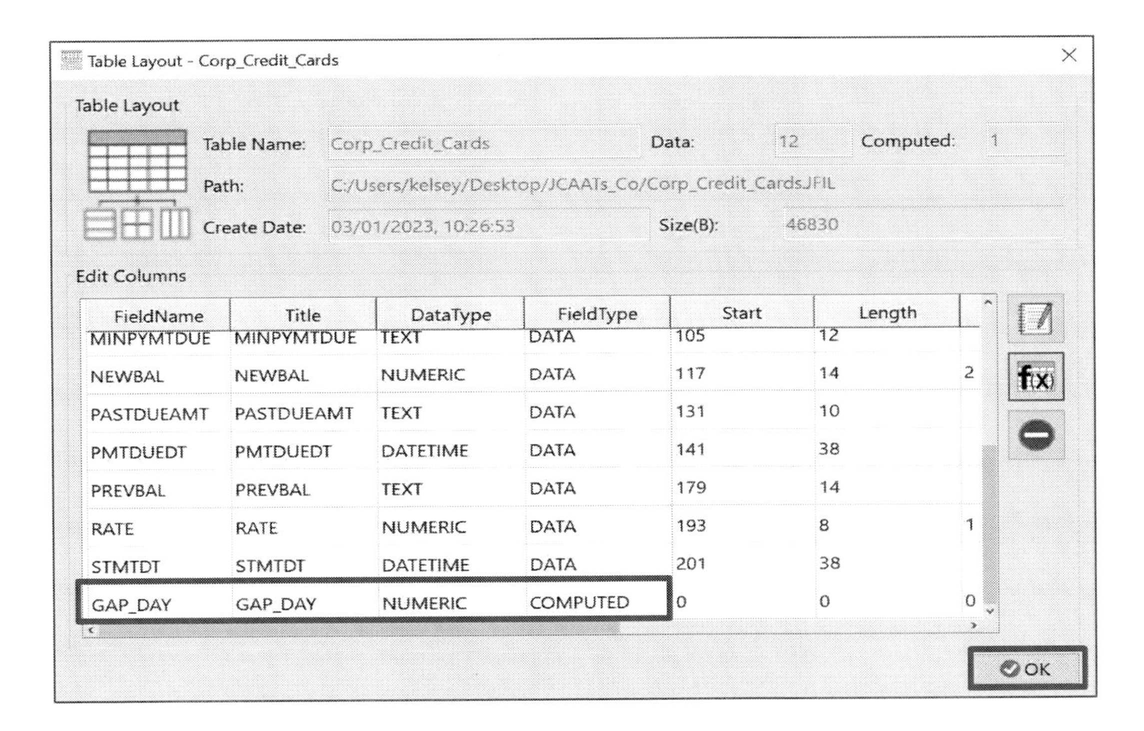

FieldName	Title	DataType	FieldType	Start	Length	
MINPYMTDUE	MINPYMTDUE	TEXT	DATA	105	12	
NEWBAL	NEWBAL	NUMERIC	DATA	117	14	2
PASTDUEAMT	PASTDUEAMT	TEXT	DATA	131	10	
PMTDUEDT	PMTDUEDT	DATETIME	DATA	141	38	
PREVBAL	PREVBAL	TEXT	DATA	179	14	
RATE	RATE	NUMERIC	DATA	193	8	1
STMTDT	STMTDT	DATETIME	DATA	201	38	
GAP_DAY	GAP_DAY	NUMERIC	COMPUTED	0	0	0

Display the newly added formula column (GAP_DAY) in the View area

Exercise 5.8
Display the newly added formula column (GAP_DAY) in the View area.

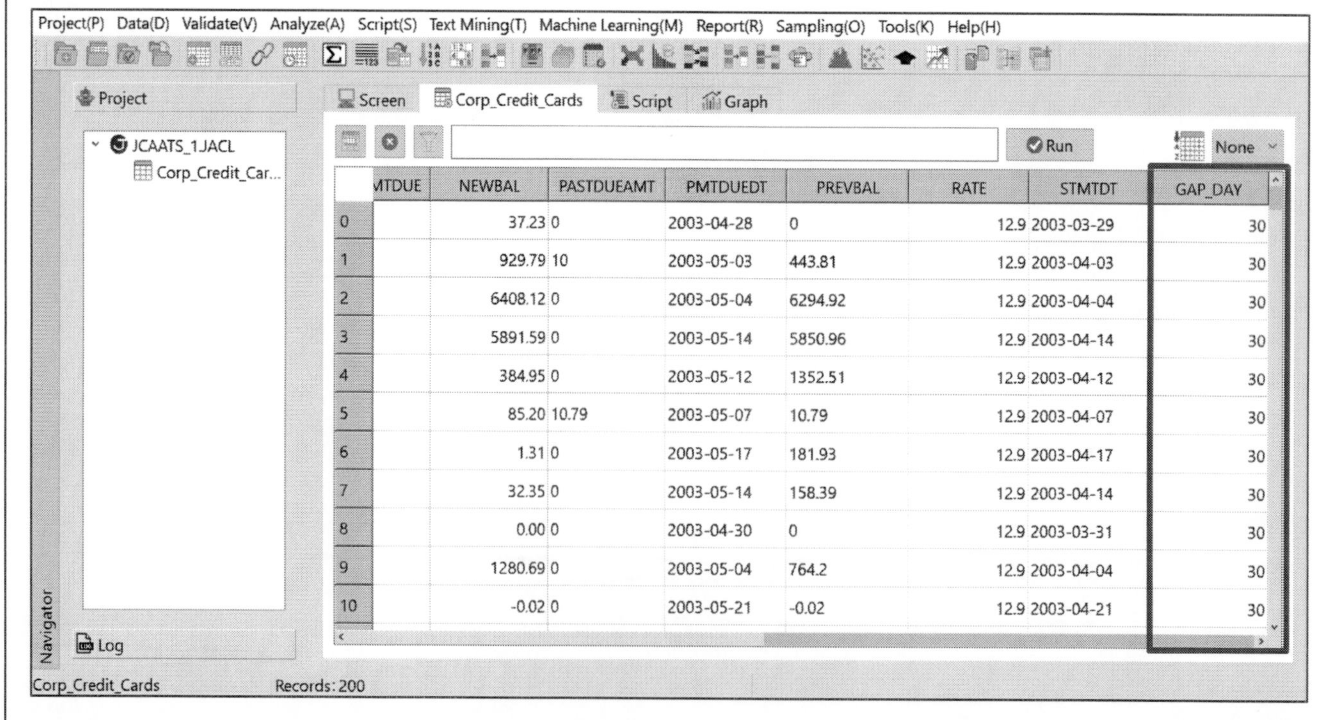

	MTDUE	NEWBAL	PASTDUEAMT	PMTDUEDT	PREVBAL	RATE	STMTDT	GAP_DAY
0		37.23	0	2003-04-28	0	12.9	2003-03-29	30
1		929.79	10	2003-05-03	443.81	12.9	2003-04-03	30
2		6408.12	0	2003-05-04	6294.92	12.9	2003-04-04	30
3		5891.59	0	2003-05-14	5850.96	12.9	2003-04-14	30
4		384.95	0	2003-05-12	1352.51	12.9	2003-04-12	30
5		85.20	10.79	2003-05-07	10.79	12.9	2003-04-07	30
6		1.31	0	2003-05-17	181.93	12.9	2003-04-17	30
7		32.35	0	2003-05-14	158.39	12.9	2003-04-14	30
8		0.00	0	2003-04-30	0	12.9	2003-03-31	30
9		1280.69	0	2003-05-04	764.2	12.9	2003-04-04	30
10		-0.02	0	2003-05-21	-0.02	12.9	2003-04-21	30

JCAATs - AI Audit Software

Chapter 6 - Exercise

JCAATs - AI Audit Software

Exercise 6.2

Please practice using the Verify command to test if there are any abnormalities in the project data table. Export the abnormal information for later audit use, and use the Locate command to understand the abnormal information.

Exercise 6.2
To verify using Badfile

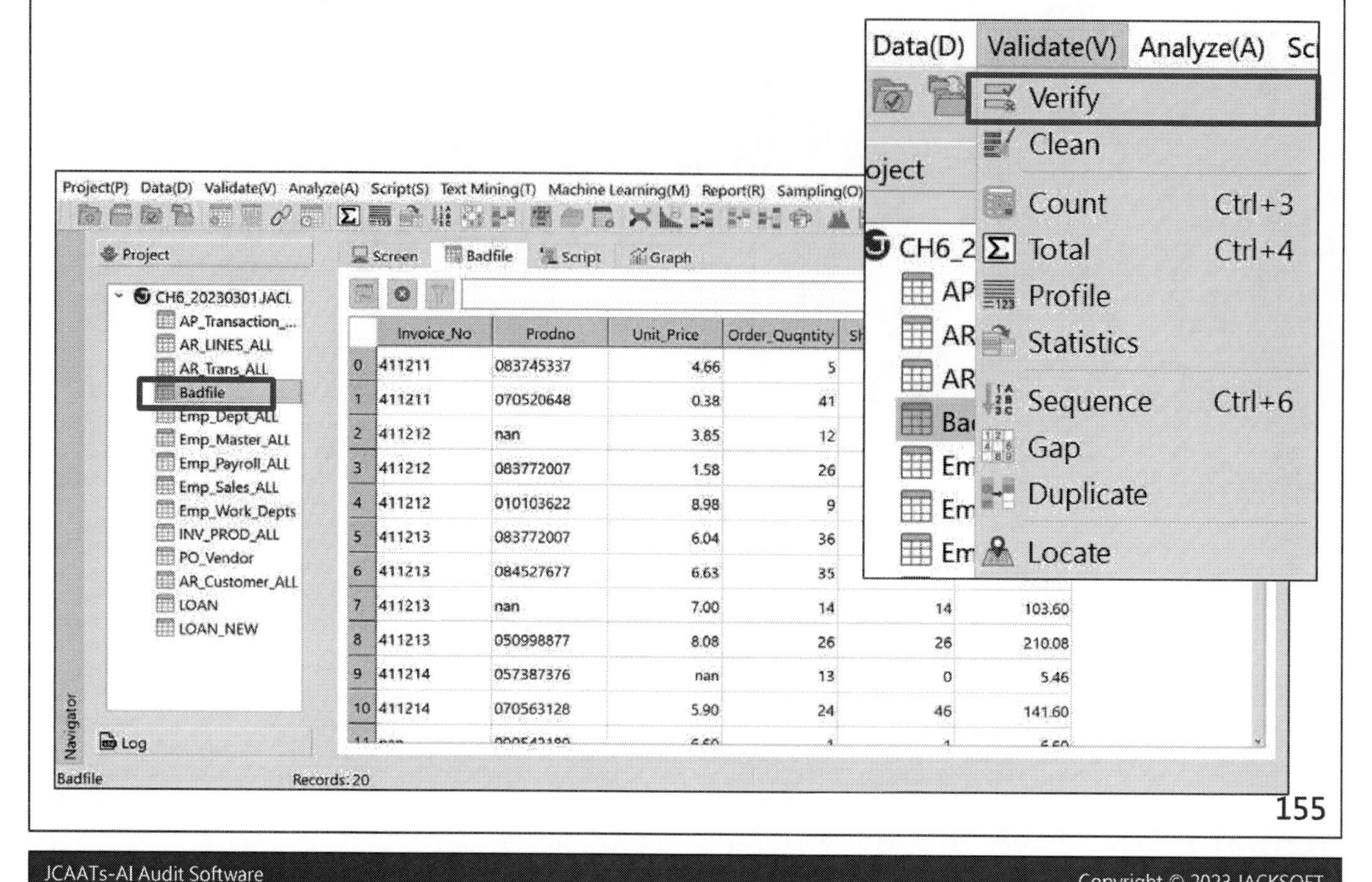

Exercise 6.2
To verify using Badfile>Select the verification field

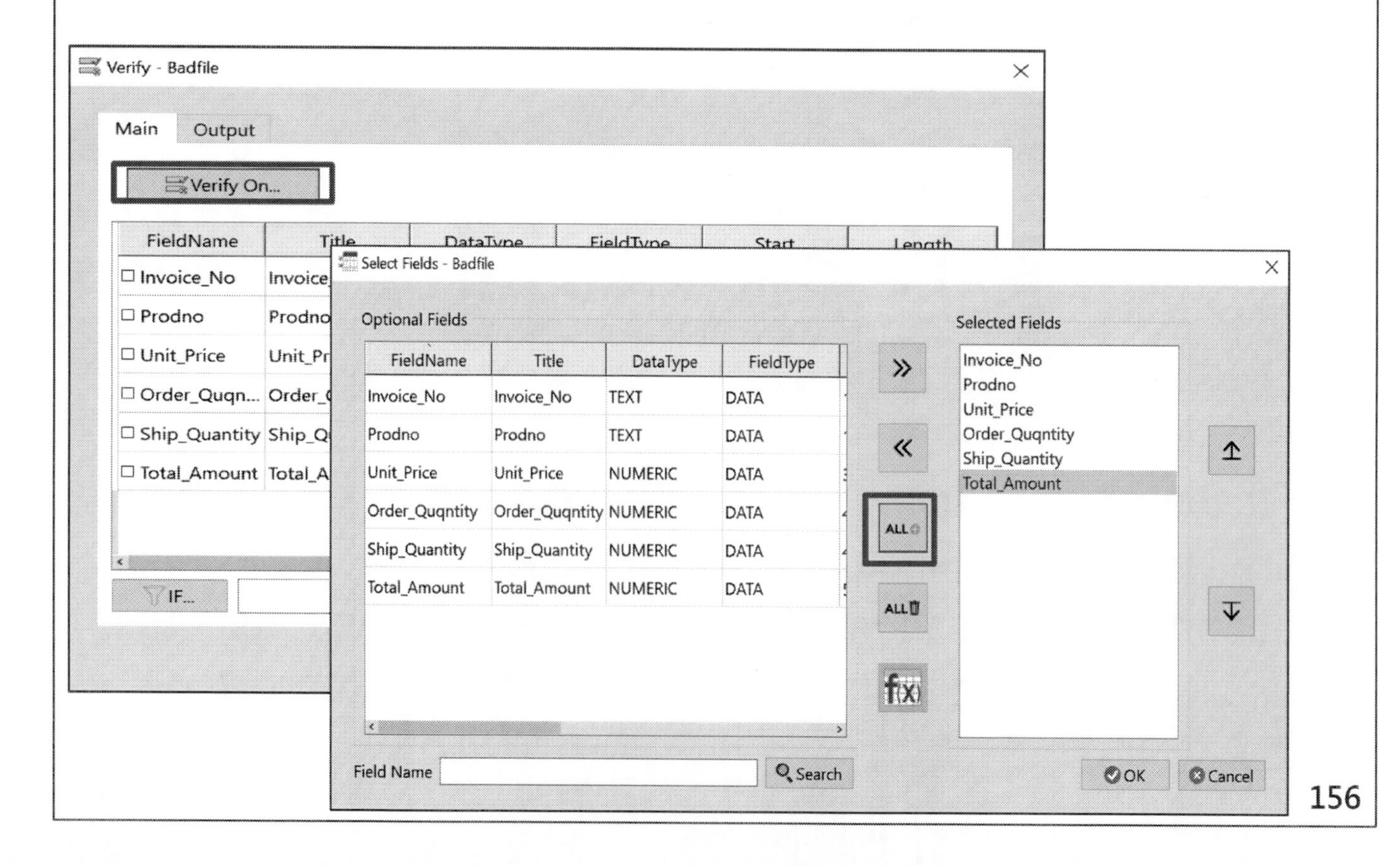

Exercise 6.2
To verify using Badfile >Export the verify command result

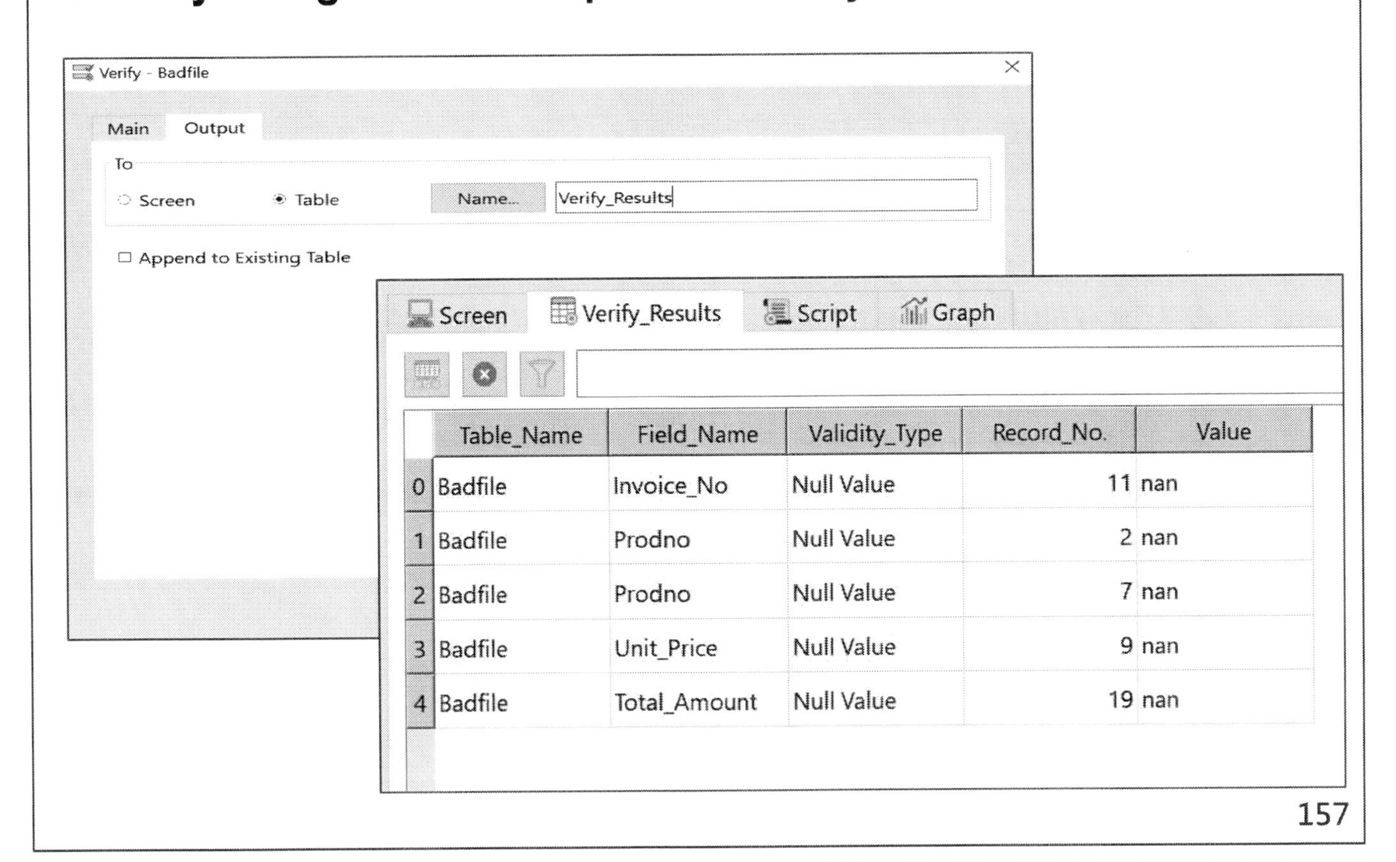

Exercise 6.2
For the verification results that have errors, use the Locate command to further understand the problem.

- Open the "Badfile" Table
- Verify→Locate
- Set the Locate data range to 2:2.

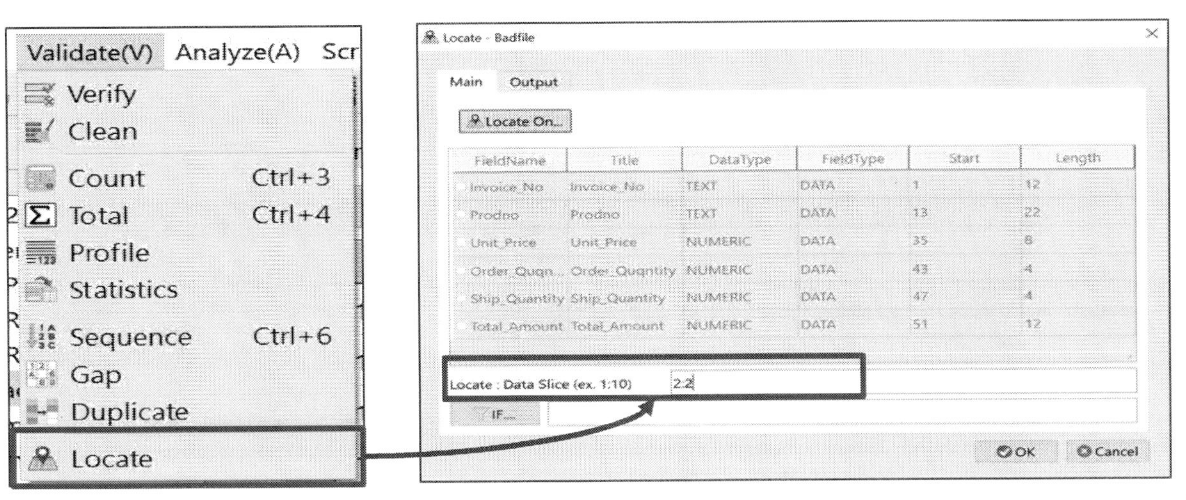

Exercise 6.2

For the verification results that have errors, use the Locate command to further understand the problem.

JCAATs >>Badfile.LOCATE(RECNO=[2:2], TO="")
Table : Badfile
Note: 2023/03/02 09:13:55
Result - Records : 1

Invoice_No	Prodno	Unit_Price	Order_Quqntity	Ship_Quantity	Total_Amount
411212	nan	3.85	12	12	46.20

JCAATs >>Badfile.LOCATE(RECNO=[7:7], TO="")
Table : Badfile
Note: 2023/03/02 09:14:15
Result - Records : 1

Invoice_No	Prodno	Unit_Price	Order_Quqntity	Ship_Quantity	Total_Amount
411213	nan	7.00	14	14	103.60

Exercise 6.2

For the verification results that have errors, use the Locate command to further understand the problem.

JCAATs >>Badfile.LOCATE(RECNO=[9:9], TO="")
Table : Badfile
Note: 2023/03/02 09:15:06
Result - Records : 1

Invoice_No	Prodno	Unit_Price	Order_Quqntity	Ship_Quantity	Total_Amount
411214	057387376	nan	13	0	5.46

JCAATs >>Badfile.LOCATE(RECNO=[11:11], TO="")
Table : Badfile
Note: 2023/03/02 09:15:10
Result - Records : 1

Invoice_No	Prodno	Unit_Price	Order_Quqntity	Ship_Quantity	Total_Amount
nan	090542189	6.60	1	1	6.60

JCAATs >>Badfile.LOCATE(RECNO=[19:19], TO="")
Table : Badfile
Note: 2023/03/02 09:15:18
Result - Records : 1

Invoice_No	Prodno	Unit_Price	Order_Quqntity	Ship_Quantity	Total_Amount
411217	340240664	0.10	49	49	nan

JCAATs Learning Note:

JCAATs - AI Audit Software

Please practice using the Clean command to purify the problematic data according to the audit requirements.

Exercise 6.3

For the verification results that have errors, use the clean command to facilitate subsequent verification.

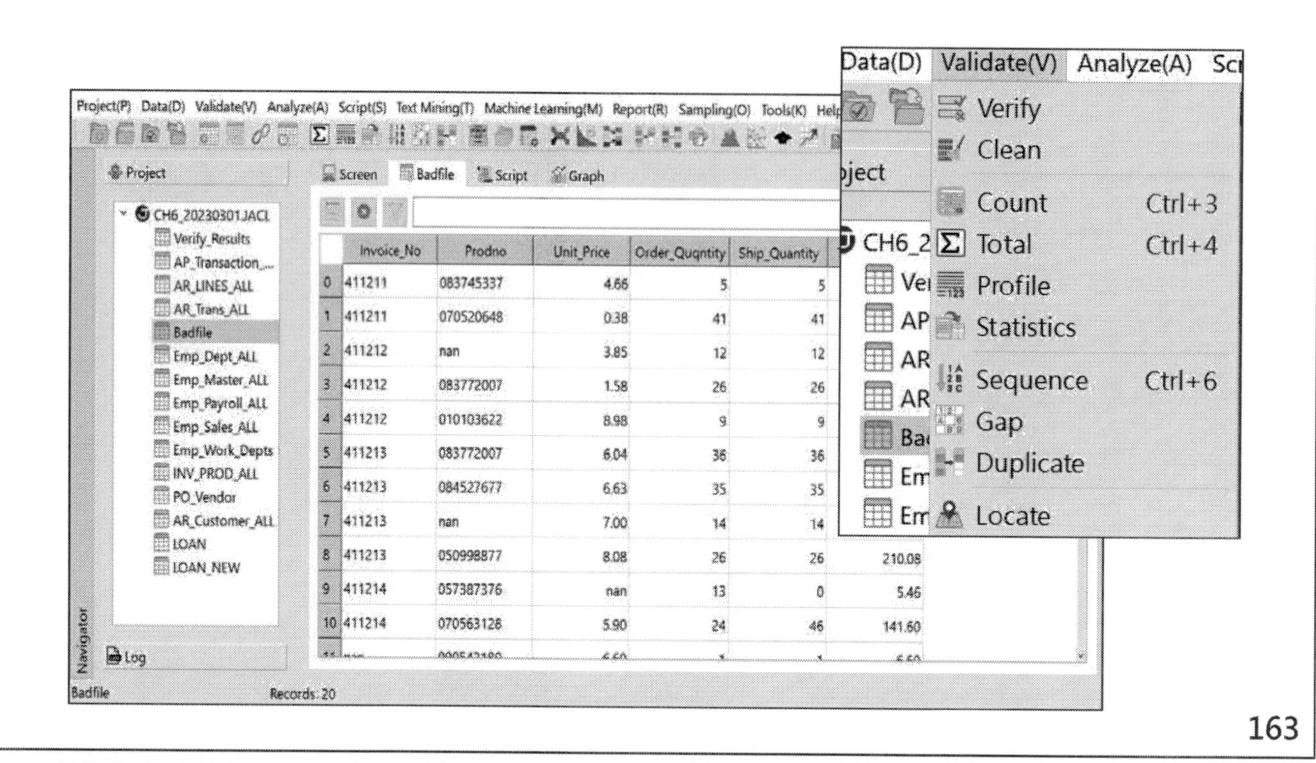

Exercise 6.3

Clean>Select the clean field.

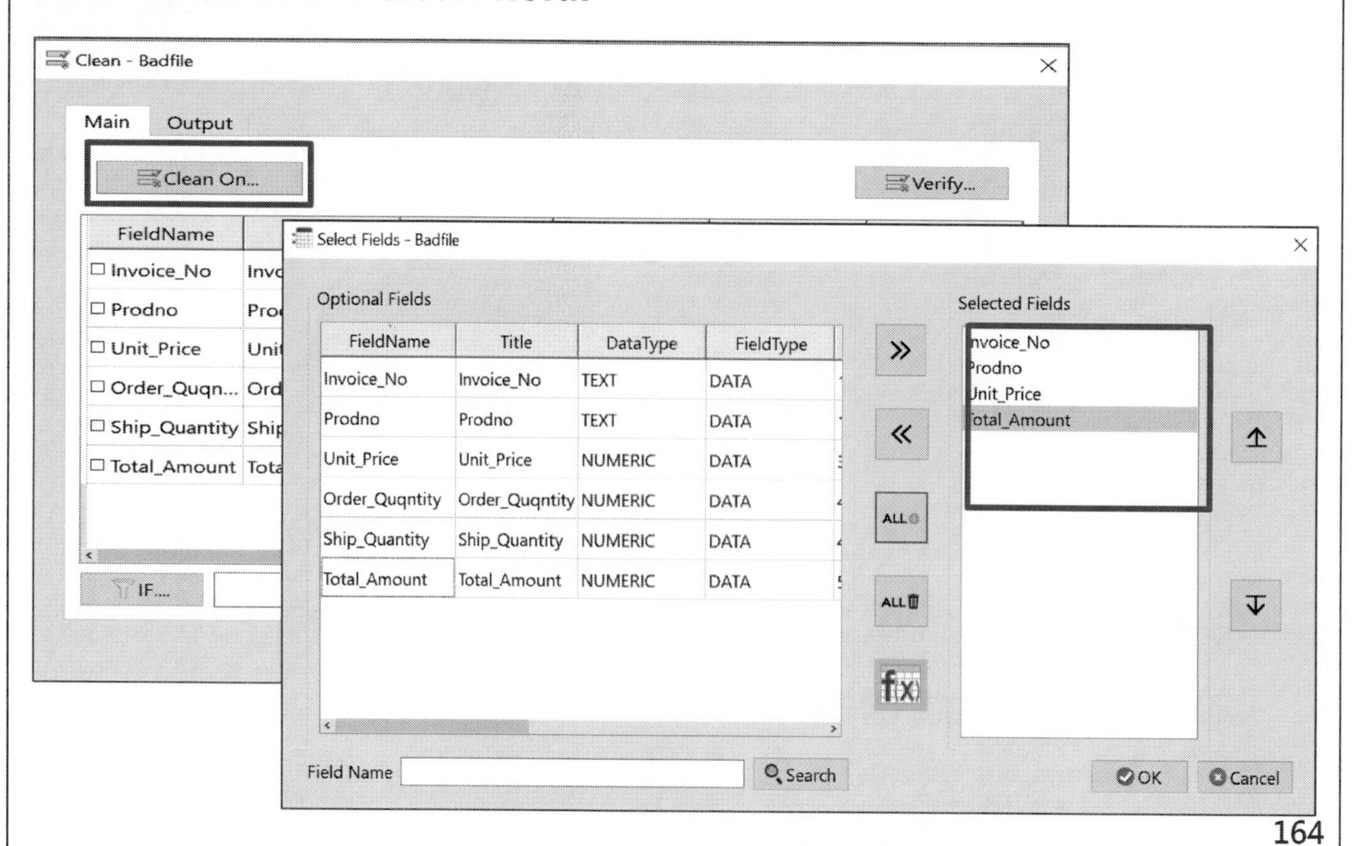

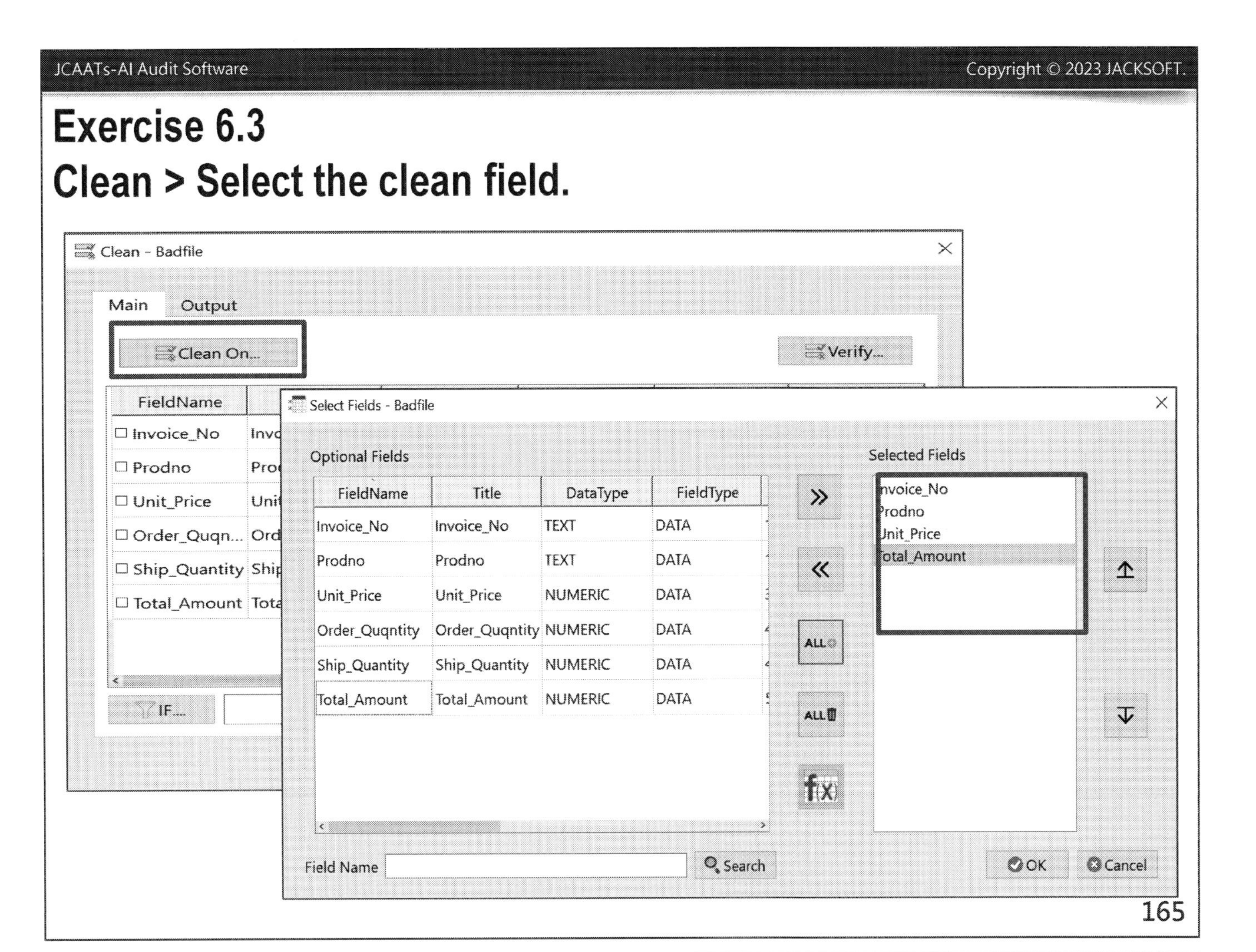

JCAATs-AI Audit Software
Copyright © 2023 JACKSOFT.
Exercise 6.3
Clean > Select the clean field.
Clean - Badfile
Main
Output
Clean On...
Verify...
FieldName
Invoice_No
Prodno
Unit_Price
Order_Quqn...
Ship_Quantity
Total_Amount
IF...
Select Fields - Badfile
Optional Fields
FieldName
Title
DataType
FieldType
Invoice_No
Invoice_No
TEXT
DATA
Prodno
Prodno
TEXT
DATA
Unit_Price
Unit_Price
NUMERIC
DATA
Order_Quqntity
Order_Quqntity
NUMERIC
DATA
Ship_Quantity
Ship_Quantity
NUMERIC
DATA
Total_Amount
Total_Amount
NUMERIC
DATA
Selected Fields
nvoice_No
rodno
Jnit_Price
otal_Amount
Field Name
Search
OK
Cancel
165

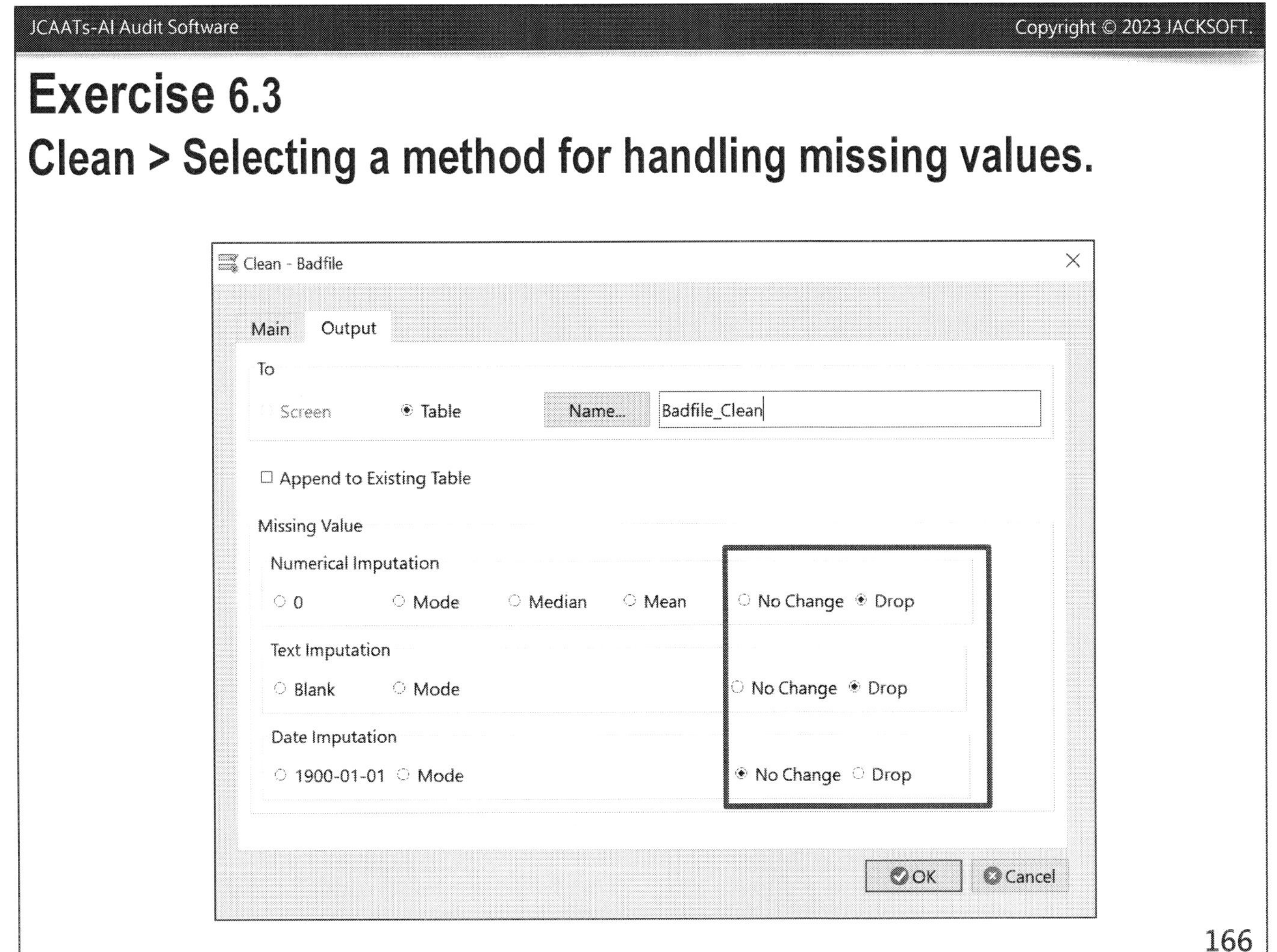

JCAATs-AI Audit Software
Copyright © 2023 JACKSOFT.
Exercise 6.3
Clean > Selecting a method for handling missing values.
Clean - Badfile
Main
Output
To
Screen
Table
Name...
Badfile_Clean
Append to Existing Table
Missing Value
Numerical Imputation
0
Mode
Median
Mean
No Change
Drop
Text Imputation
Blank
Mode
No Change
Drop
Date Imputation
1900-01-01
Mode
No Change
Drop
OK
Cancel
166

Exercise 6.3

After completing the Clean, there are a total of 15 records in the table.

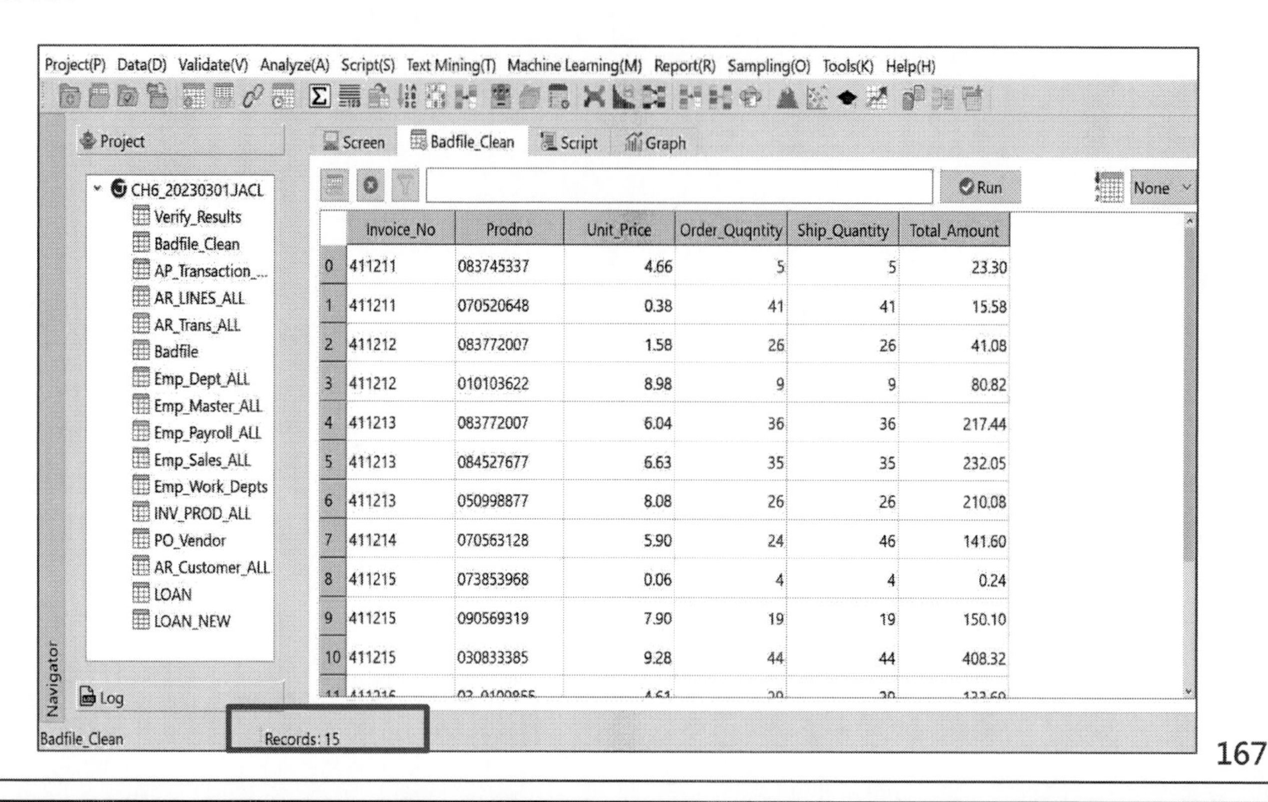

	Invoice_No	Prodno	Unit_Price	Order_Quqntity	Ship_Quantity	Total_Amount
0	411211	083745337	4.66	5	5	23.30
1	411211	070520648	0.38	41	41	15.58
2	411212	083772007	1.58	26	26	41.08
3	411212	010103622	8.98	9	9	80.82
4	411213	083772007	6.04	36	36	217.44
5	411213	084527677	6.63	35	35	232.05
6	411213	050998877	8.08	26	26	210.08
7	411214	070563128	5.90	24	46	141.60
8	411215	073853968	0.06	4	4	0.24
9	411215	090569319	7.90	19	19	150.10
10	411215	030833385	9.28	44	44	408.32

Chapter 6 - Exercise

- Exercise 6.4: Please practice the **Count command** to obtain the number of records in the data table to facilitate integrity verification.
- Exercise 6.5: Please practice the **Total command** to obtain the numeric sum of the fields to be verified for integrity verification.

JCAATs - AI Audit Software

Please practice the Count command to obtain the number of records in the data table to facilitate integrity verification.

AR_LINES_ALL, AR_Trans_ALL , AR_Customer_ALL

Exercise 6.4

Check the AR-related data table, count the number of cases, and generate a statistical file of case counts.

- Open the "AR_LINES_ALL" file
- Validate→Count

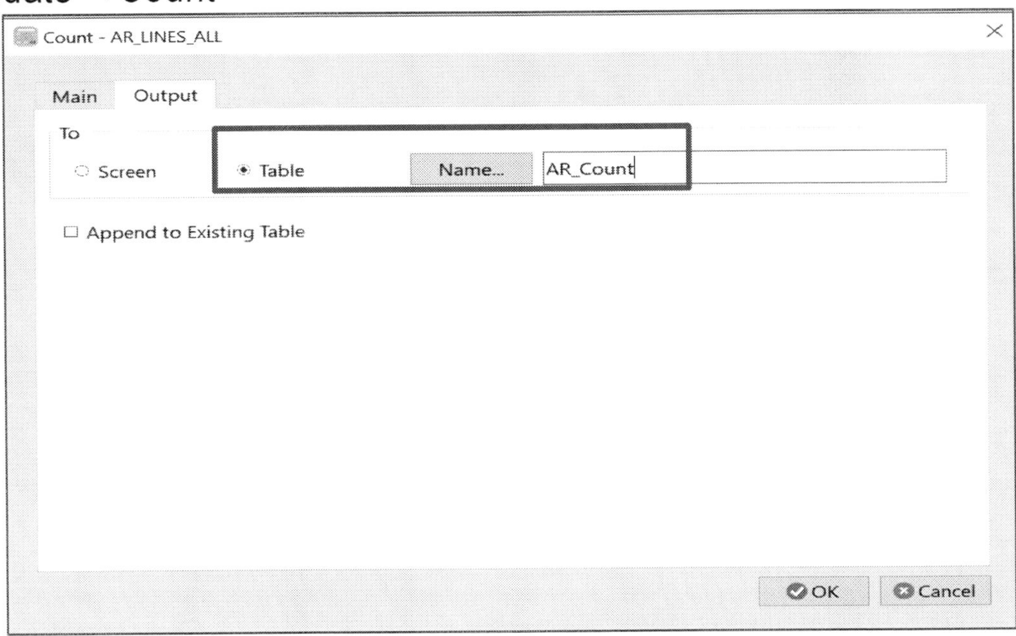

Exercise 6.4
Perform a record count and append it to the existing data table.

- Open the " AR_Trans_ALL " file
- Validate→Count
- Append to the existing table

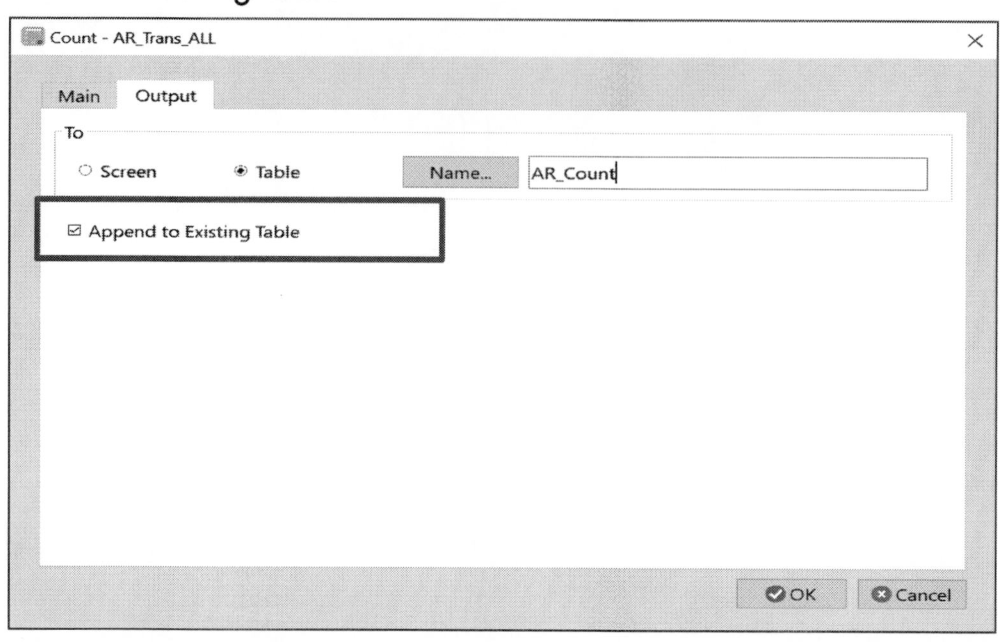

JCAATs-AI Audit Software

Exercise 6.4
Perform a record count and append it to the existing data table.

- Open the "AR_Customer_ALL" file
- Validate→Count
- Append to the existing table

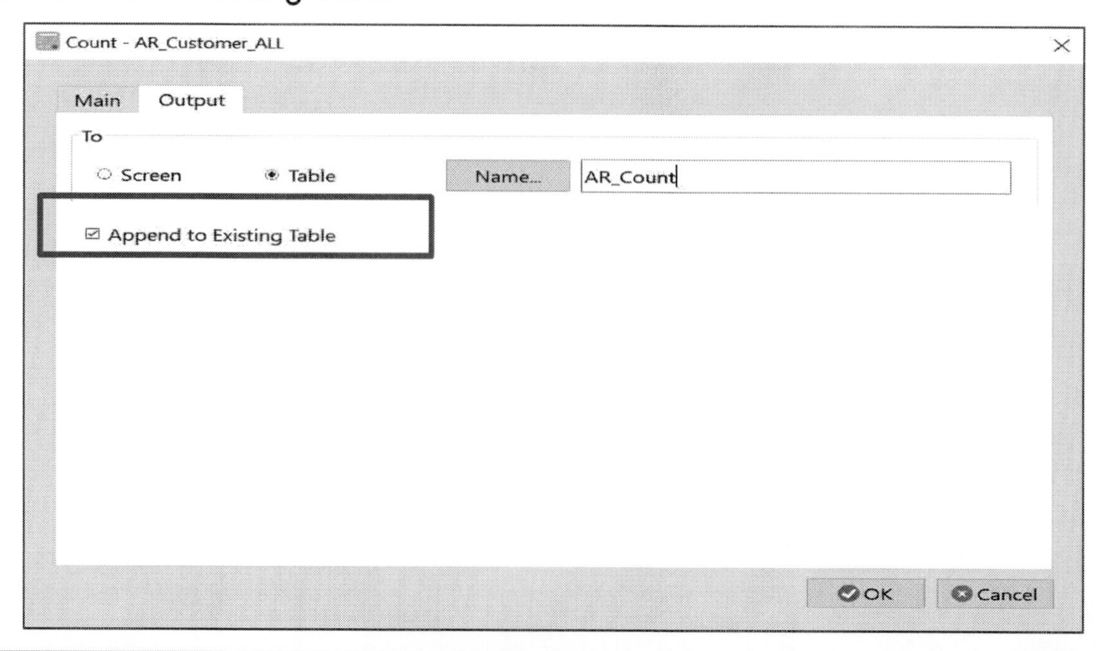

Exercise 6.4
Statistical file of the results from the check of AR-related data table.
AR_LINES_ALL、AR_Trans_ALL、AR_Customer_ALL

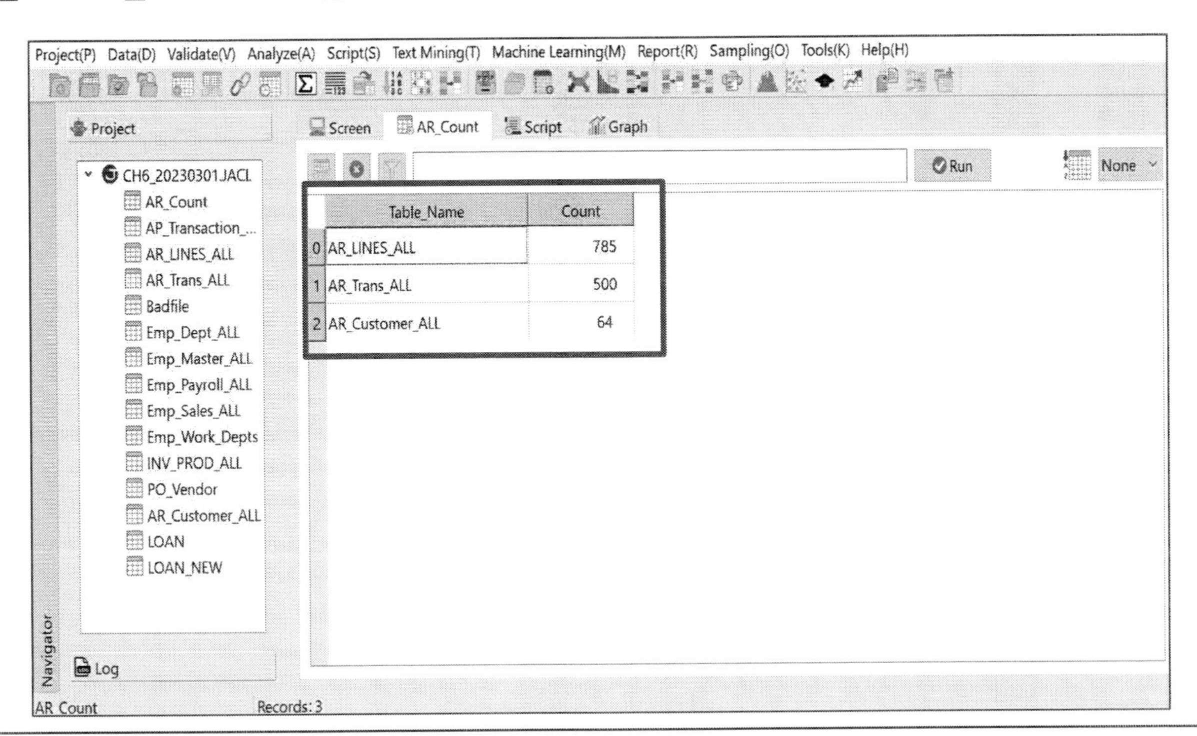

JCAATs Learning Note:

JCAATs - AI Audit Software

Please practice the Total command to obtain the numeric sum of the fields to be verified for integrity verification.

Exercise 6.5

Check the AP-related data file, perform a summation, and generate a summation table.

AP_Transaction_ALL

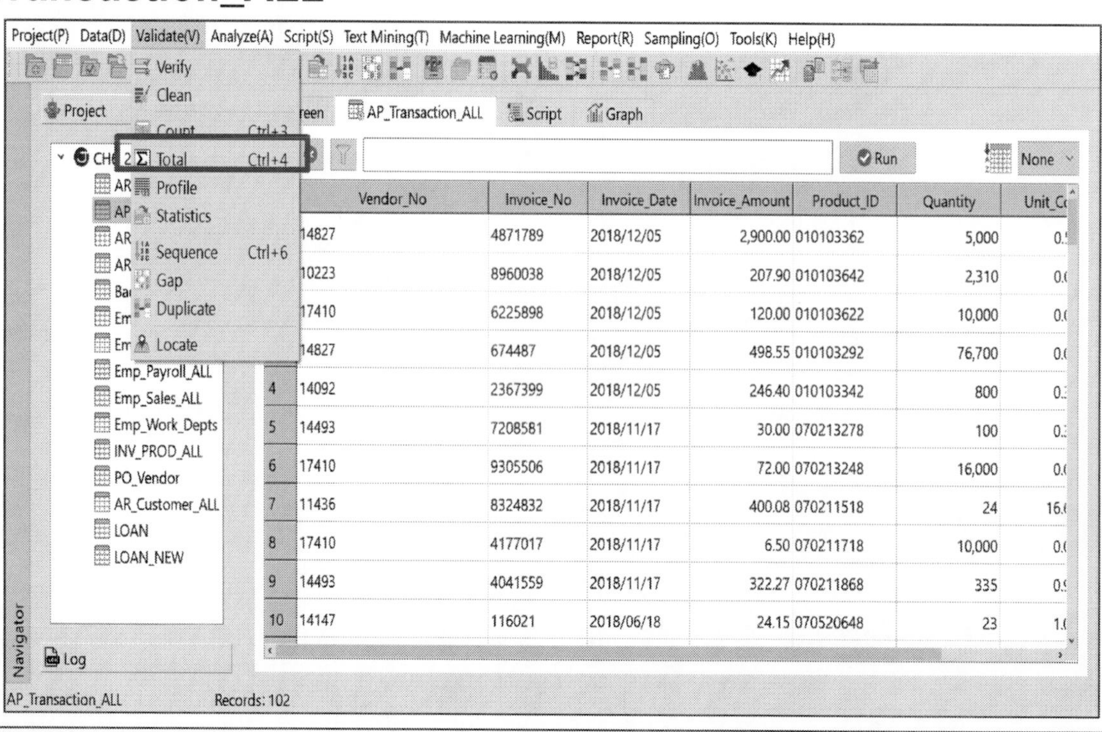

Exercise 6.5 Total>Select the Total column.
AP_Transaction_ALL

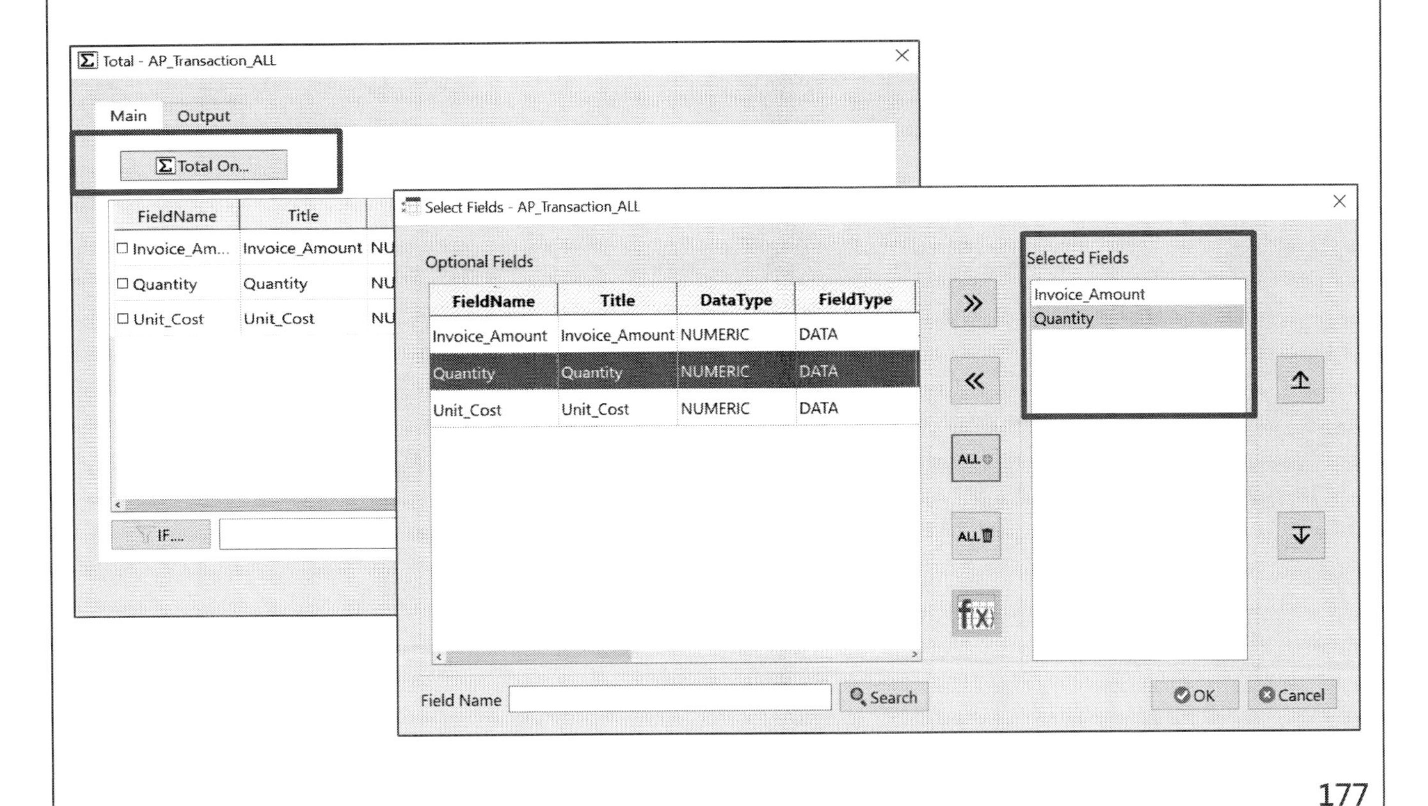

Exercise 6.5
AP related audit summary table
AP_Transaction_ALL

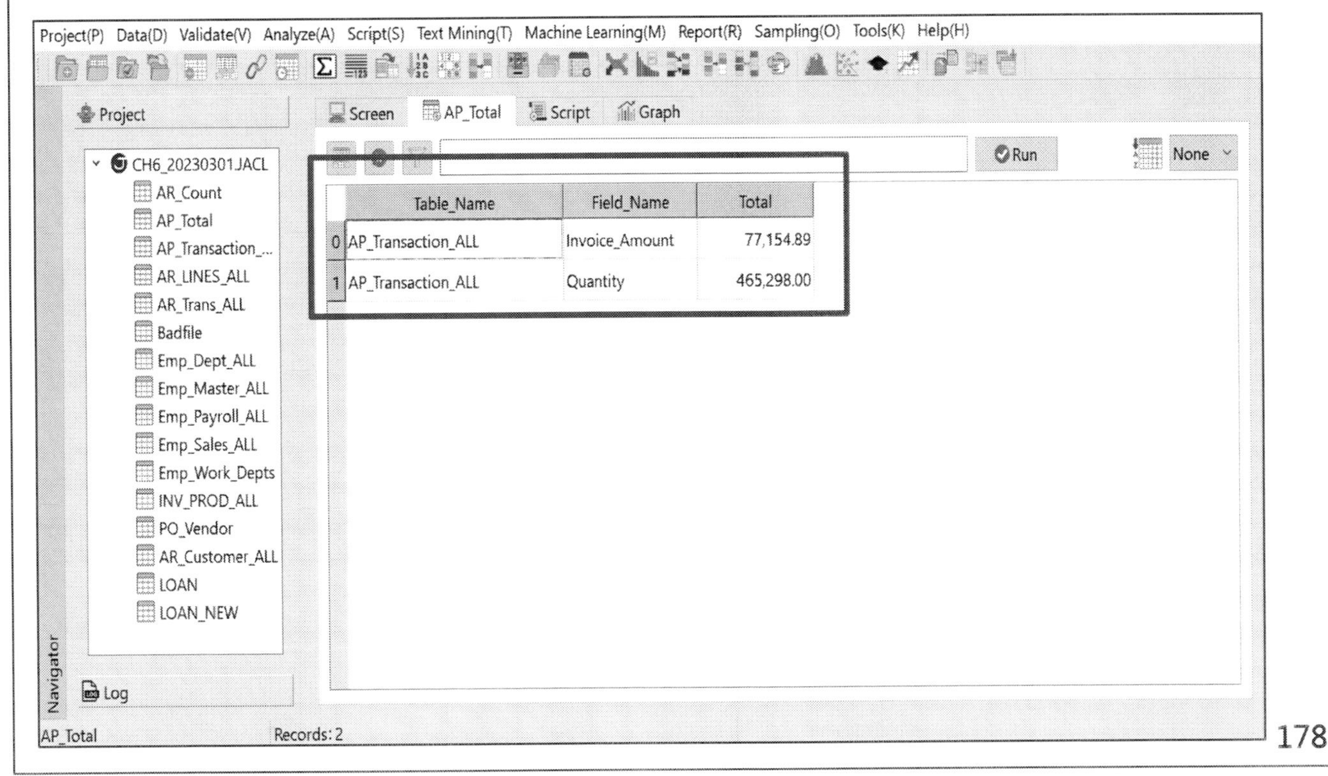

	Table_Name	Field_Name	Total
0	AP_Transaction_ALL	Invoice_Amount	77,154.89
1	AP_Transaction_ALL	Quantity	465,298.00

Chapter 6 - Exercise

- Exercise 6.6: Please practice **Profile** command to obtain the numerical basic statistical information of the fields that need to be verified for integrity validation.

- Exercise 6.7: Please practice **Statistics** command to obtain the interval and statistical information of the fields that need to be verified for integrity validation.

JCAATs - AI Audit Software

Please practice the Profile command to obtain basic statistical information on the numerical fields that need to be verified for integrity verification.

Exercise 6.6

Check the data file related to AP, and after executing the Profile, generate a table of the resulting data.

AP_Transaction_ALL

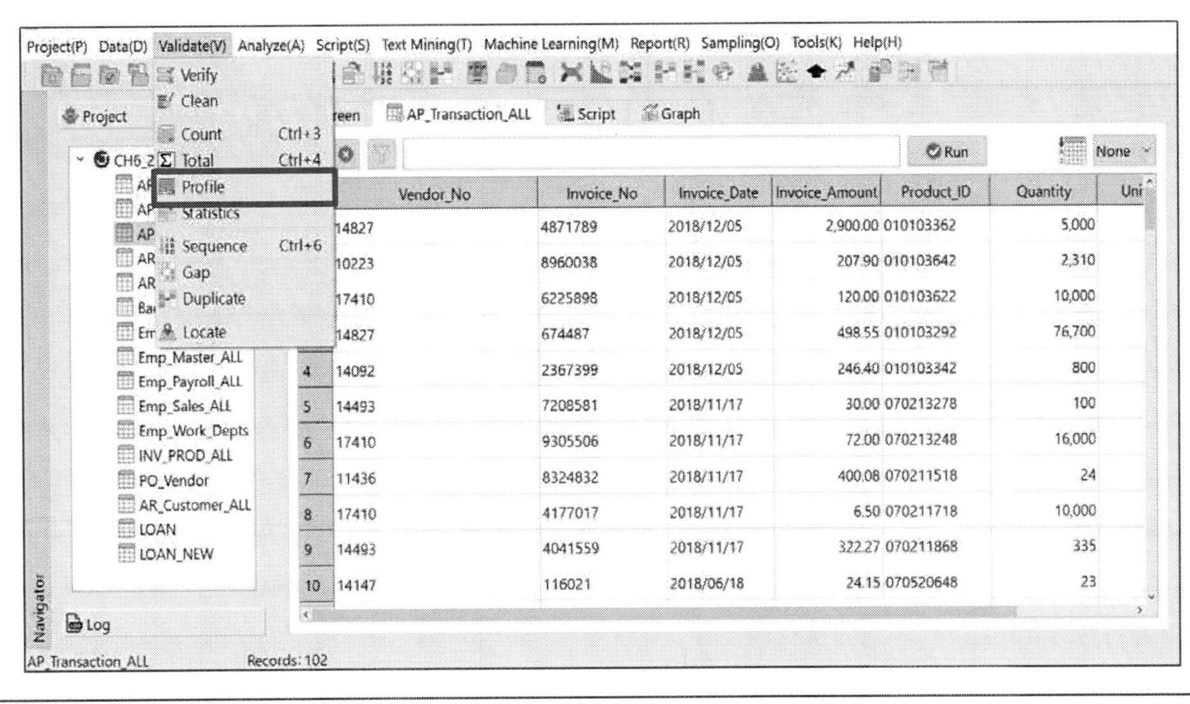

Exercise 6.6

Validate>Profile>Select Profile fields.

AP_Transaction_ALL

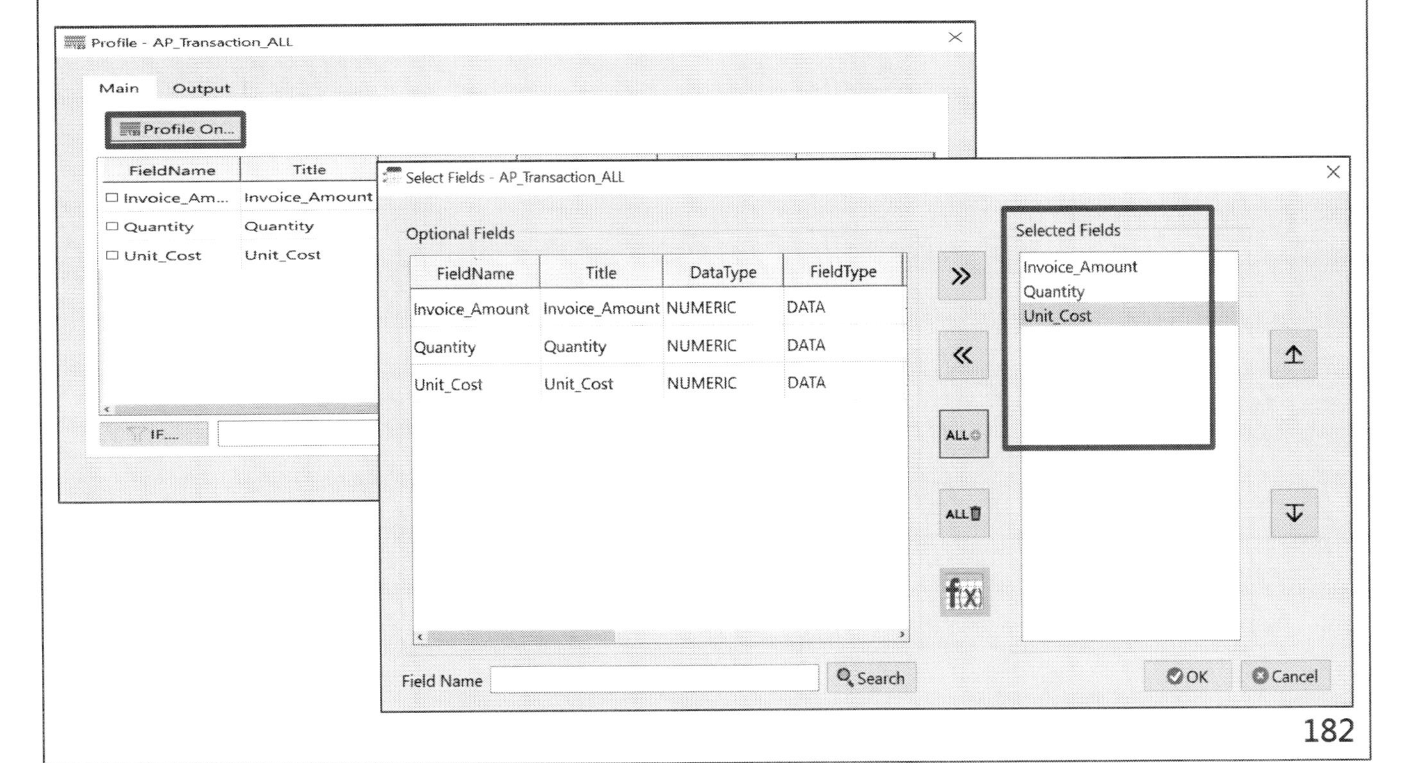

Exercise 6.6
AP-related data file Check the Profile command results.
AP_Transaction_ALL

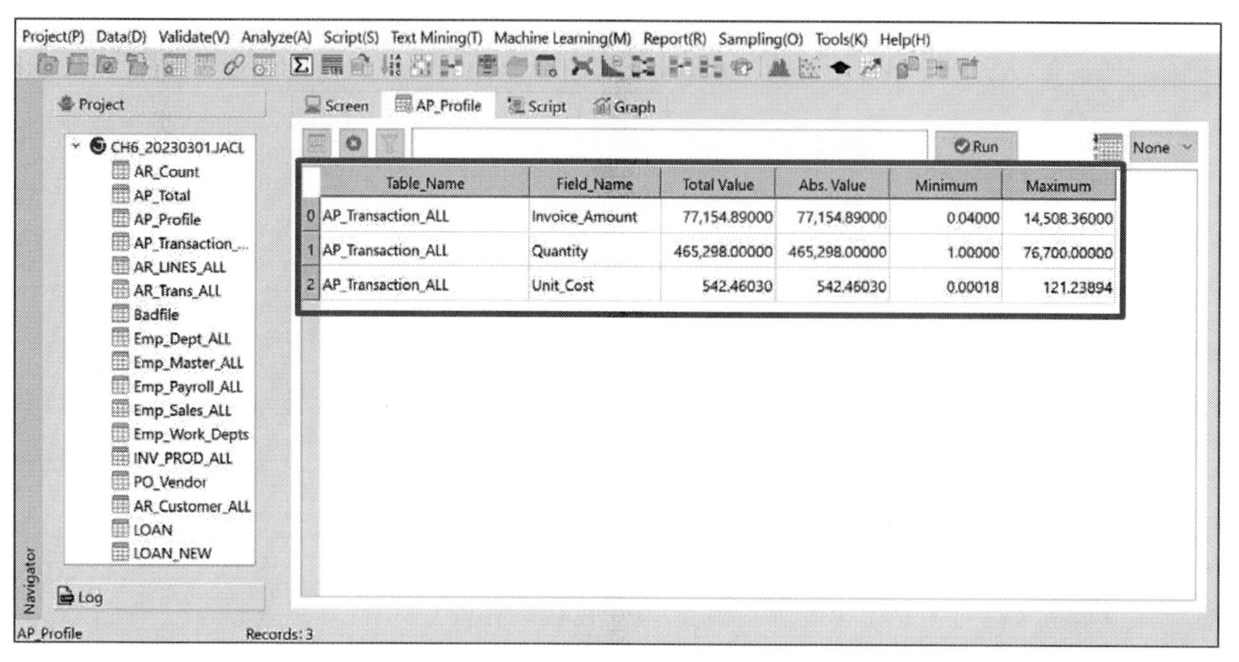

	Table_Name	Field_Name	Total Value	Abs. Value	Minimum	Maximum
0	AP_Transaction_ALL	Invoice_Amount	77,154.89000	77,154.89000	0.04000	14,508.36000
1	AP_Transaction_ALL	Quantity	465,298.00000	465,298.00000	1.00000	76,700.00000
2	AP_Transaction_ALL	Unit_Cost	542.46030	542.46030	0.00018	121.23894

Exercise 6.6
Perform a verification check on inventory, execute the Profile command, and find abnormal data.
INV_PROD_ALL

- Open "INV_PROD_ALL" file
- Validate→Profile
- Select "UnCst" field

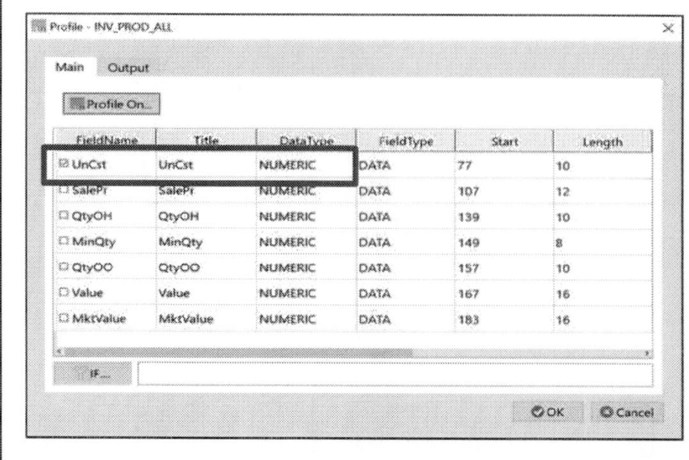

FieldName	Title	DataType	FieldType	Start	Length
☑ UnCst	UnCst	NUMERIC	DATA	77	10
☐ SalePr	SalePr	NUMERIC	DATA	107	12
☐ QtyOH	QtyOH	NUMERIC	DATA	139	10
☐ MinQty	MinQty	NUMERIC	DATA	149	8
☐ QtyOO	QtyOO	NUMERIC	DATA	157	10
☐ Value	Value	NUMERIC	DATA	167	16
☐ MktValue	MktValue	NUMERIC	DATA	183	16

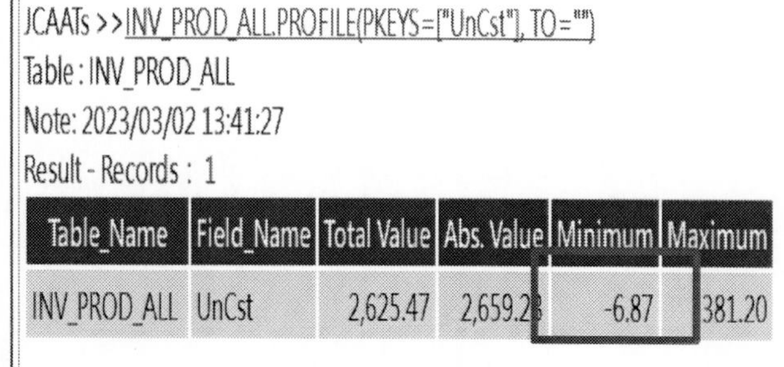

JCAATs >>INV_PROD_ALL.PROFILE(PKEYS=["UnCst"], TO="")

Table : INV_PROD_ALL

Note: 2023/03/02 13:41:27

Result - Records : 1

Table_Name	Field_Name	Total Value	Abs. Value	Minimum	Maximum
INV_PROD_ALL	UnCst	2,625.47	2,659.23	-6.87	381.20

Exercise 6.6
Perform a verification check on inventory, execute the Profile command, and find abnormal data.
INV_PROD_ALL

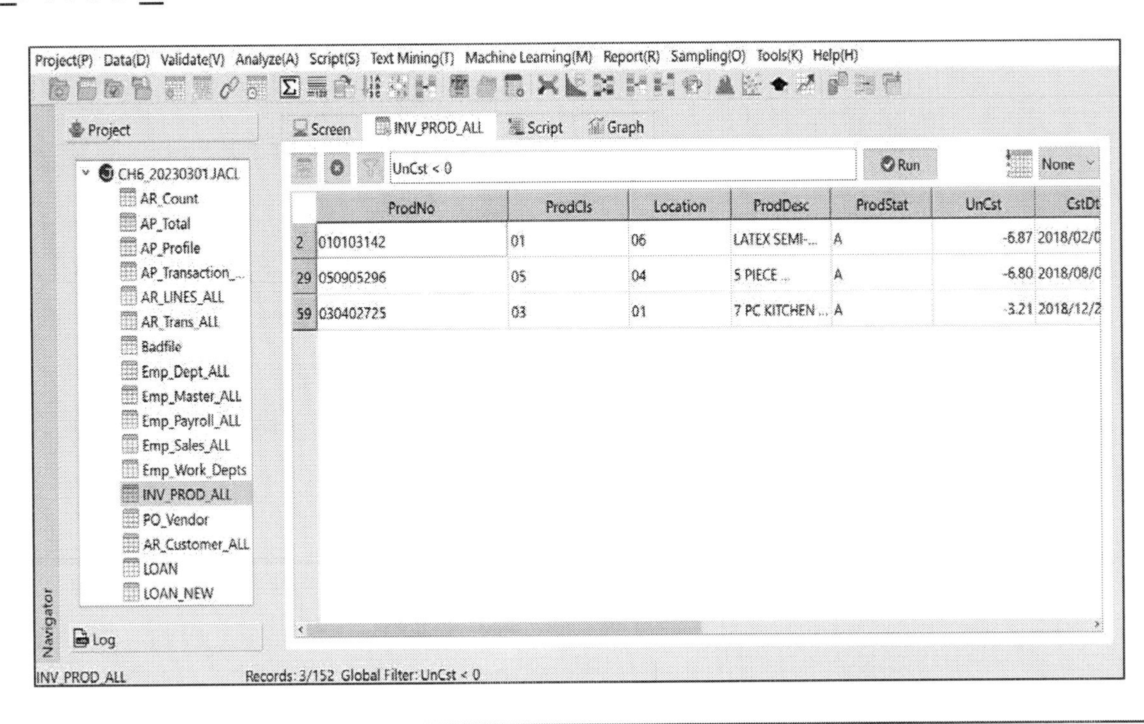

	ProdNo	ProdCls	Location	ProdDesc	ProdStat	UnCst	CstDt
2	010103142	01	06	LATEX SEMI-...	A	-6.87	2018/02/0
29	050905296	05	04	5 PIECE ...	A	-6.80	2018/08/0
59	030402725	03	01	7 PC KITCHEN ...	A	-3.21	2018/12/2

JCAATs Learning Note:

JCAATs - AI Audit Software

Please practice the Statistics command to obtain interval and statistical information on the fields that need to be verified for integrity verification.

Exercise 6.7

Check the data file related to AR, execute the Statistics command, and generate a statistical data table.

AR_Trans_ALL

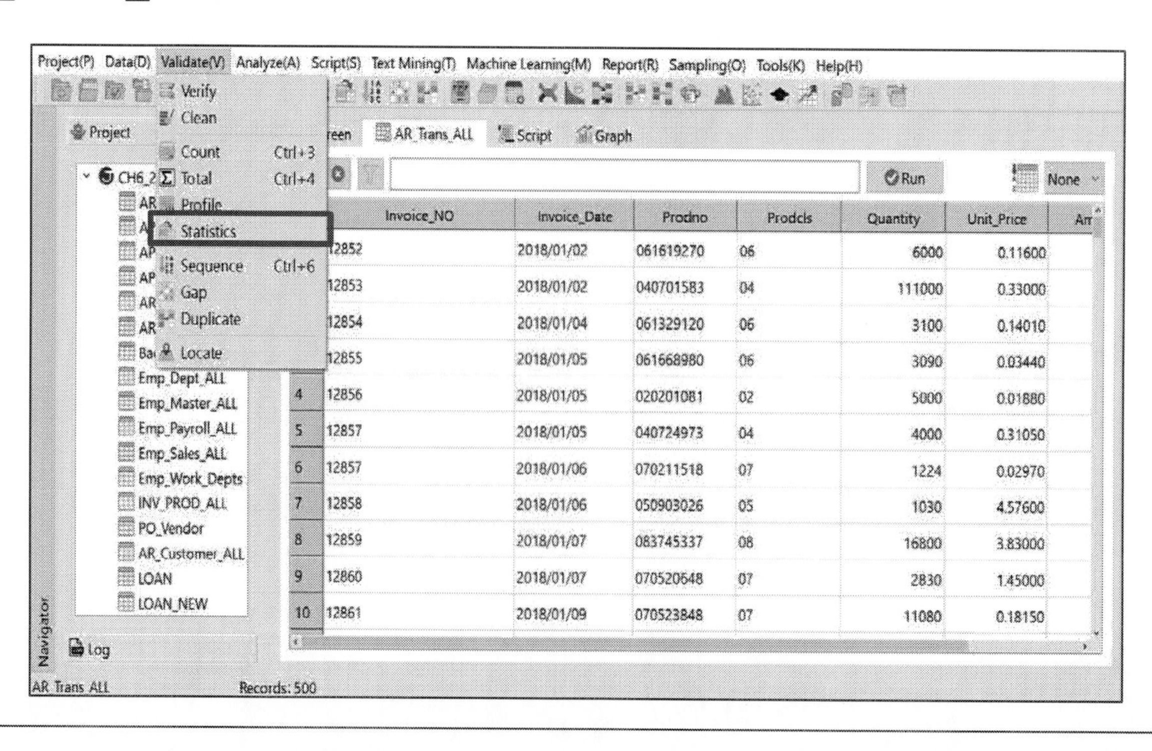

Exercise 6.7

Validate > Statistics >Select Statistics fields and set the number of records.

AR_Trans_ALL

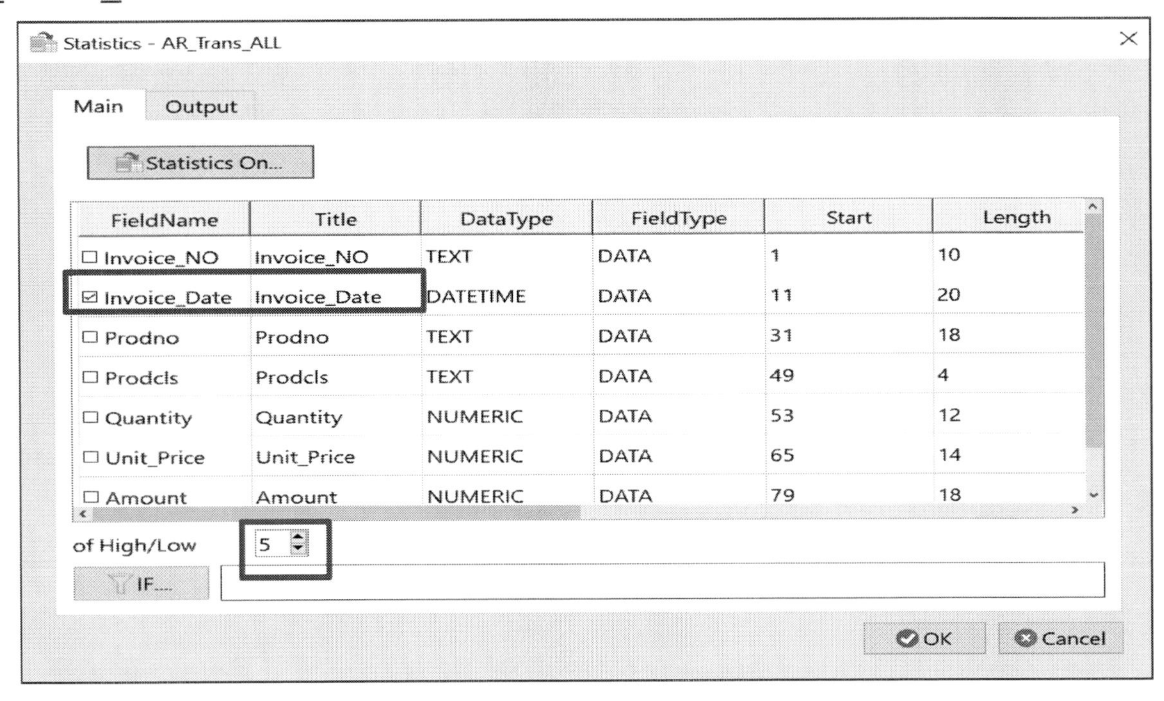

FieldName	Title	DataType	FieldType	Start	Length
☐ Invoice_NO	Invoice_NO	TEXT	DATA	1	10
☑ Invoice_Date	Invoice_Date	DATETIME	DATA	11	20
☐ Prodno	Prodno	TEXT	DATA	31	18
☐ Prodcls	Prodcls	TEXT	DATA	49	4
☐ Quantity	Quantity	NUMERIC	DATA	53	12
☐ Unit_Price	Unit_Price	NUMERIC	DATA	65	14
☐ Amount	Amount	NUMERIC	DATA	79	18

Exercise 6.7

Statistical results of the data file related to AR

AR_Trans_ALL

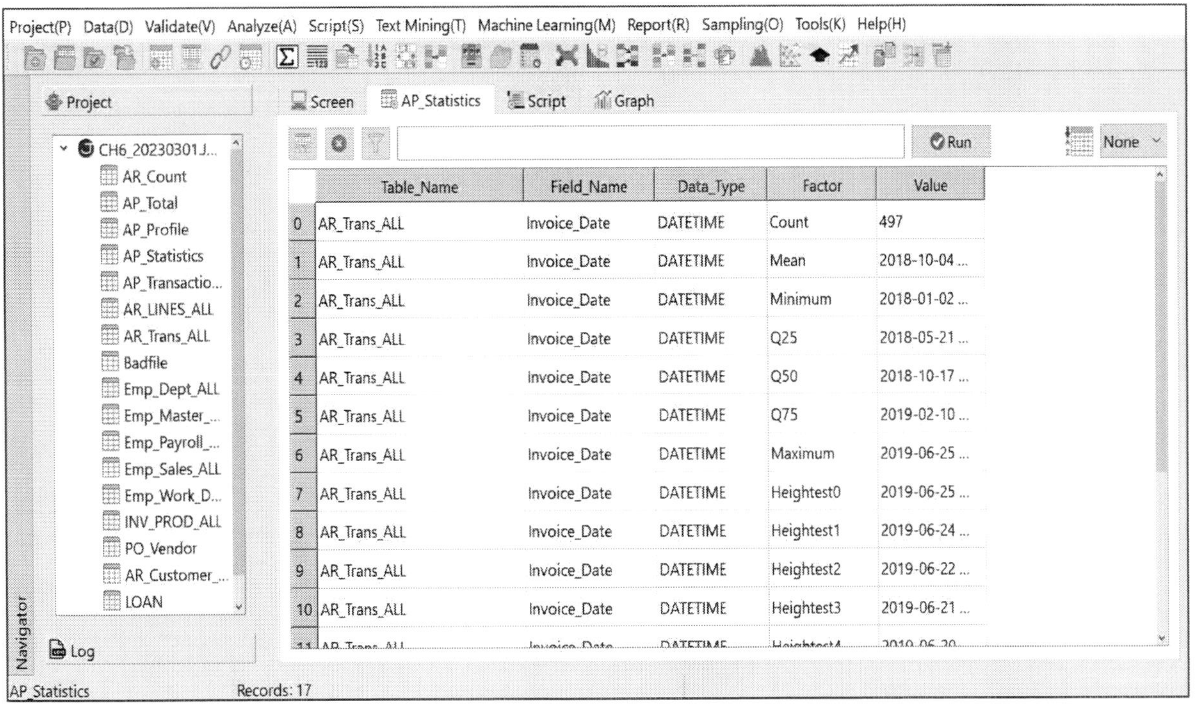

	Table_Name	Field_Name	Data_Type	Factor	Value
0	AR_Trans_ALL	Invoice_Date	DATETIME	Count	497
1	AR_Trans_ALL	Invoice_Date	DATETIME	Mean	2018-10-04 ...
2	AR_Trans_ALL	Invoice_Date	DATETIME	Minimum	2018-01-02 ...
3	AR_Trans_ALL	Invoice_Date	DATETIME	Q25	2018-05-21 ...
4	AR_Trans_ALL	Invoice_Date	DATETIME	Q50	2018-10-17 ...
5	AR_Trans_ALL	Invoice_Date	DATETIME	Q75	2019-02-10 ...
6	AR_Trans_ALL	Invoice_Date	DATETIME	Maximum	2019-06-25 ...
7	AR_Trans_ALL	Invoice_Date	DATETIME	Heightest0	2019-06-25 ...
8	AR_Trans_ALL	Invoice_Date	DATETIME	Heightest1	2019-06-24 ...
9	AR_Trans_ALL	Invoice_Date	DATETIME	Heightest2	2019-06-22 ...
10	AR_Trans_ALL	Invoice_Date	DATETIME	Heightest3	2019-06-21 ...
11	AR_Trans_ALL	Invoice_Date	DATETIME	Heightest4	2019-06-20

Chapter 6 - Exercise

- Exercise 6.8: Please practice **Sequence command** to check if the required field has data that is not sorted in order, and generate an abnormal data table to facilitate further investigation.

- Exercise 6.9: Please practice **Gap command** to check if the required field has data with missing numbers, and generate an abnormal data table to facilitate further investigation.

- Exercise 6.10: Please practice **Duplicate command** to check if the required field has duplicate data, and generate an abnormal data table to facilitate further investigation.

JCAATs-AI Audit Software
Copyright © 2023 JACKSOFT.

JCAATs Learning Note:

JCAATs - AI Audit Software

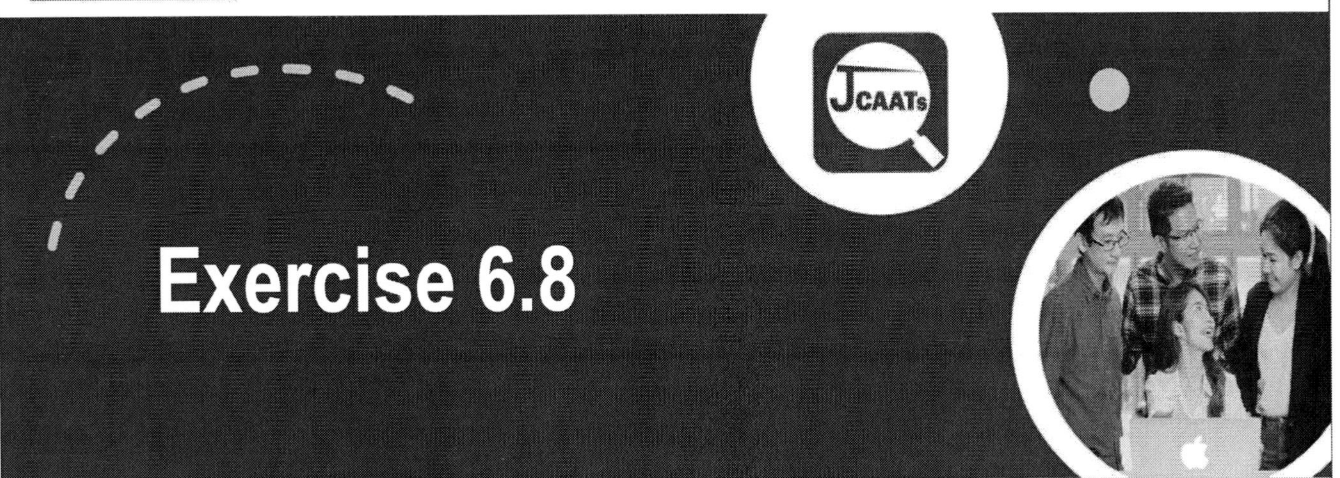

Exercise 6.8

Please practice the Sequence command to determine if the fields that need to be verified are not arranged in order, generate an abnormal data table, and facilitate further investigation.

Exercise 6.8

Check the data file related to AR, execute the Sequence command, and generate a sequence data table.

AR_Trans_ALL

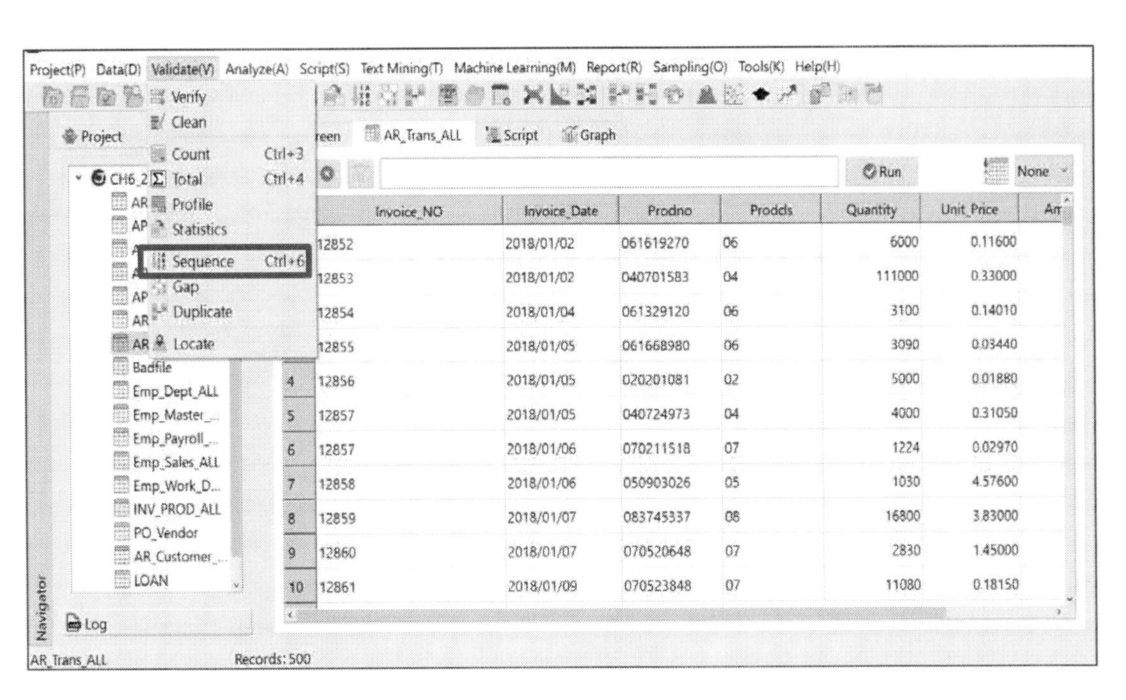

Exercise 6.8
Validate > Sequence >Select the values and fields to be verified AR_Trans_ALL

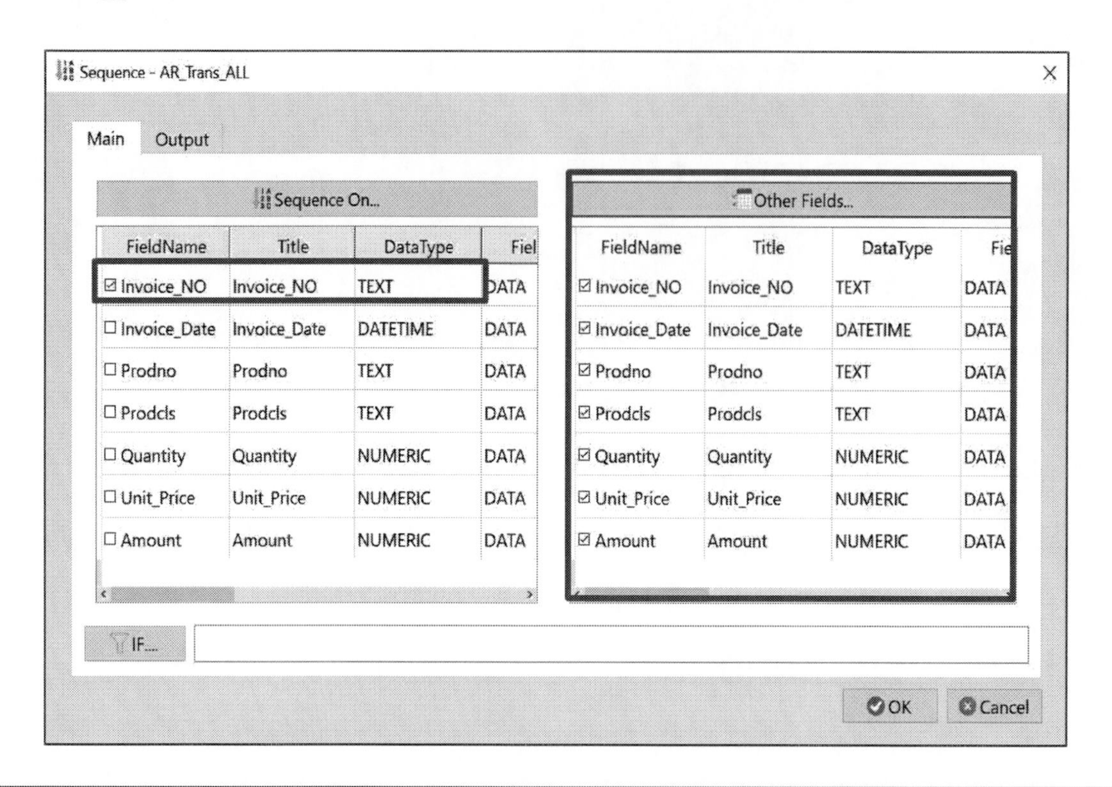

Exercise 6.8
Check the data file related to AR , execute the Sequence command, and generate a sequence data table. AR_Trans_ALL

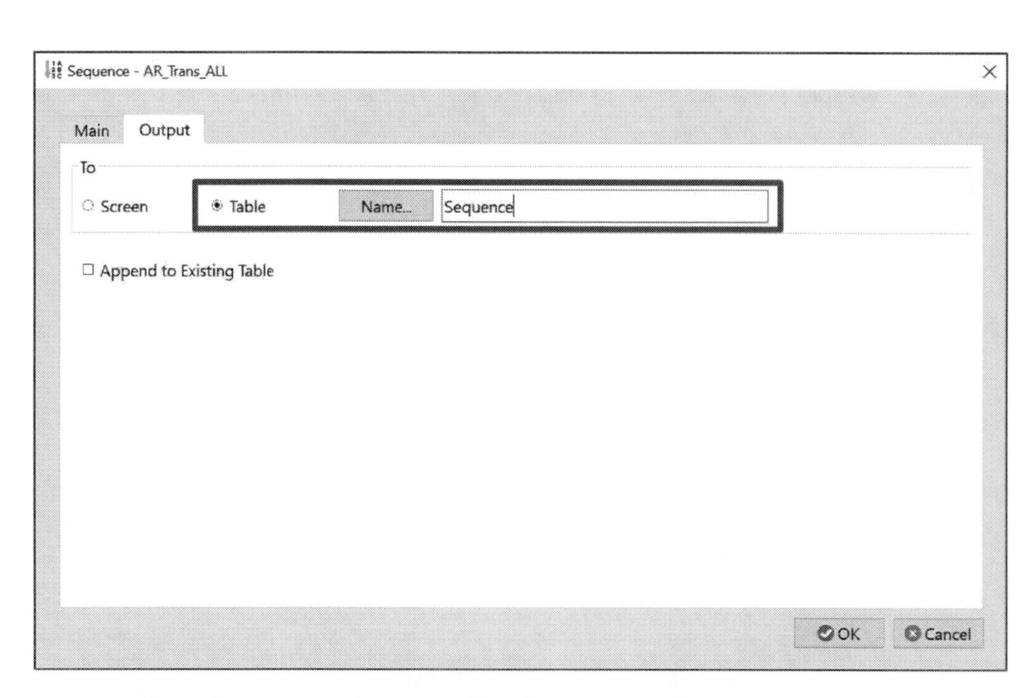

Exercise 6.8
Check the data file related to AR , execute the Sequence command, and generate a sequence data table.
AR_Trans_ALL

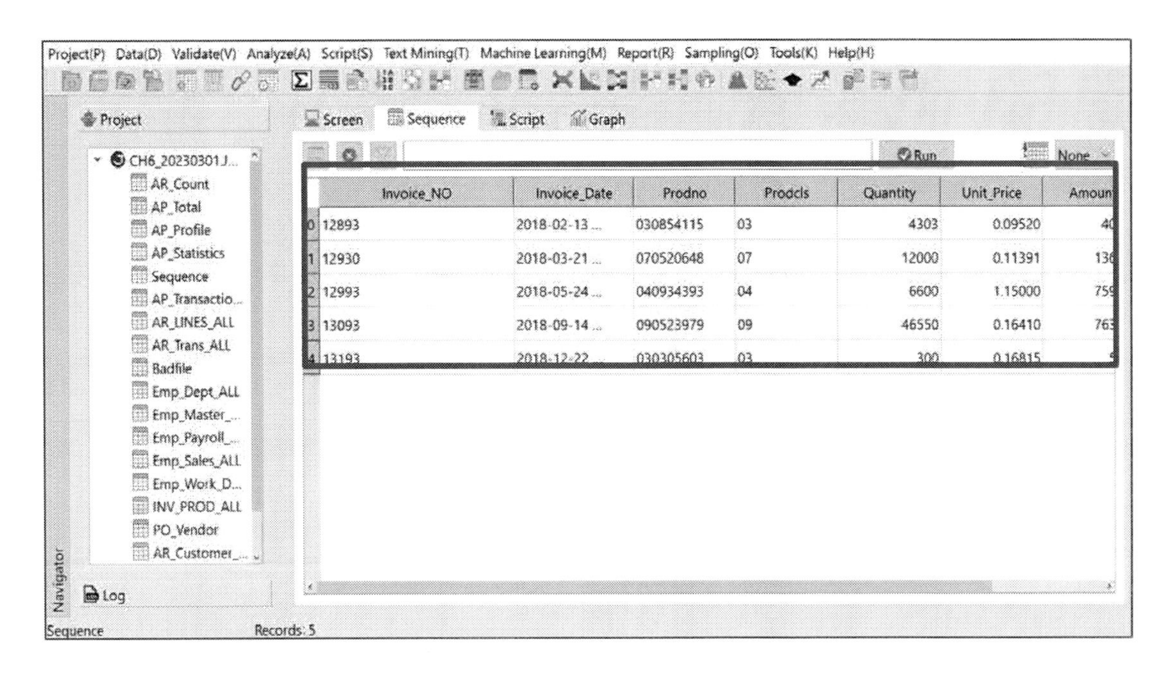

JCAATs Learning Note:

JCAATs - AI Audit Software

Please practice the Gap command to determine if there are any missing numbers in the fields that need to be verified, generate an abnormal data table, and facilitate further investigation.

Exercise 6.9

Check the data file related to employees, execute the Gap command

Emp_Payroll_ALL

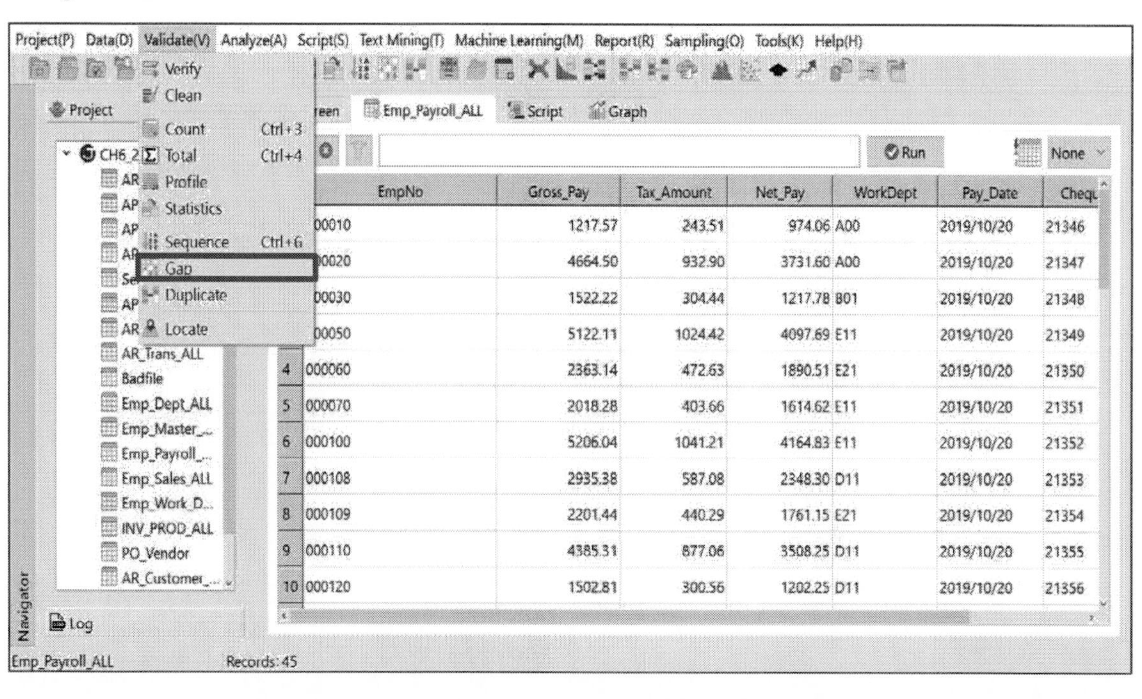

Exercise 6.9

Validate > Gap >Select the values to be verified and set the display record count.

Emp_Payroll_ALL

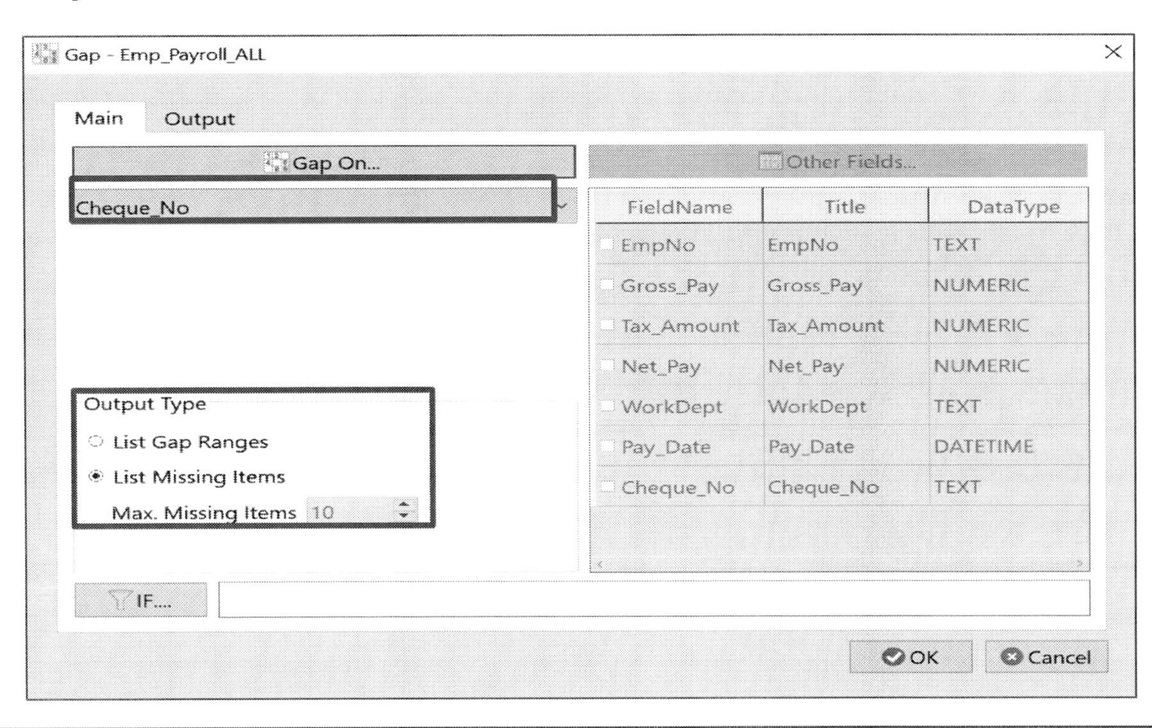

Exercise 6.9

Check the data file related to employees, execute the Gap command

Emp_Payroll_ALL

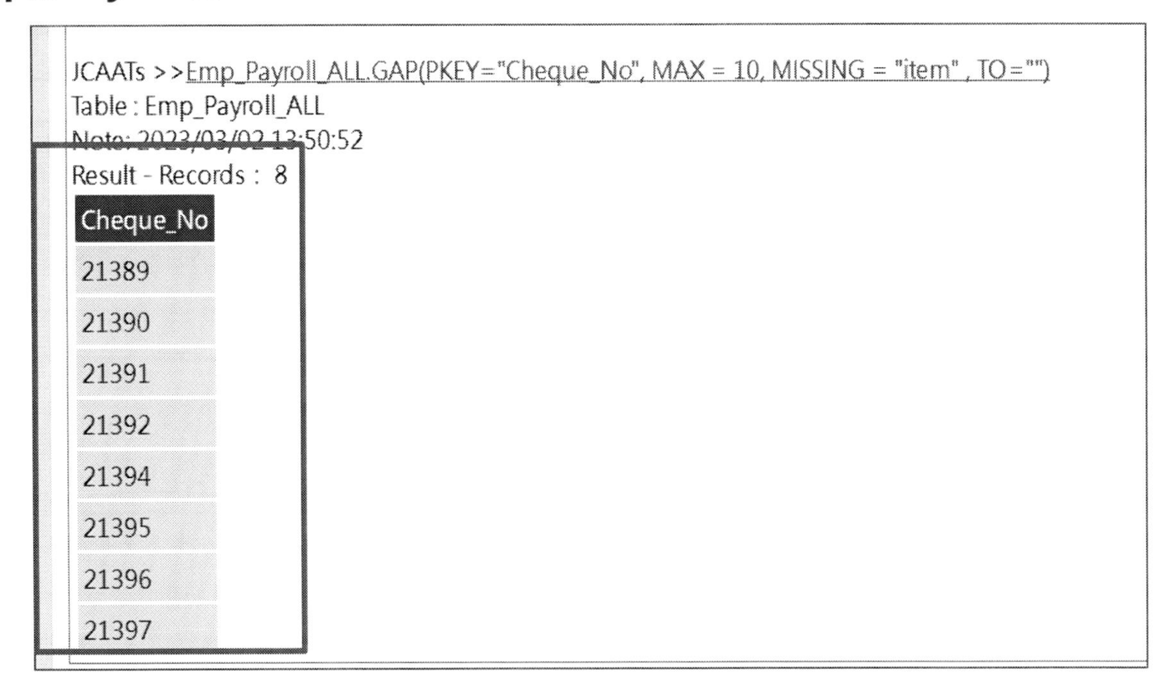

JCAATs - AI Audit Software

Please practice the Duplicate command to determine if there are any duplicate data in the fields that need to be verified, generate an abnormal data table, and facilitate further investigation.

Exercise 6.10

Check the data file related to employees, execute the Duplicate command, and generate a table of duplicate data.

Emp_Payroll_ALL

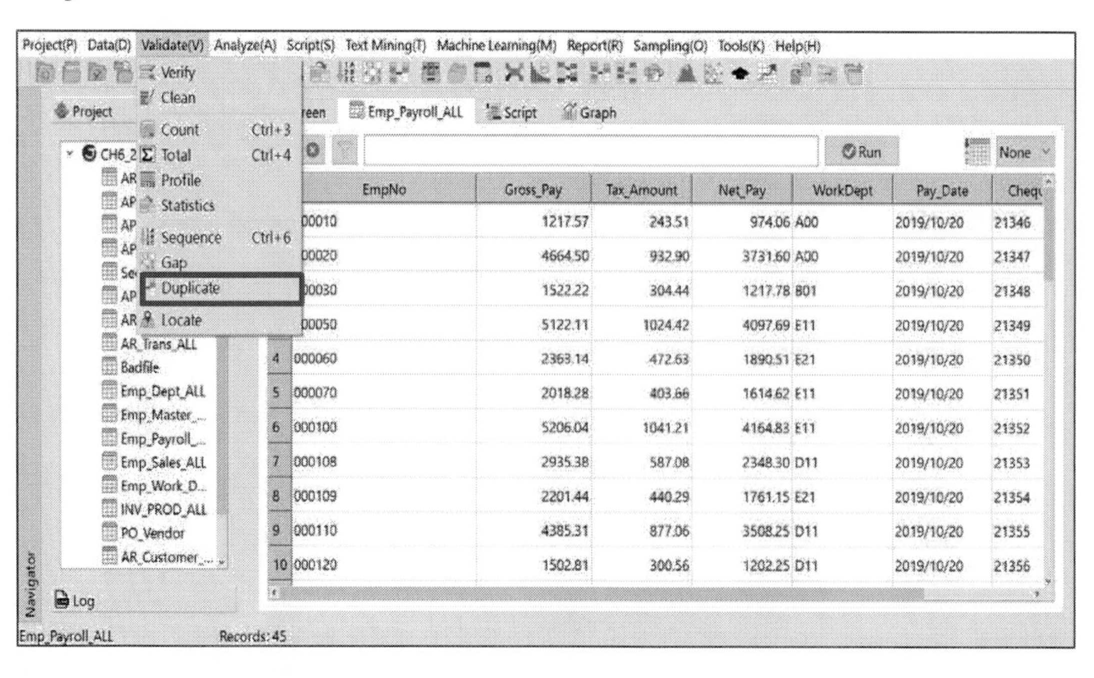

Exercise 6.10

Validate> Duplicate >Select the values and fields to be verified

Emp_Payroll_ALL

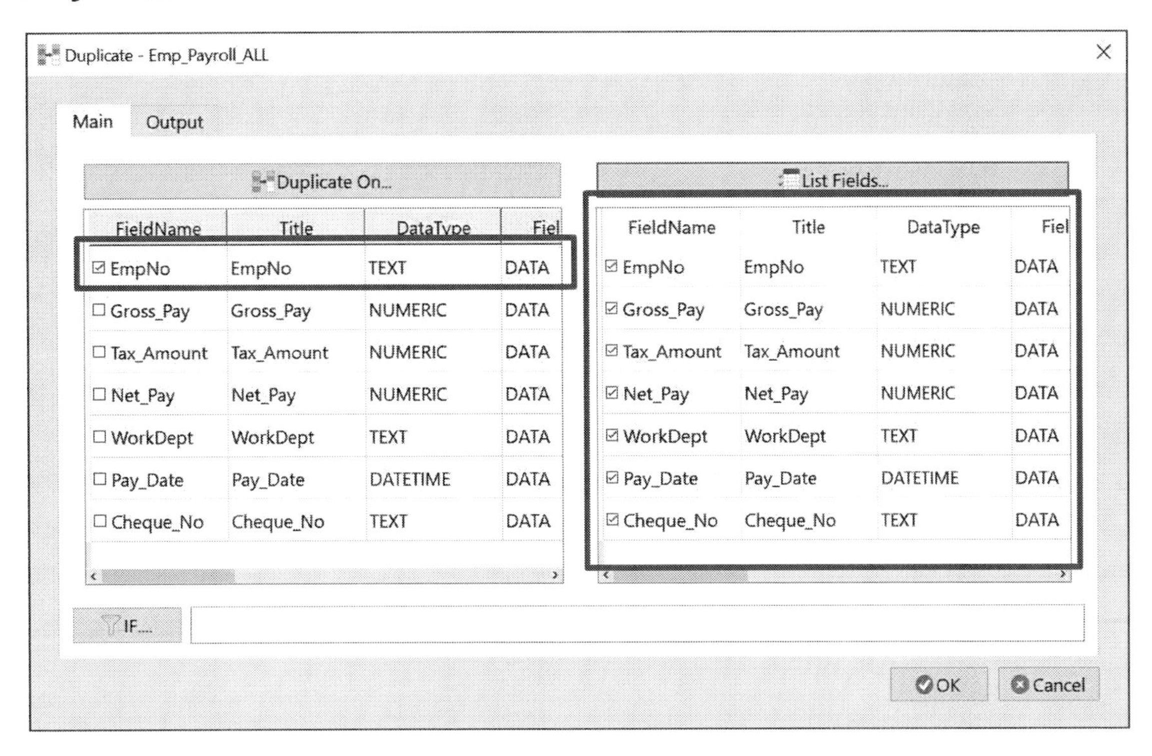

JCAATs-AI Audit Software Copyright © 2023 JACKSOFT.

Exercise 6.10

Check the data file related to employees, execute the Duplicate command, and generate a table of duplicate data.

Emp_Payroll_ALL

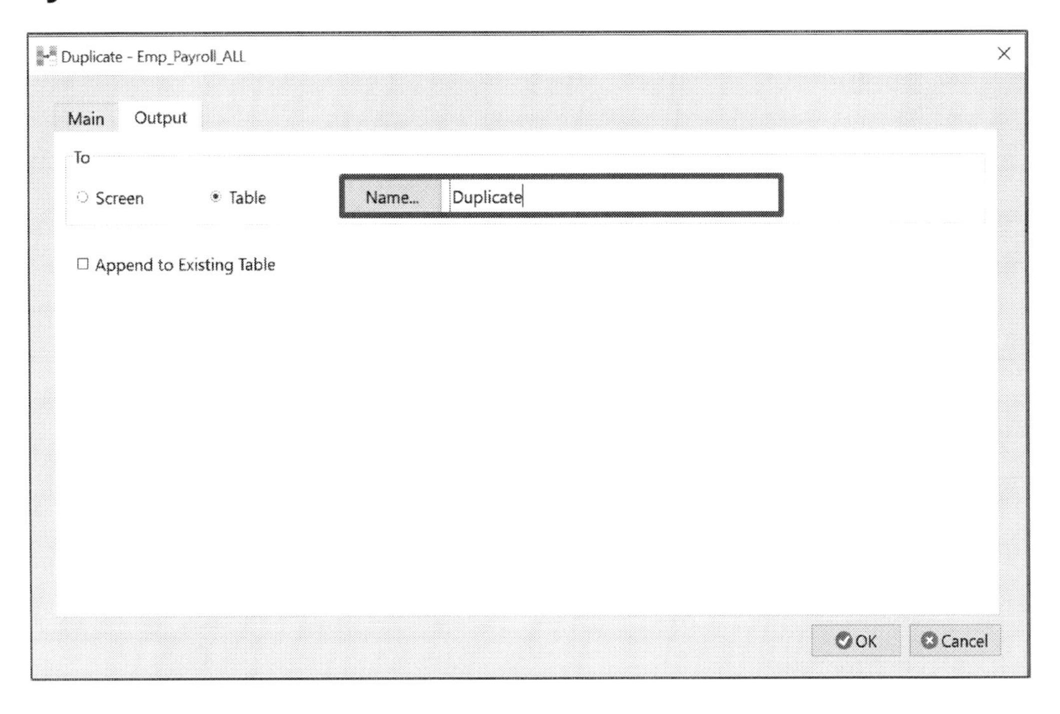

Exercise 6.10
Check the table of duplicate data for the related employee data file.
Emp_Payroll_ALL

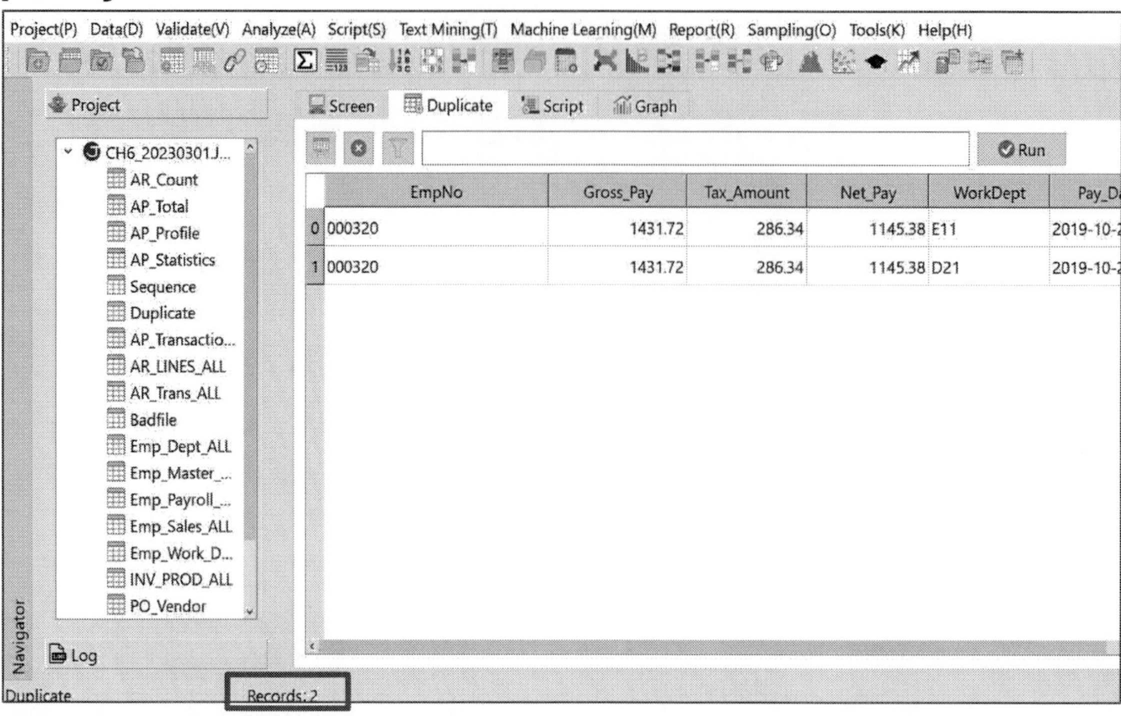

Exercise 6.11 A Case Study

1. Please create a new project file with the project name "Audit". (5 points)
2. Please import Credit_Cards transaction data according to the provided file format (schema) below. (15 points)

Length	Field Name	Note	Type
16	CARDNUM	Card No.	C
10	CREDLIM	Credit Limit	N
6	CUSTNO	Customer No.	C
10	EXPDT	Expired Date	D
10	AMT	Amount	N

3. Verify if the data is incorrect (10 points)

   □ No, there are no errors in the data.

   □ Yes, error in row , column .

Exercise 6.11 A Case Study

4. Number of records (15 points)
   i. The total number of records in Credit_Cards is .
   ii. In Credit_Cards, there are records with a credit limit of 9000.
   iii. In Credit_Cards, there are records with a credit limit greater than 9000.

Exercise 6.11 A Case Study

5. Summation Calculation (25 points)
   i. The total transaction amount of credit card transaction data is
   ii. The highest credit limit in the credit card transaction data belongs to the customer with the ID number

      ____

   iii. The lowest credit limit in the credit card transaction data belongs to the customer with the ID number

      ____

Exercise 6.11 A Case Study

6. Please confirm if there are identical credit card numbers in the credit card transaction data (10 points)
   □ No, there is no such situation in the data.
   □ Yes, the duplicated credit card number is

7. Following the previous question, are there identical credit card numbers but different customer ID numbers in the data? (10 points)
   □ No, there is no such situation in the data.
   □ Yes, the duplicated credit card number is , and the customer ID numbers are respectively."

Exercise 6.11
Please create a new project file named "AR_Audit".

1. Create a new folder.
2. Click JCAATs-AI audit software.
3. Click "Project" > "Select New Project."
4. Define a project name.
5. Save.

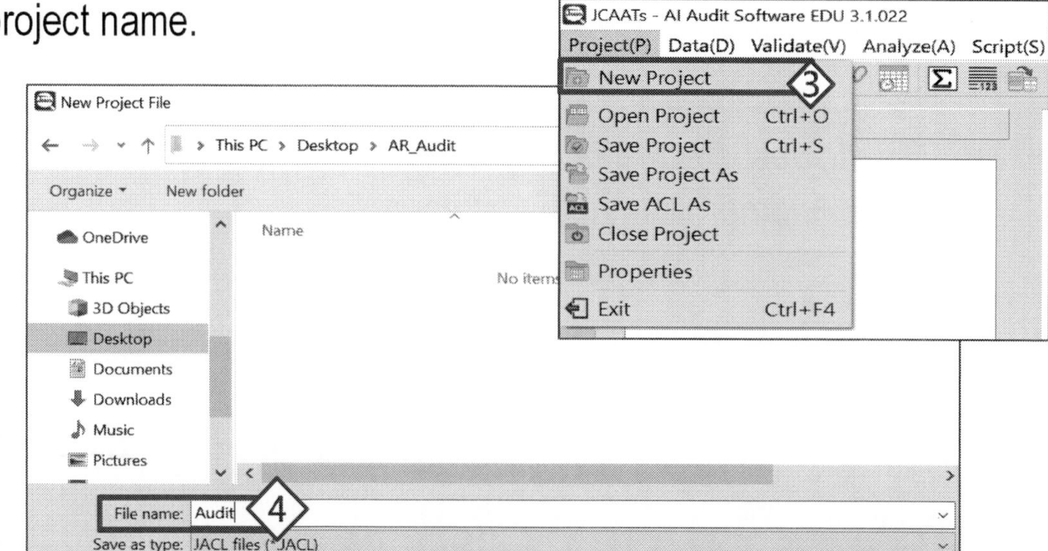

Exercise 6.11

Please import Credit_Cards transaction data according to the provided file format (schema) below.

- Select "Data" > "New Table.
- After selecting the data source platform as "File, " click "Next".

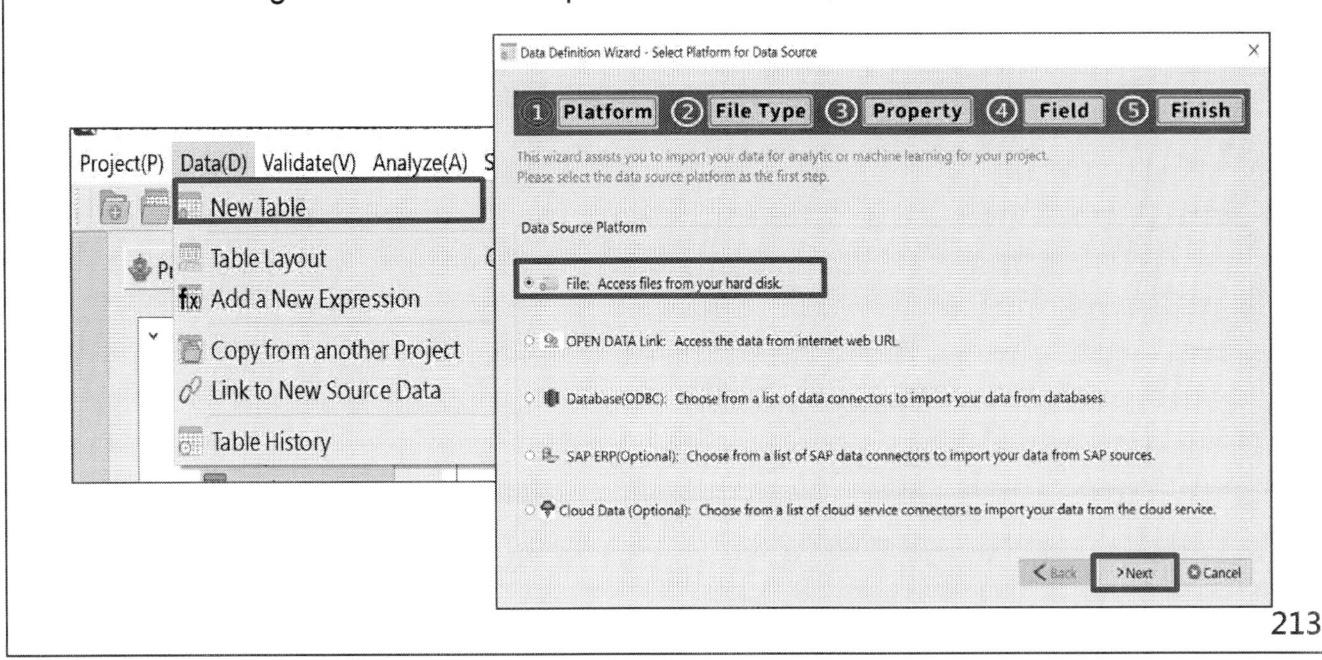

JCAATs-AI Audit Software Copyright © 2023 JACKSOFT.

Exercise 6.11

Please import Credit_Cards transaction data according to the provided file format (schema) below.

- Select Credit_Cards to import.
- Click "Open" to proceed.

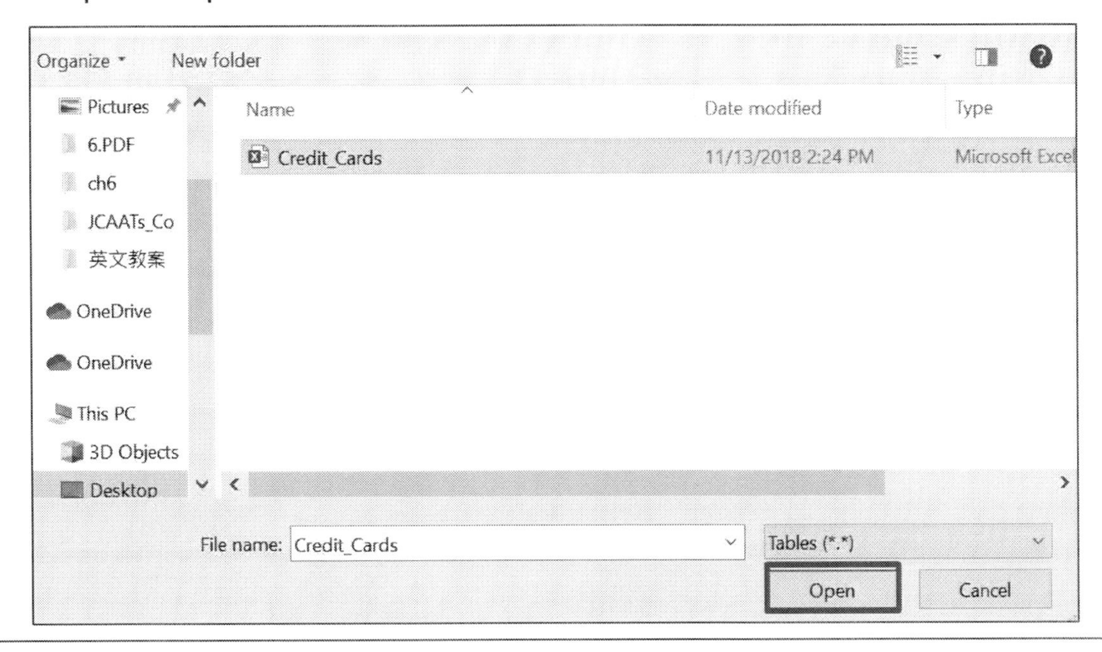

Exercise 6.11

Please import Credit_Cards transaction data according to the provided file format (schema) below.

- JCAATs will detect the file format automatically.
- If there are no errors with the detected format, select "Next" to proceed.

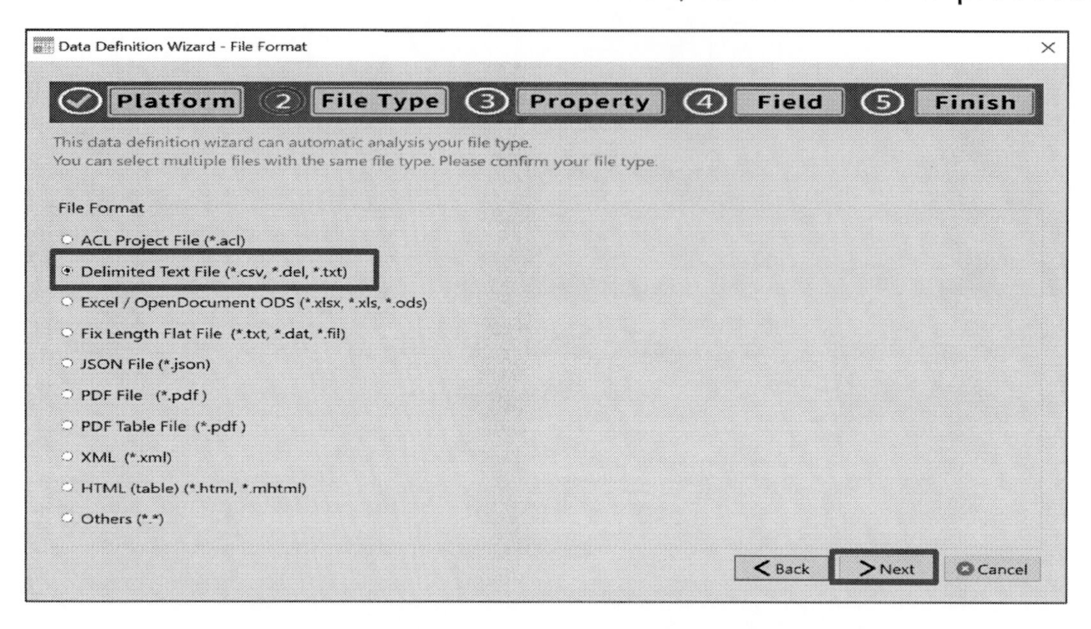

Exercise 6.11

Please import Credit_Cards transaction data according to the provided file format (schema) below.

- Set the data file separator to "Tab" one by one, check the first line field name, set the starting line number to 0, and click "Next" after completing the settings.

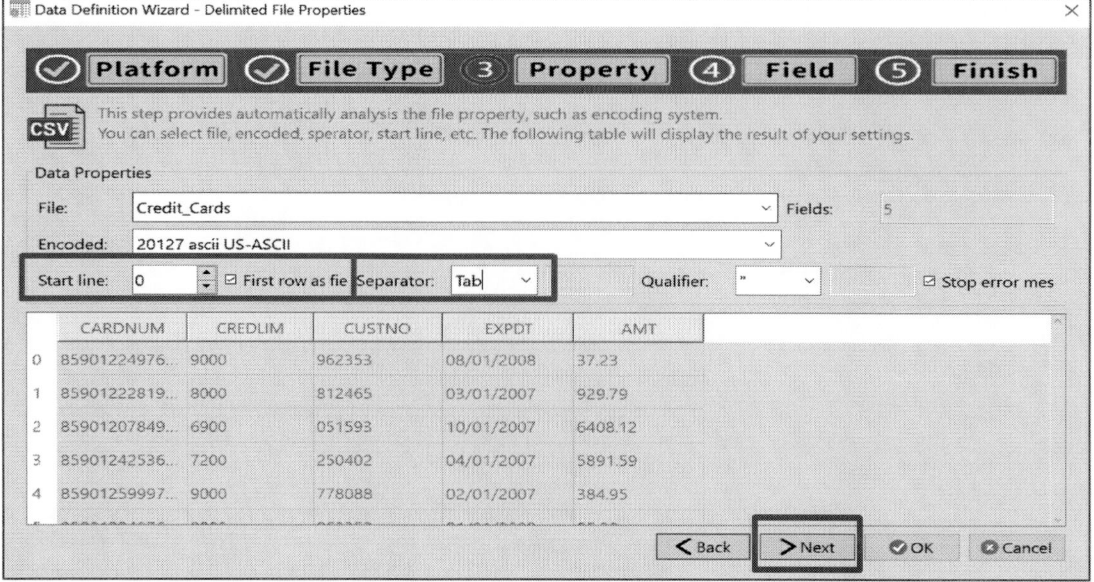

Exercise 6.11

Please import Credit_Cards transaction data according to the provided file format (schema) below.

- Define the field name , display name , data type and data format of each data table field one by one. After setting, select "Next".

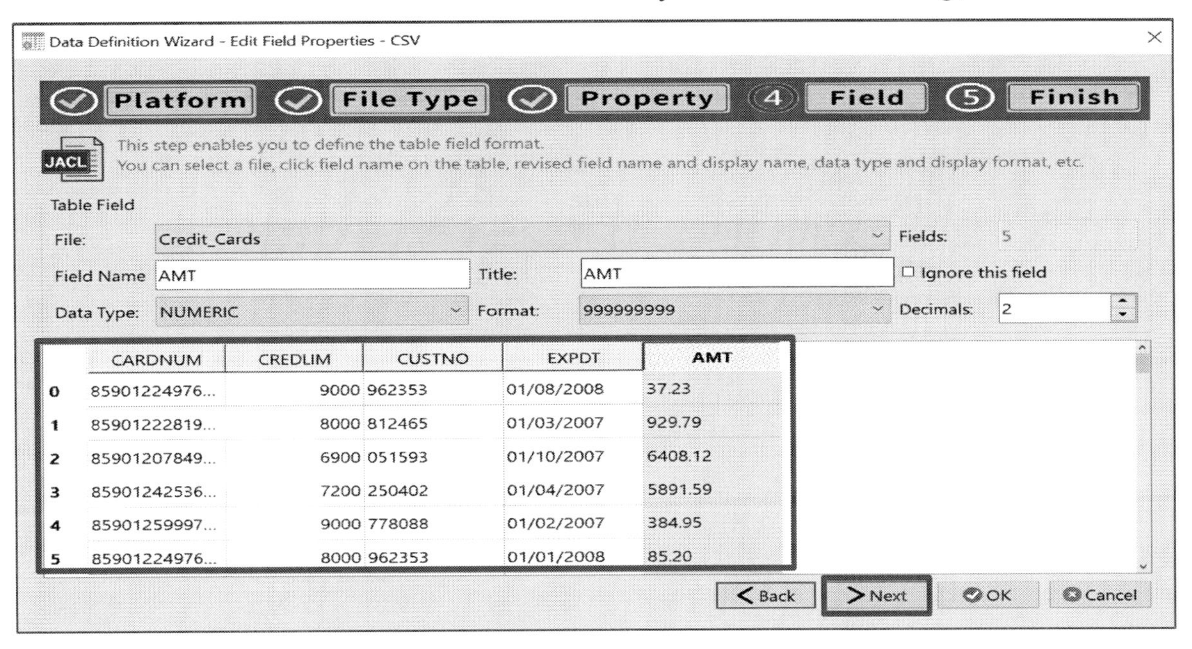

JCAATs-AI Audit Software
Copyright © 2023 JACKSOFT.

Exercise 6.11

Please import Credit_Cards transaction data according to the provided file format (schema) below.

- The default data file path is the project folder, but it can be modified as needed. After ensuring that the path and information are correct, select "Done".

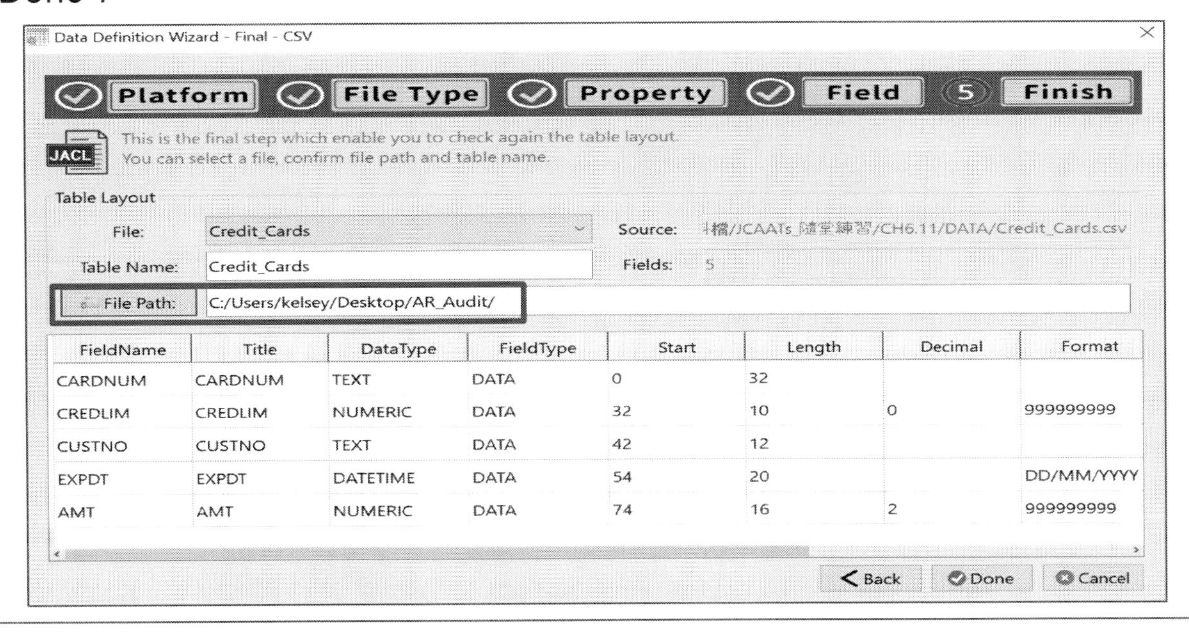

Exercise 6.11
Please import Credit_Cards transaction data according to the provided file format (schema) below.

- After the import progress is completed, we can see that there are two data tables successfully imported.

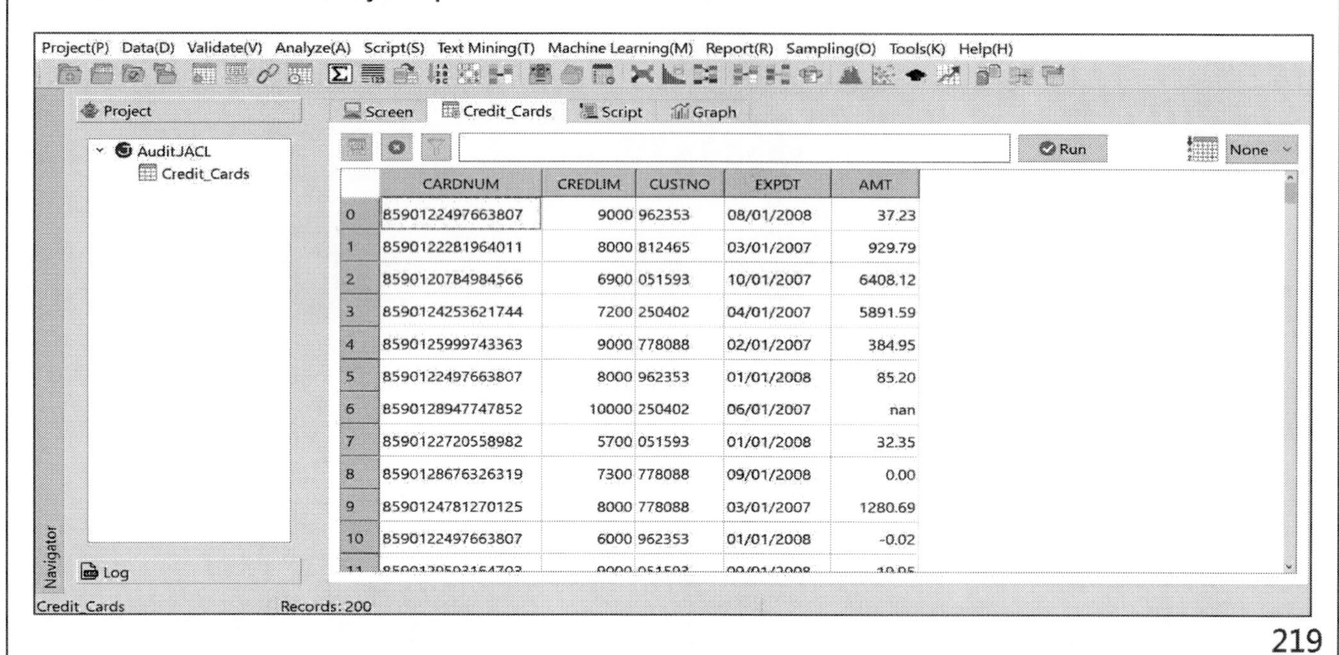

Exercise 6.11
Verify if the data is incorrect

- Click on "Validate>Verify "
- Select "Verify On" and choose "Add All" for selected fields.

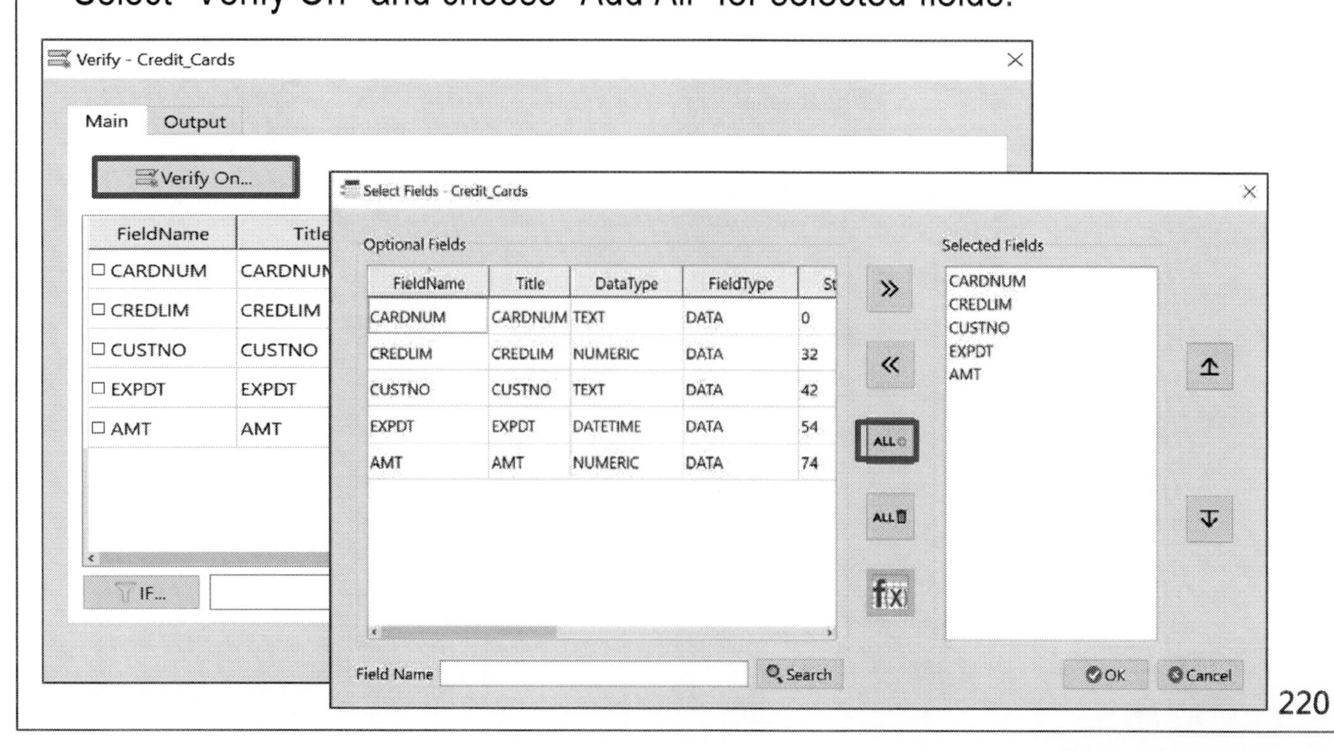

Exercise 6.11
Verify if the data is incorrect

JCAATs >>Credit_Cards.VERIFY(PKEYS=["CARDNUM","CREDLIM","CUSTNO","EXPDT","AMT"], TO="")
Table : Credit_Cards
Note: 2023/03/02 09:40:46
Result - Records : 2

Table_Name	Field_Name	Validity_Type	Record_No.	Value
Credit_Cards	EXPDT	Null Value	14	nan
Credit_Cards	AMT	Null Value	6	nan

Exercise 6.11
The total number of records in Credit_Cards is

- Open the " Credit_Cards" Table
- Click on " Validate>Total "

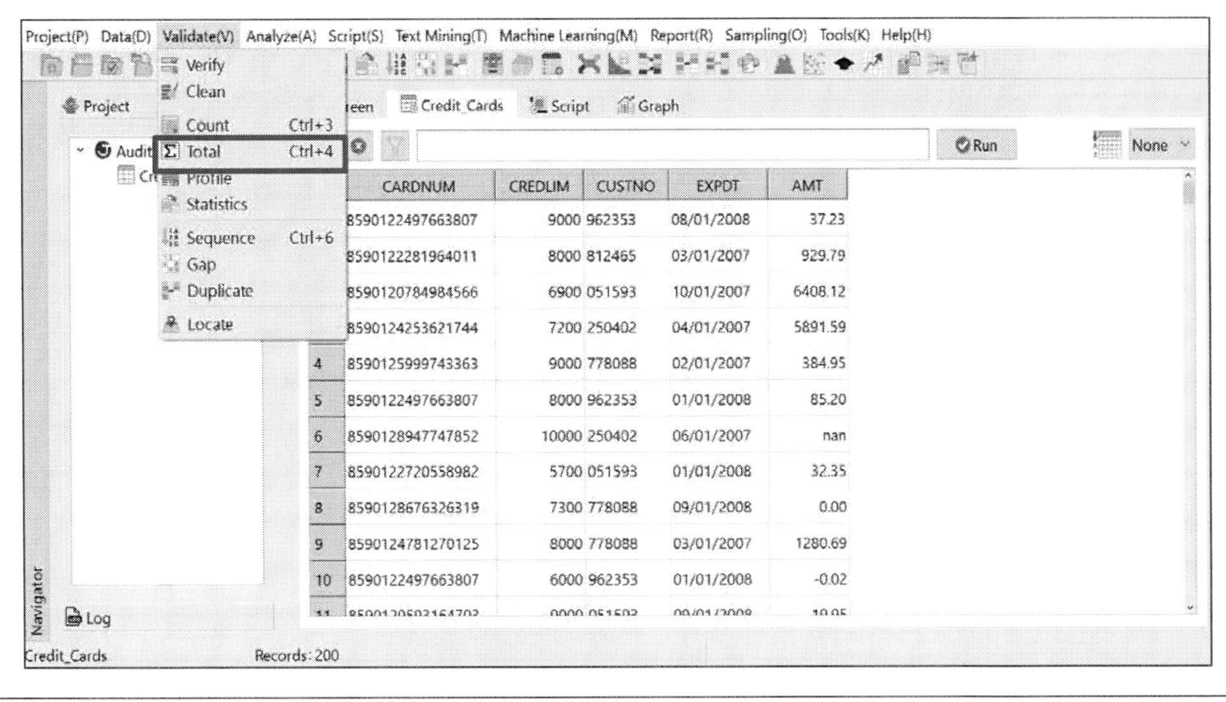

Exercise 6.11
The total number of records in Credit_Cards is

- Select the fields to total. (AMT)

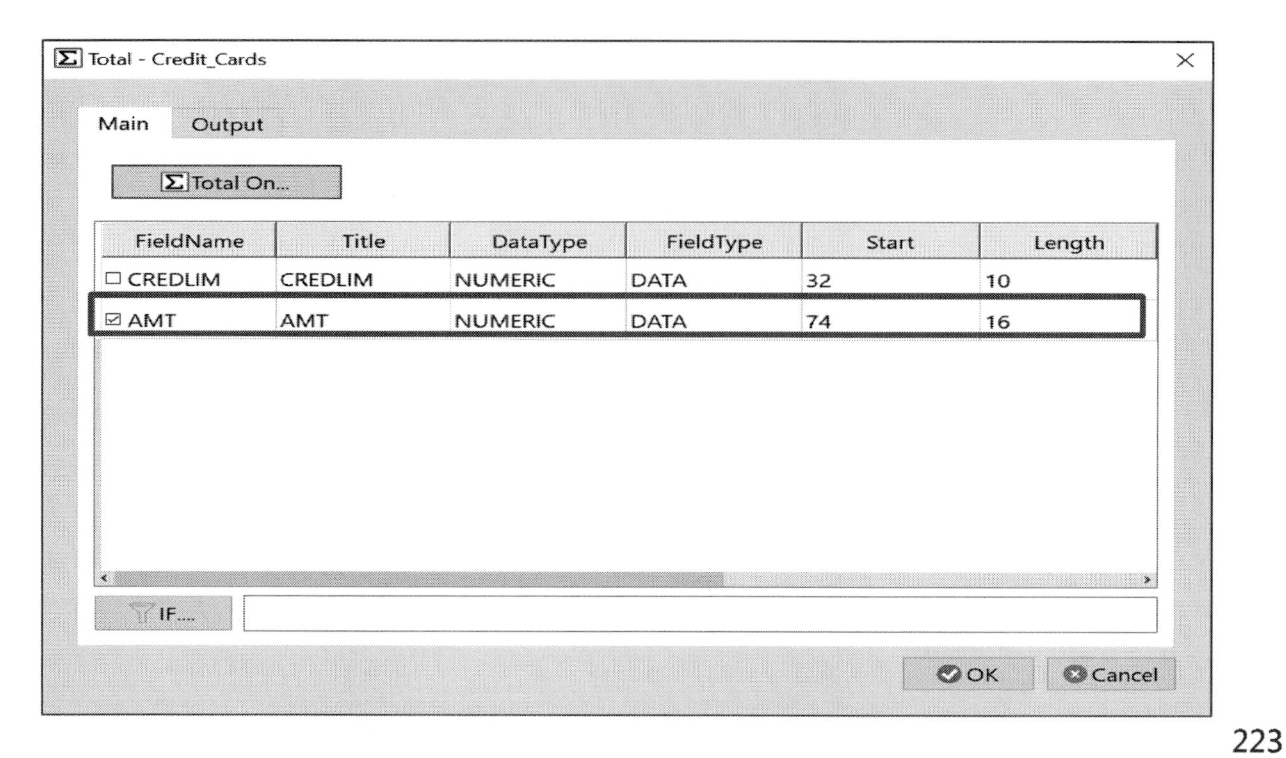

JCAATs-AI Audit Software
Copyright © 2023 JACKSOFT.

Exercise 6.11
The total number of records in Credit_Cards is

- Output to Screen

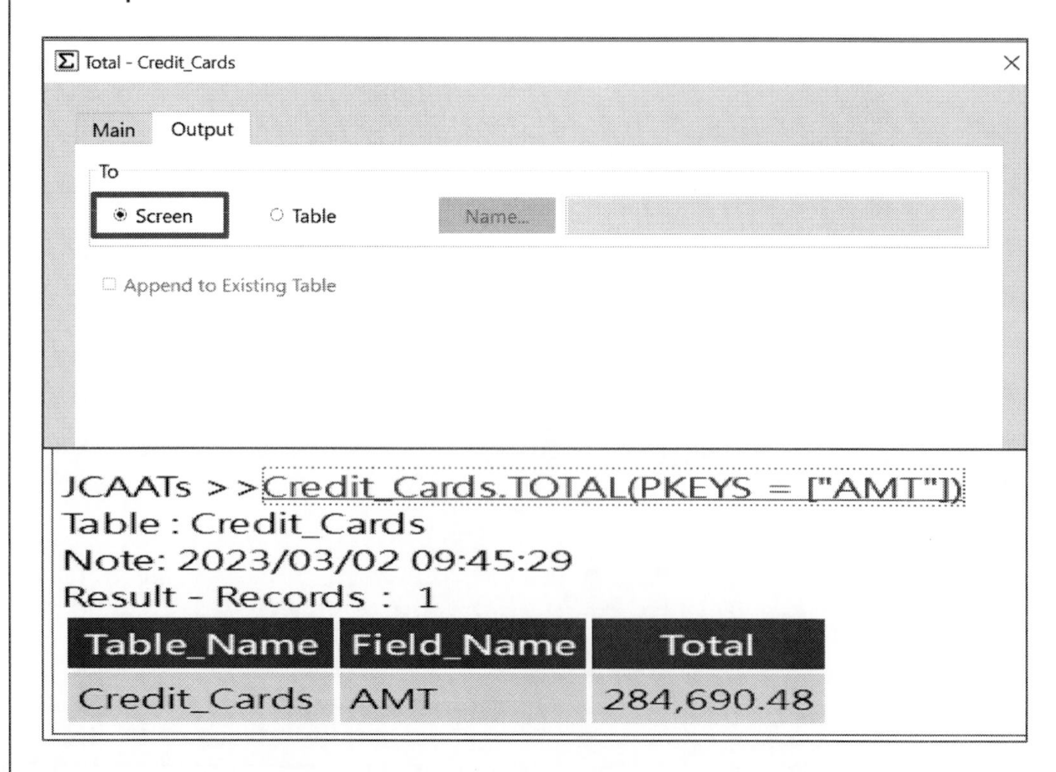

Exercise 6.11
In Credit_Cards, there are records with a credit limit of 9000.

- Select "Set Filter"

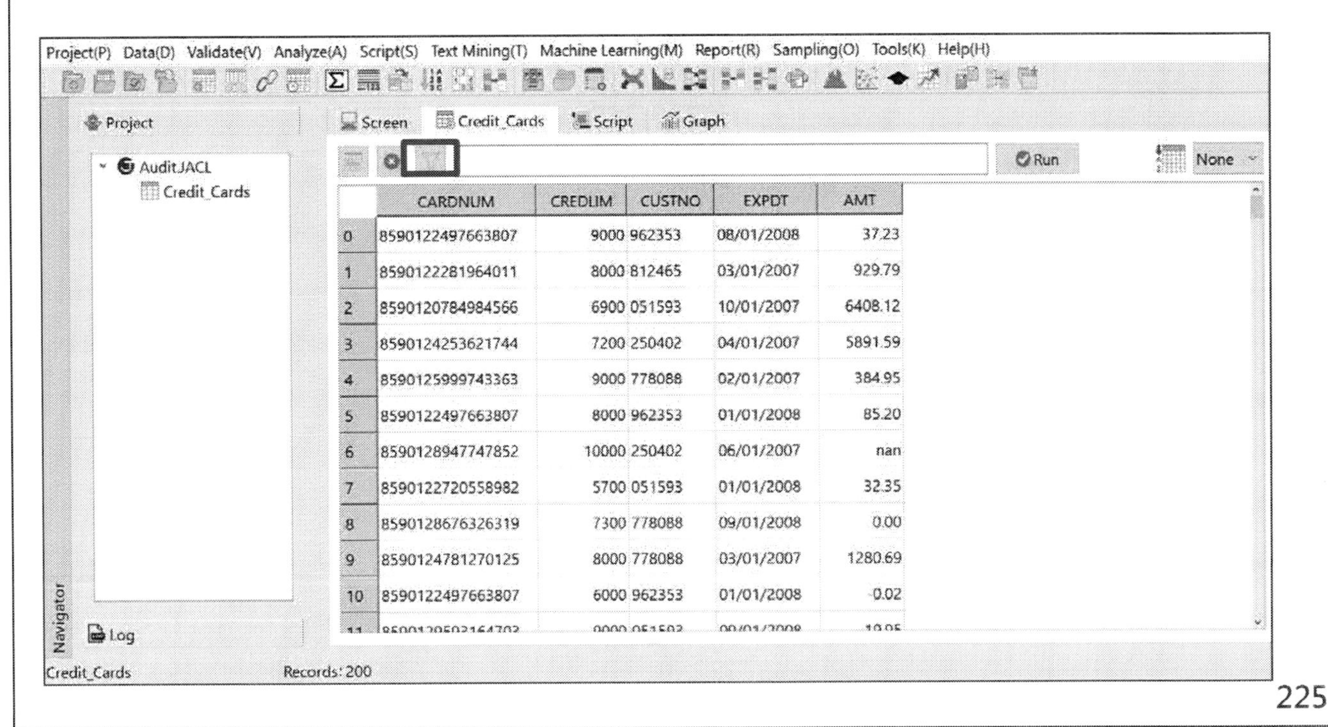

Exercise 6.11
In Credit_Cards, there are records with a credit limit of 9000.

- Add a filter condition:
  CREDLM == 9000

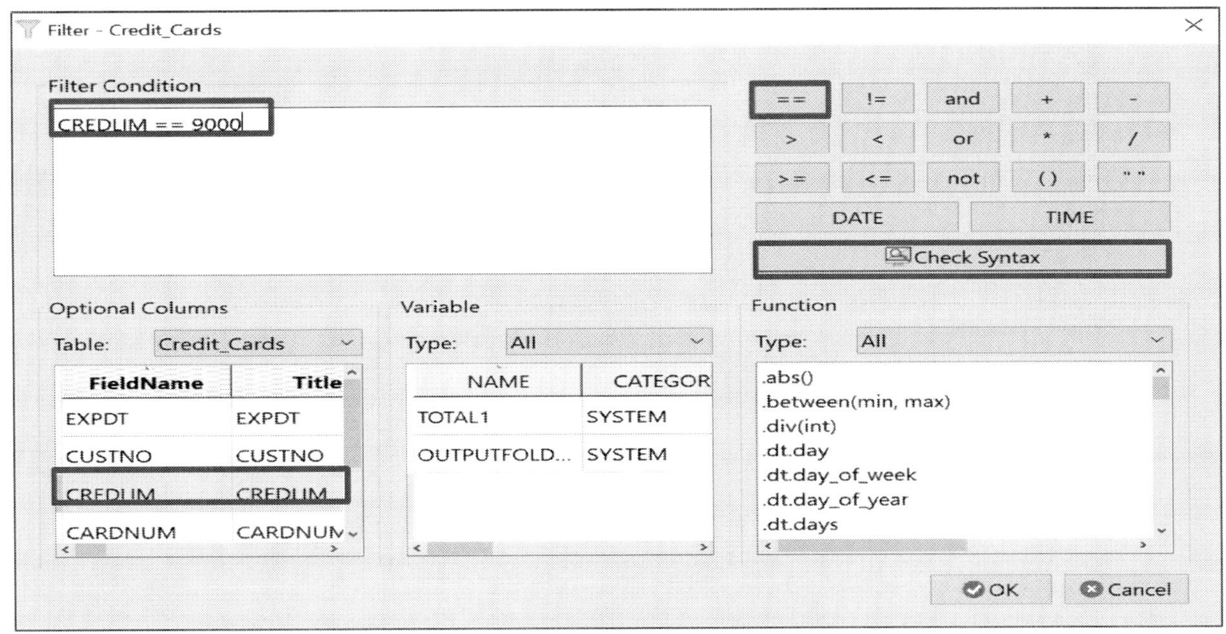

Exercise 6.11

In Credit_Cards, there are records with a credit limit of 9000.

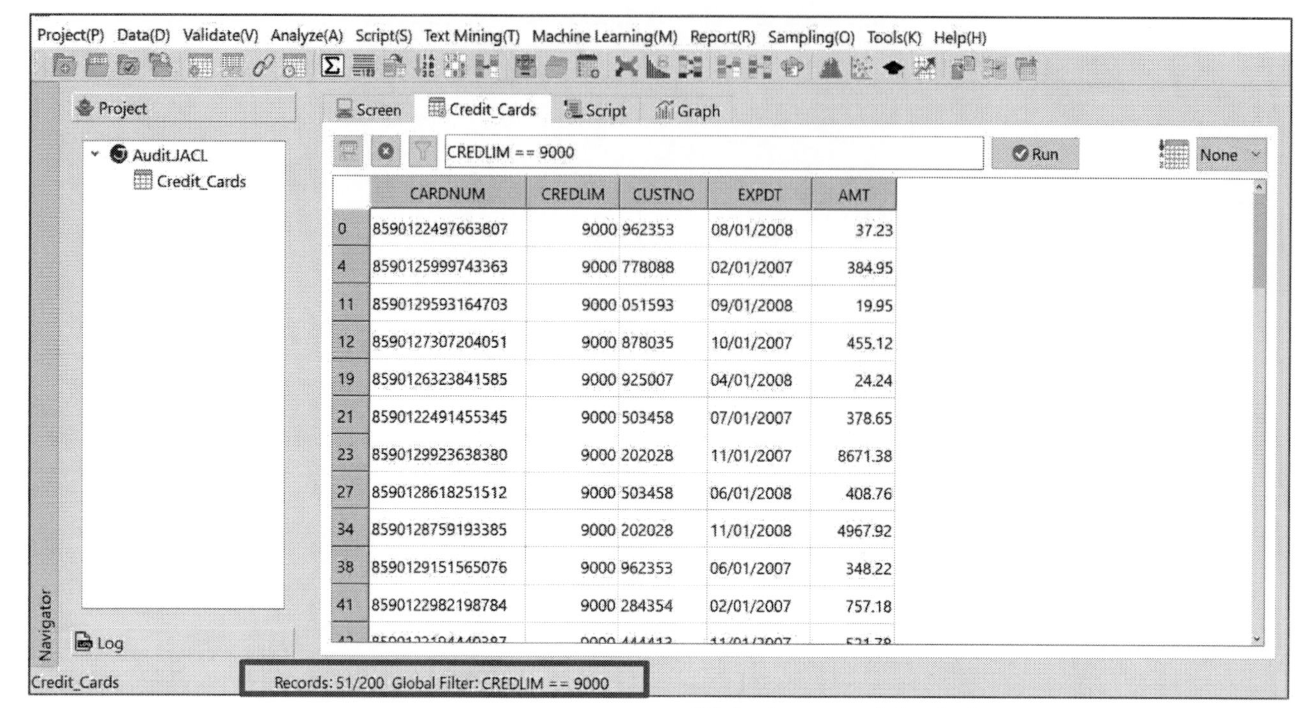

	CARDNUM	CREDLIM	CUSTNO	EXPDT	AMT
0	8590122497663807	9000	962353	08/01/2008	37.23
4	8590125999743363	9000	778088	02/01/2007	384.95
11	8590129593164703	9000	051593	09/01/2008	19.95
12	8590127307204051	9000	878035	10/01/2007	455.12
19	8590126323841585	9000	925007	04/01/2008	24.24
21	8590122491455345	9000	503458	07/01/2007	378.65
23	8590129923638380	9000	202028	11/01/2007	8671.38
27	8590128618251512	9000	503458	06/01/2008	408.76
34	8590128759193385	9000	202028	11/01/2008	4967.92
38	8590129151565076	9000	962353	06/01/2007	348.22
41	8590122982198784	9000	284354	02/01/2007	757.18

51 instances detected

JCAATs-AI Audit Software
Copyright © 2023 JACKSOFT.

Exercise 6.11

In Credit_Cards, there are records with a credit limit greater than 9000.

- Add a filter condition: CREDLM > 9000

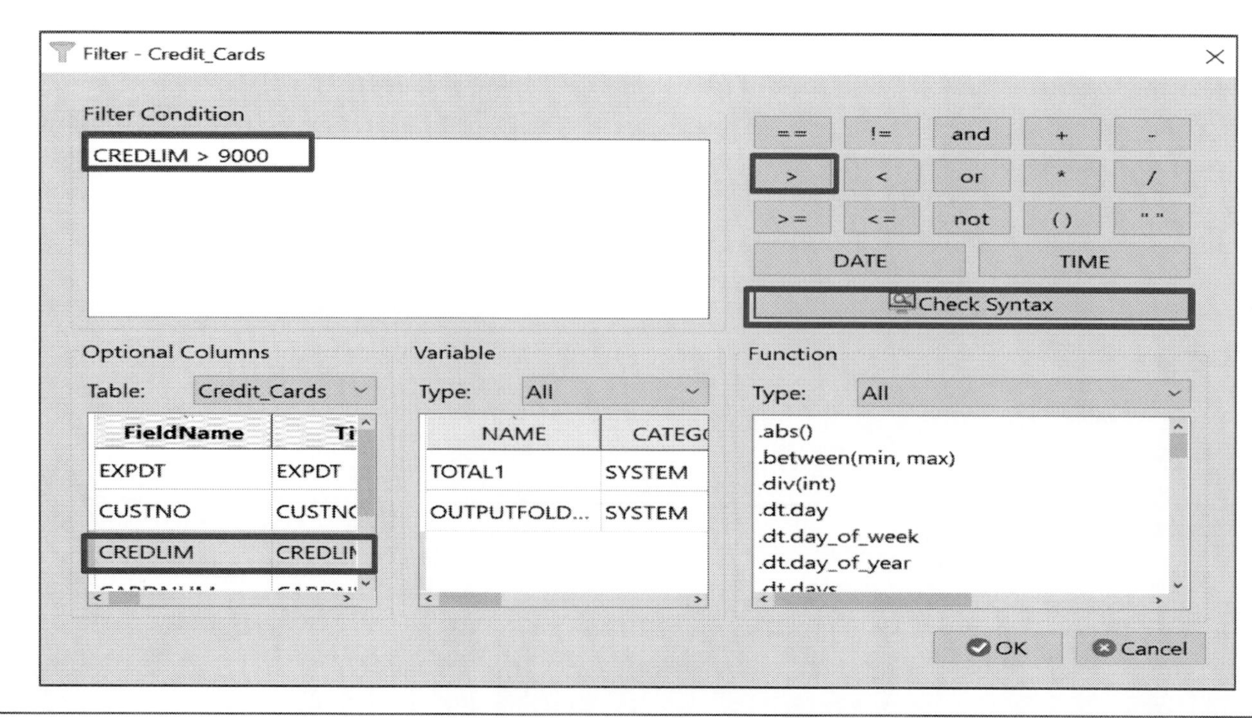

Exercise 6.11

In Credit_Cards, there are records with a credit limit greater than 9000.

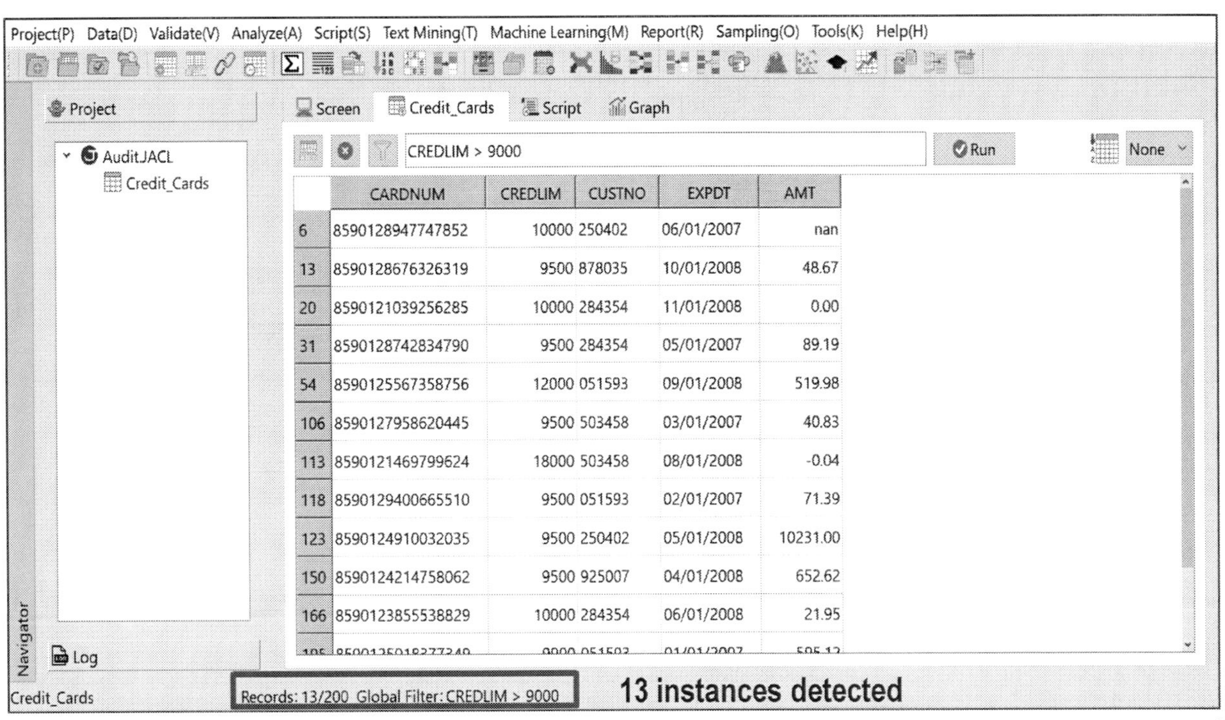

Exercise 6.11

The highest credit limit in the credit card transaction data belongs to the customer with the ID number

- Open the " Credit_Cards" Table
- Click on " Analyze>Summarize "

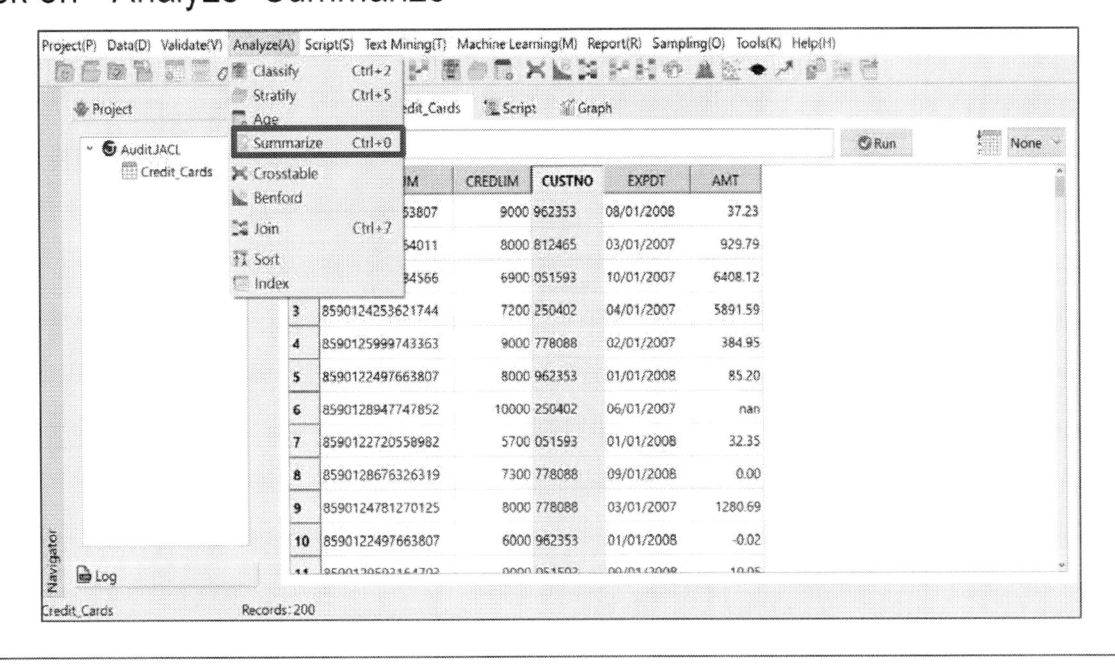

Exercise 6.11
The highest credit limit in the credit card transaction data belongs to the customer with the ID number

- Select the field to Summarize. (CUSTNO)
- Subtotal CREDLIM, choose CARDNUM for other fields.

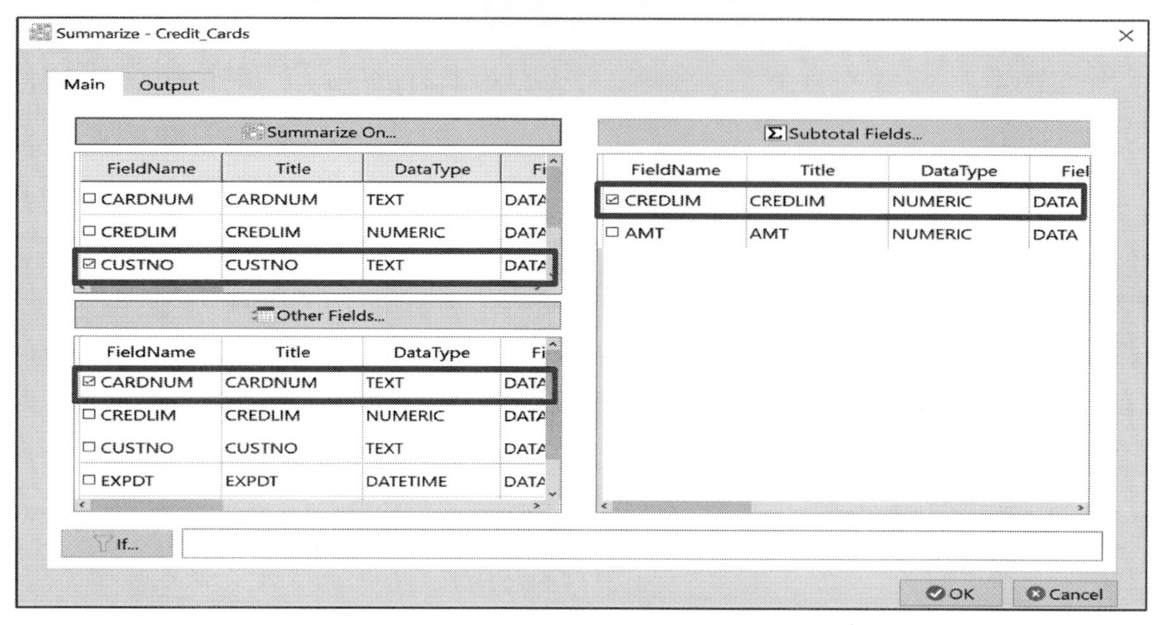

JCAATs-AI Audit Software
Copyright © 2023 JACKSOFT.

Exercise 6.11
The highest credit limit in the credit card transaction data belongs to the customer with the ID number

- Output to a new table

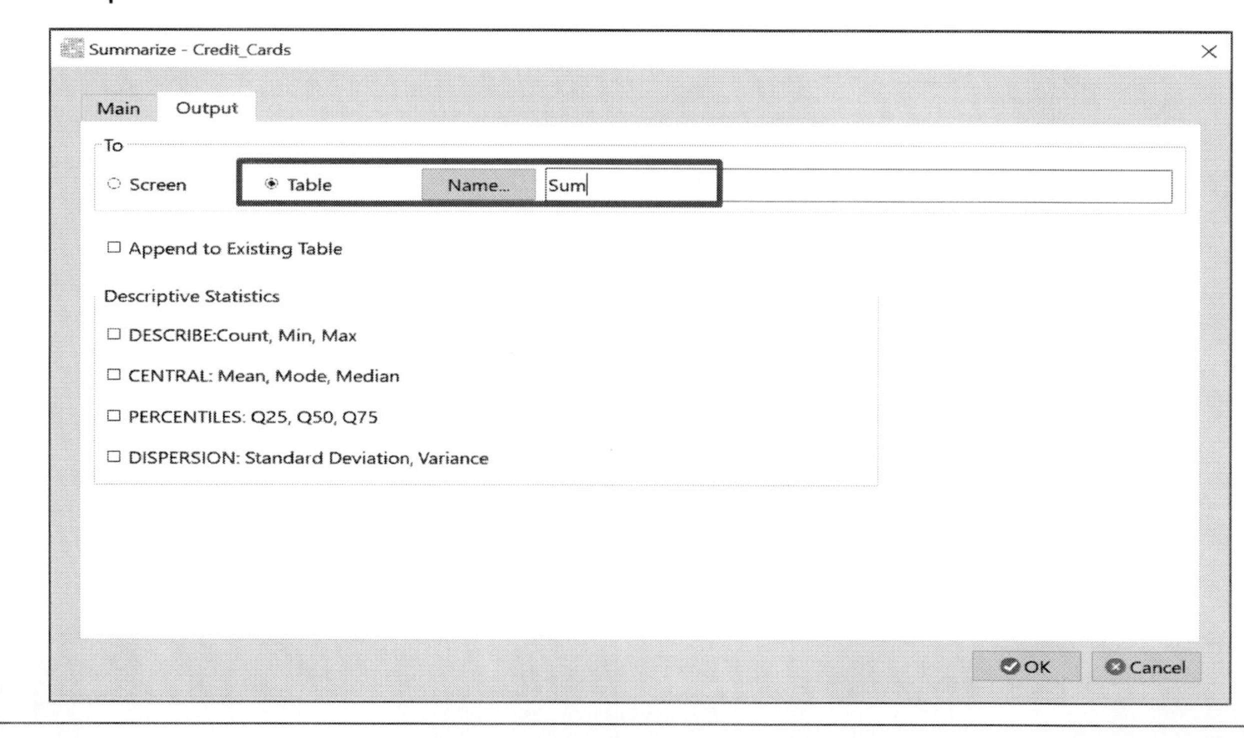

Exercise 6.11
The highest credit limit in the credit card transaction data belongs to the customer with the ID number

- Right-click on the field to sort the column in descending order.

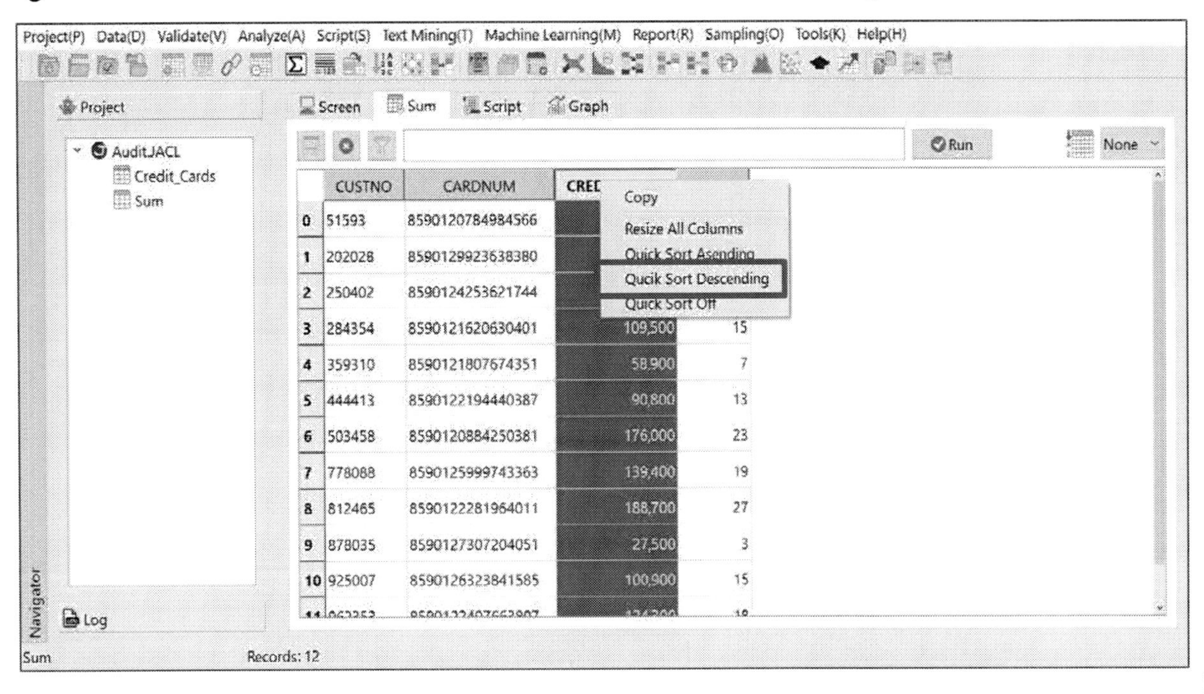

	CUSTNO	CARDNUM	CRED	
0	51593	8590120784984566		
1	202028	8590129923638380		
2	250402	8590124253621744		
3	284354	8590121620630401	109,500	15
4	359310	8590121807674351	58,900	7
5	444413	8590122194440387	90,800	13
6	503458	8590120884250381	176,000	23
7	778088	8590125999743363	139,400	19
8	812465	8590122281964011	188,700	27
9	878035	8590127307204051	27,500	3
10	925007	8590126323841585	100,900	15

JCAATs-AI Audit Software Copyright © 2023 JACKSOFT.

Exercise 6.11
The highest credit limit in the credit card transaction data belongs to the customer with the ID number

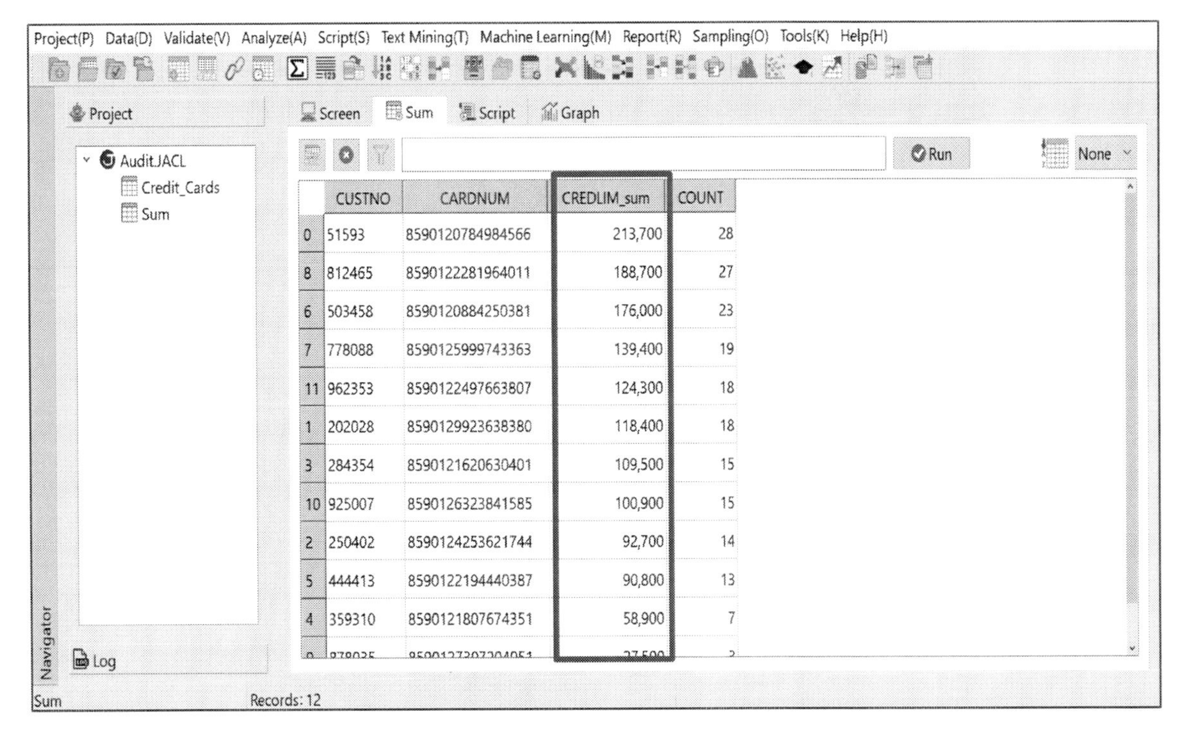

	CUSTNO	CARDNUM	CREDLIM_sum	COUNT
0	51593	8590120784984566	213,700	28
8	812465	8590122281964011	188,700	27
6	503458	8590120884250381	176,000	23
7	778088	8590125999743363	139,400	19
11	962353	8590122497663807	124,300	18
1	202028	8590129923638380	118,400	18
3	284354	8590121620630401	109,500	15
10	925007	8590126323841585	100,900	15
2	250402	8590124253621744	92,700	14
5	444413	8590122194440387	90,800	13
4	359310	8590121807674351	58,900	7

Exercise 6.11

The lowest credit limit in the credit card transaction data belongs to the customer with the ID number

- Right-click on the field to sort the column in ascending order.

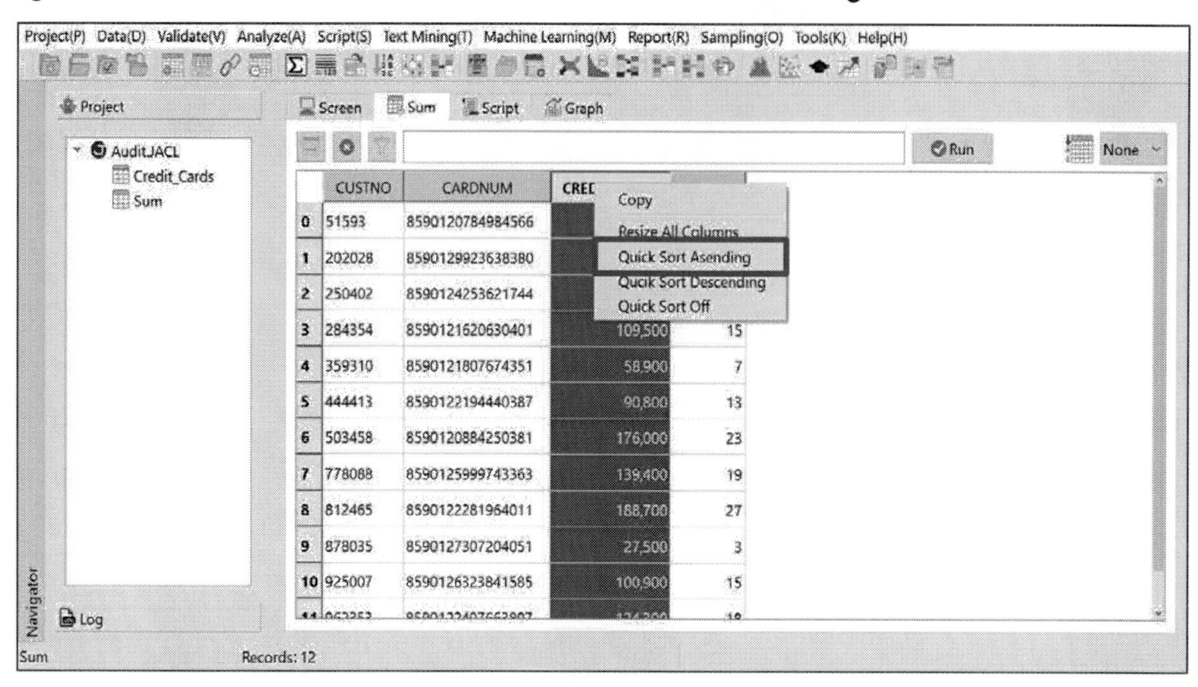

JCAATs-AI Audit Software

Copyright © 2023 JACKSOFT.

Exercise 6.11

The lowest credit limit in the credit card transaction data belongs to the customer with the ID number

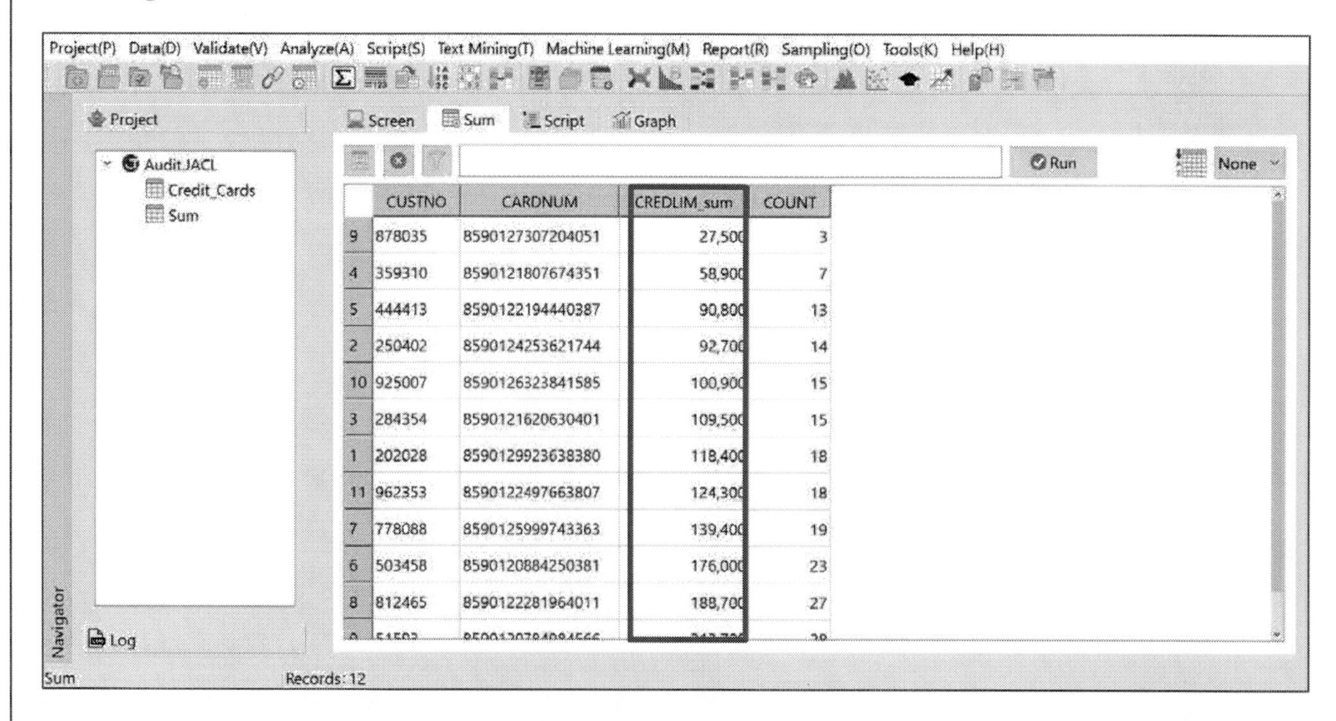

	CUSTNO	CARDNUM	CREDLIM_sum	COUNT
9	878035	8590127307204051	27,500	3
4	359310	8590121807674351	58,900	7
5	444413	8590122194440387	90,800	13
2	250402	8590124253621744	92,700	14
10	925007	8590126323841585	100,900	15
3	284354	8590121620630401	109,500	15
1	202028	8590129923638380	118,400	18
11	962353	8590122497663807	124,300	18
7	778088	8590125999743363	139,400	19
6	503458	8590120884250381	176,000	23
8	812465	8590122281964011	188,700	27

Exercise 6.11

Please confirm if there are identical credit card numbers in the credit card transaction data

- Click on " Validate>Duplicate "

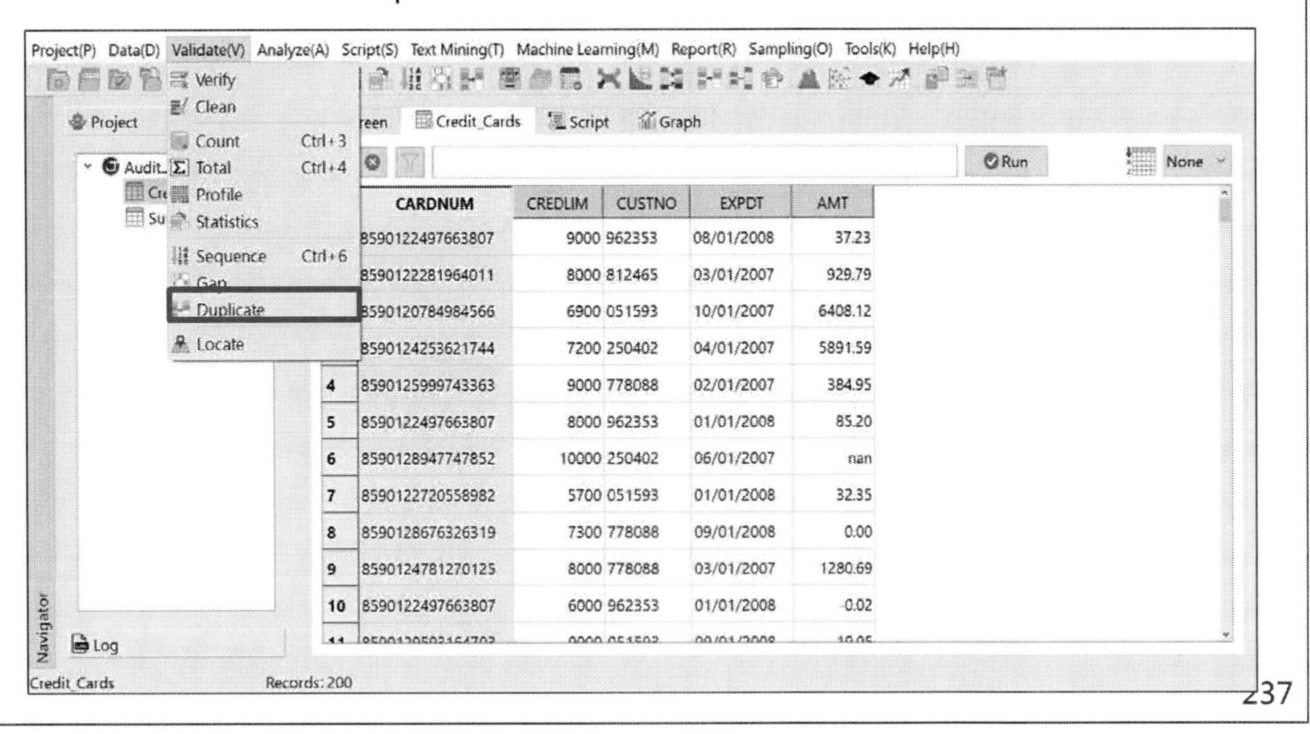

Exercise 6.11

Please confirm if there are identical credit card numbers in the credit card transaction data

- Select the field to Duplicate. (CARDNUM)

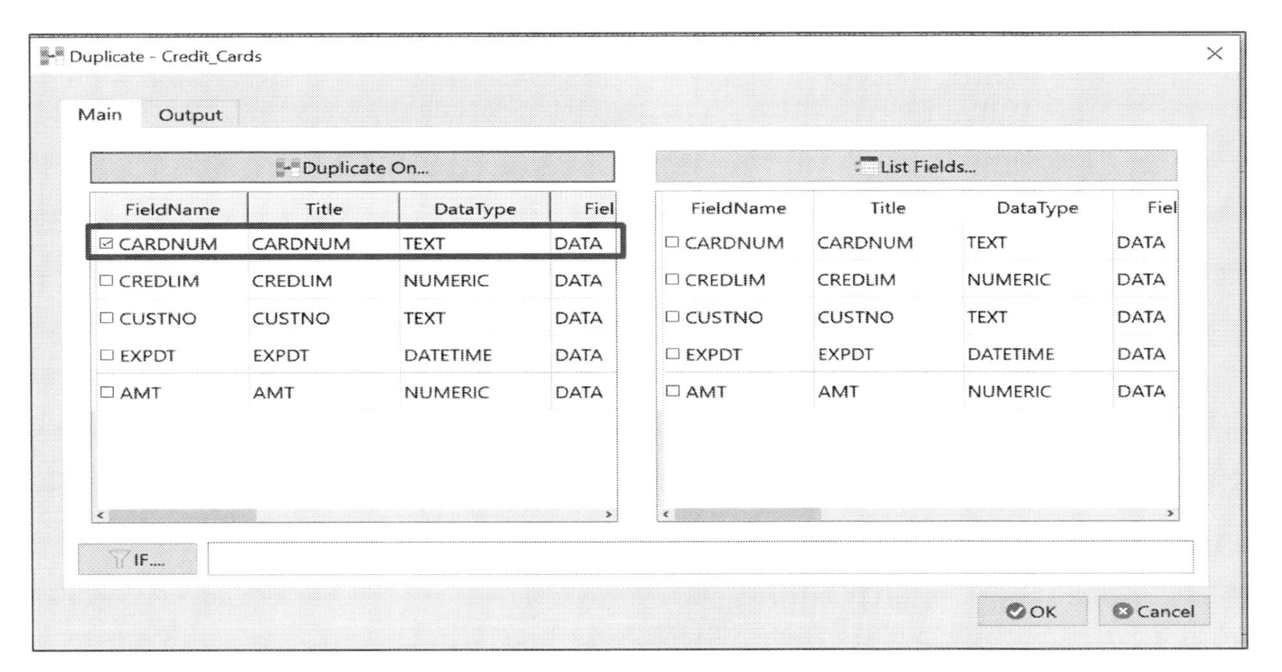

Exercise 6.11

Please confirm if there are identical credit card numbers in the credit card transaction data

- Output to Screen

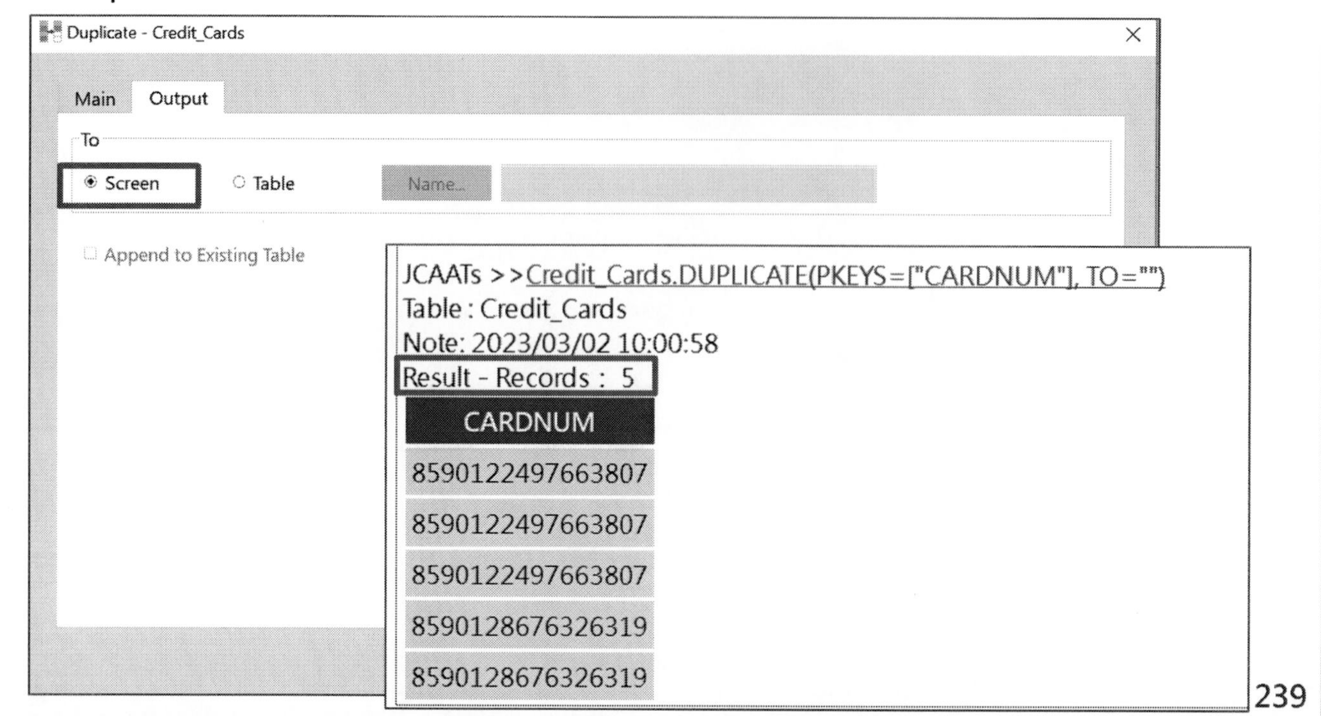

JCAATs-AI Audit Software Copyright © 2023 JACKSOFT.

Exercise 6.11

Following the previous question, are there identical credit card numbers but different customer ID numbers in the data?

- Select the field to Duplicate. (CARDNUM)
- Choose the fields to include in the duplicate table. (CUSTNO)

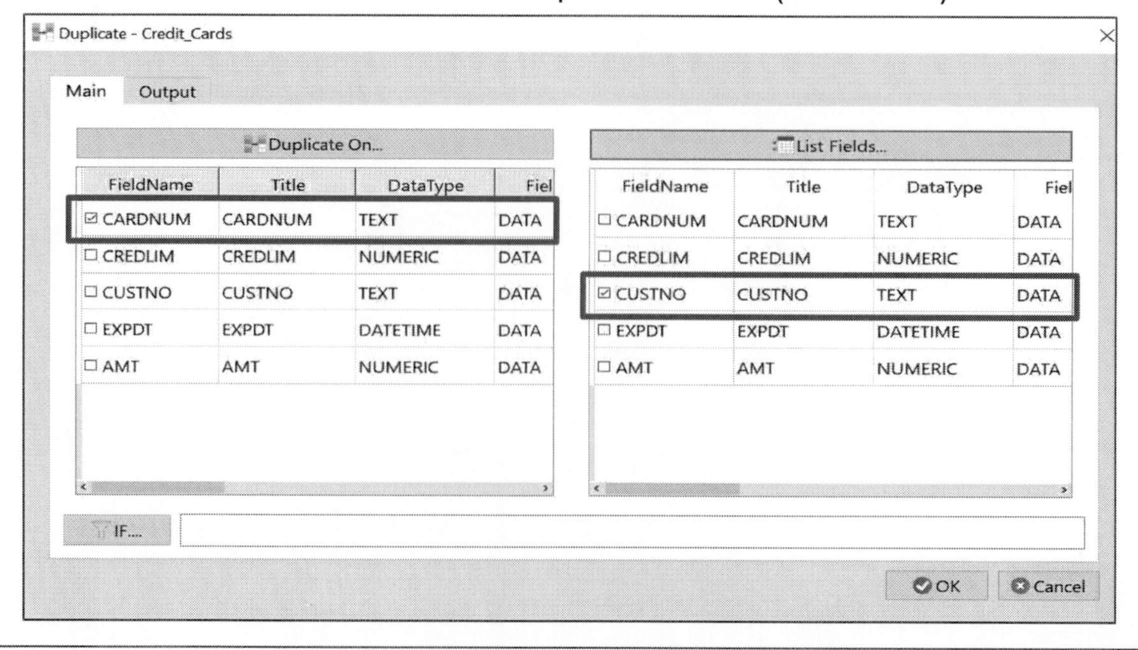

Exercise 6.11
Following the previous question, are there identical credit card numbers but different customer ID numbers in the data?

- Output to Screen

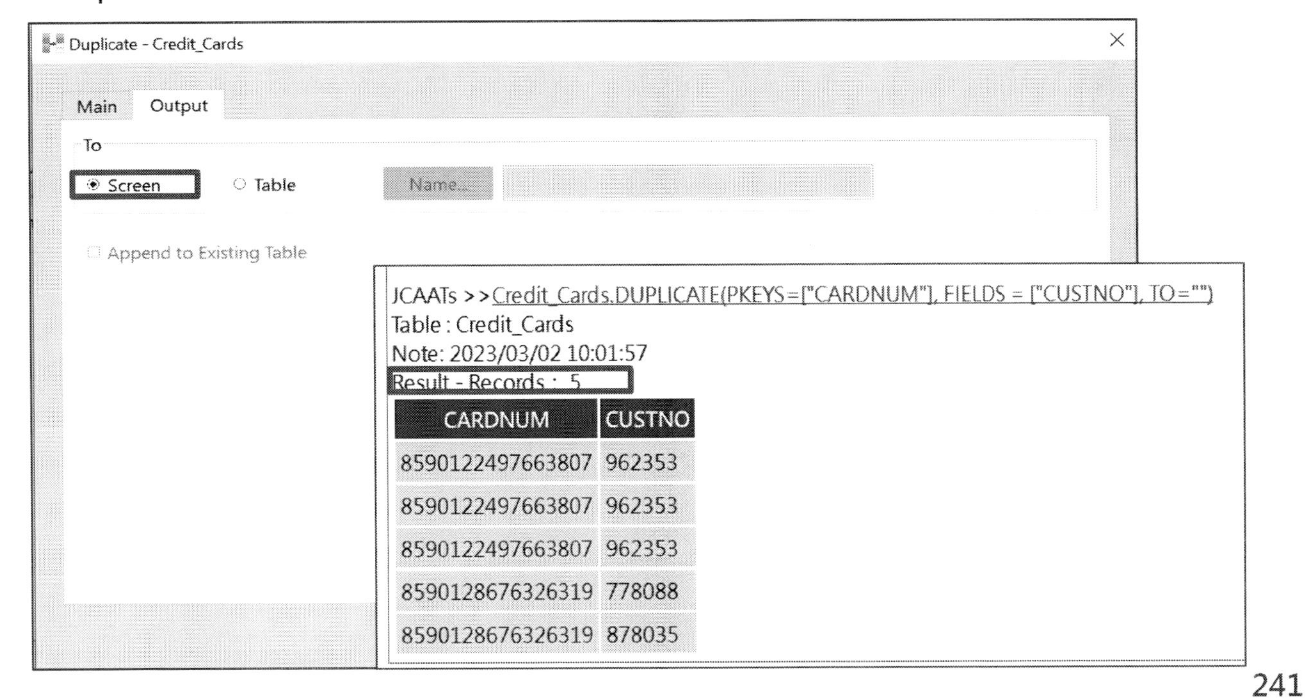

CARDNUM	CUSTNO
8590122497663807	962353
8590122497663807	962353
8590122497663807	962353
8590128676326319	778088
8590128676326319	878035

JCAATs Learning Note:

JCAATs - AI Audit Software

Chapter 7 - Exercise

Chapter 7 - Exercise

- Practice 7.11: Please use the **Join** command. Use the employee data table (Emp_Master_ALL) as the primary table and the employee payroll table (Emp_Payroll_ALL) as the secondary table. Match the two tables using the employee number as the matching field. Select and display the employee number, work department, payment period (EmpNo, WorkDept, Pay_Per_Period) columns from the primary table and the payment amount (Gross_Pay) column from the secondary table. Save the results as the "Emp_Pay_Dtl" data table.

- Practice 7.12: Please use the **Join** command. Use the employee data table (Emp_Master_ALL) as the primary table and the employee payroll table (Emp_Payroll_ALL) as the secondary table. Match the two tables using the employee number as the matching column. Select and display all columns from the primary table, and save all data of employees who have not been paid as the unpaid employee data table.

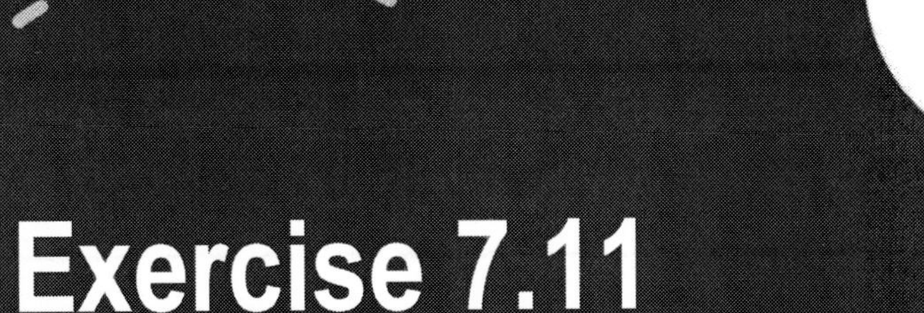

Exercise 7.11

Exercise 7.11

Use the Join command to perform a reconciliation check by using the employee data table (Emp_Master_ALL) as the primary table and the employee payroll table (Emp_Payroll_ALL) as the secondary table.

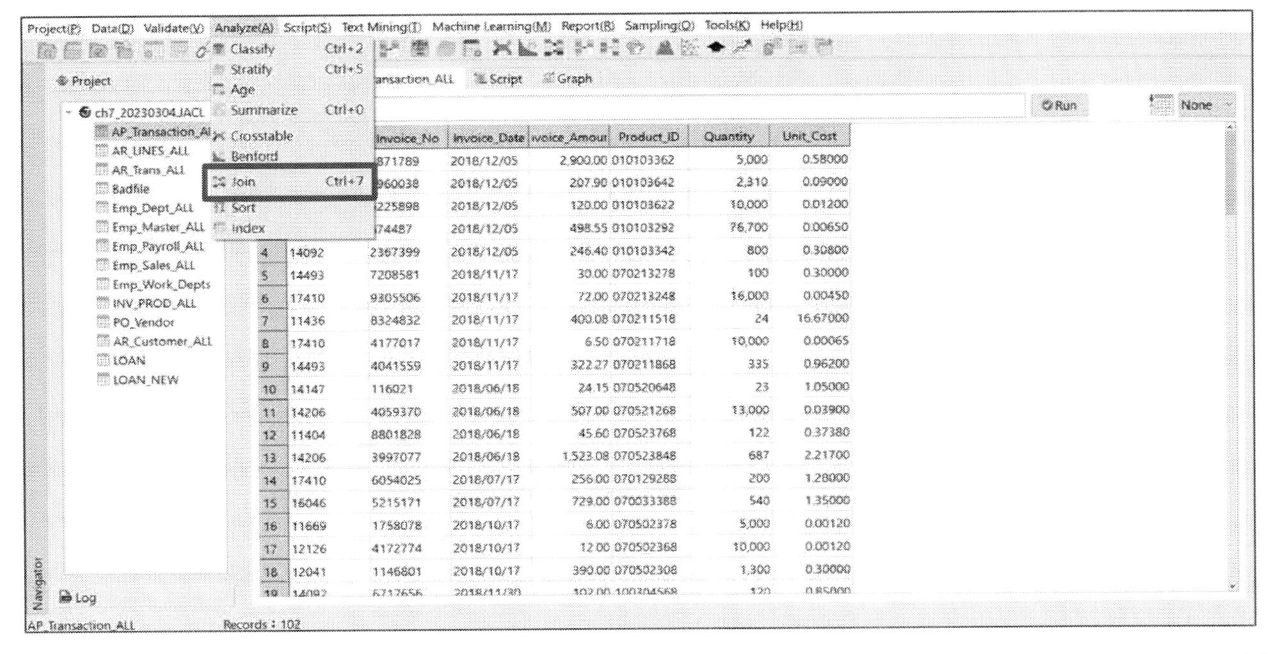

Exercise 7.11

Select and display the employee number, work department, payment period (EmpNo, WorkDept, Pay_Per_Period) columns from the primary table and the payment amount (Gross_Pay) column from the secondary table.

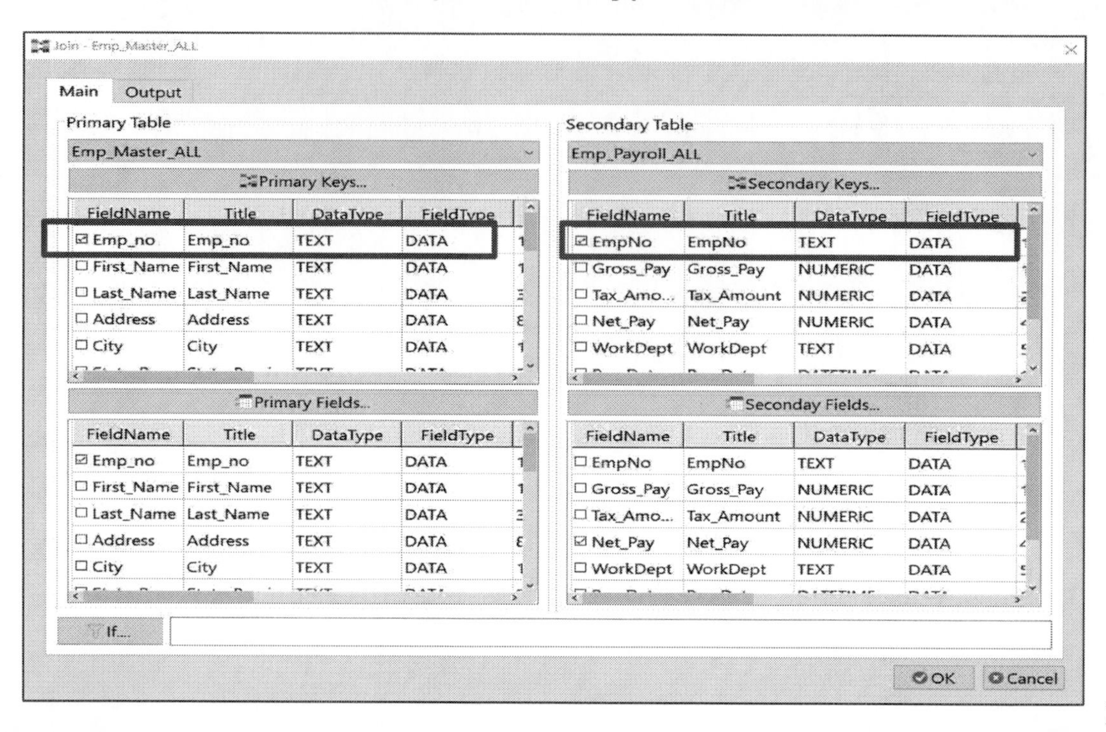

JCAATs-AI Audit Software
Copyright © 2023 JACKSOFT.

Exercise 7.11

Select the Join Type as "Matched Primary with the first Secondary"

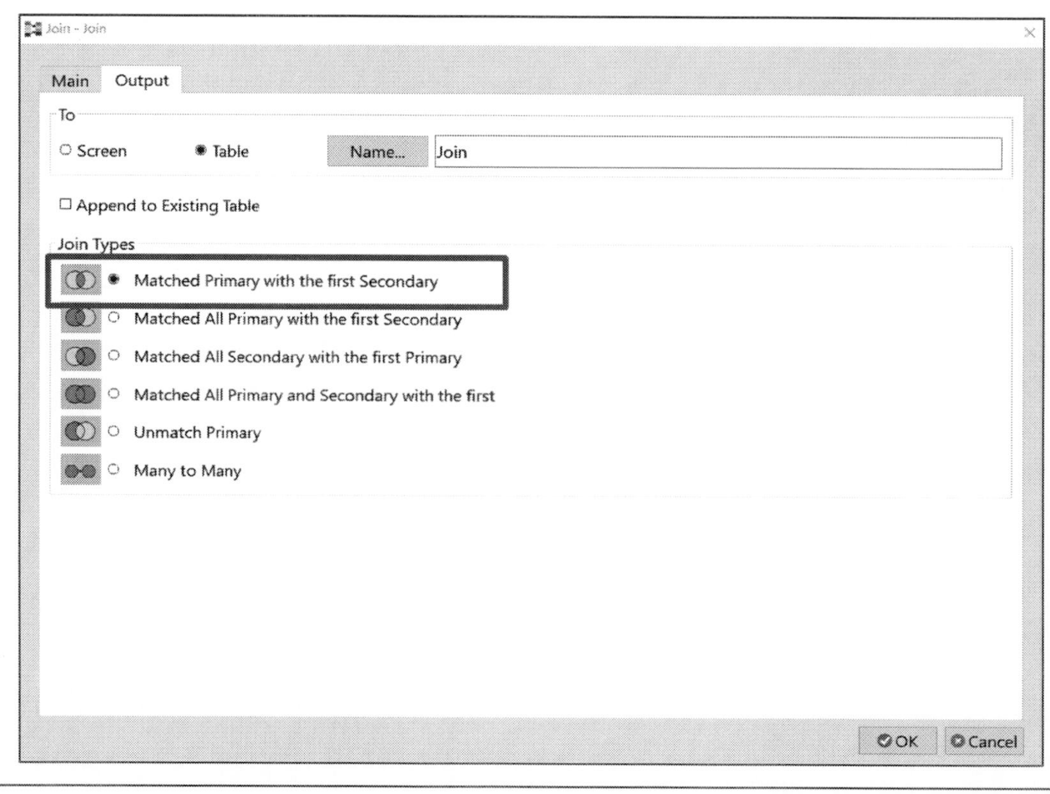

Exercise 7.11

We can view the results in the display area.

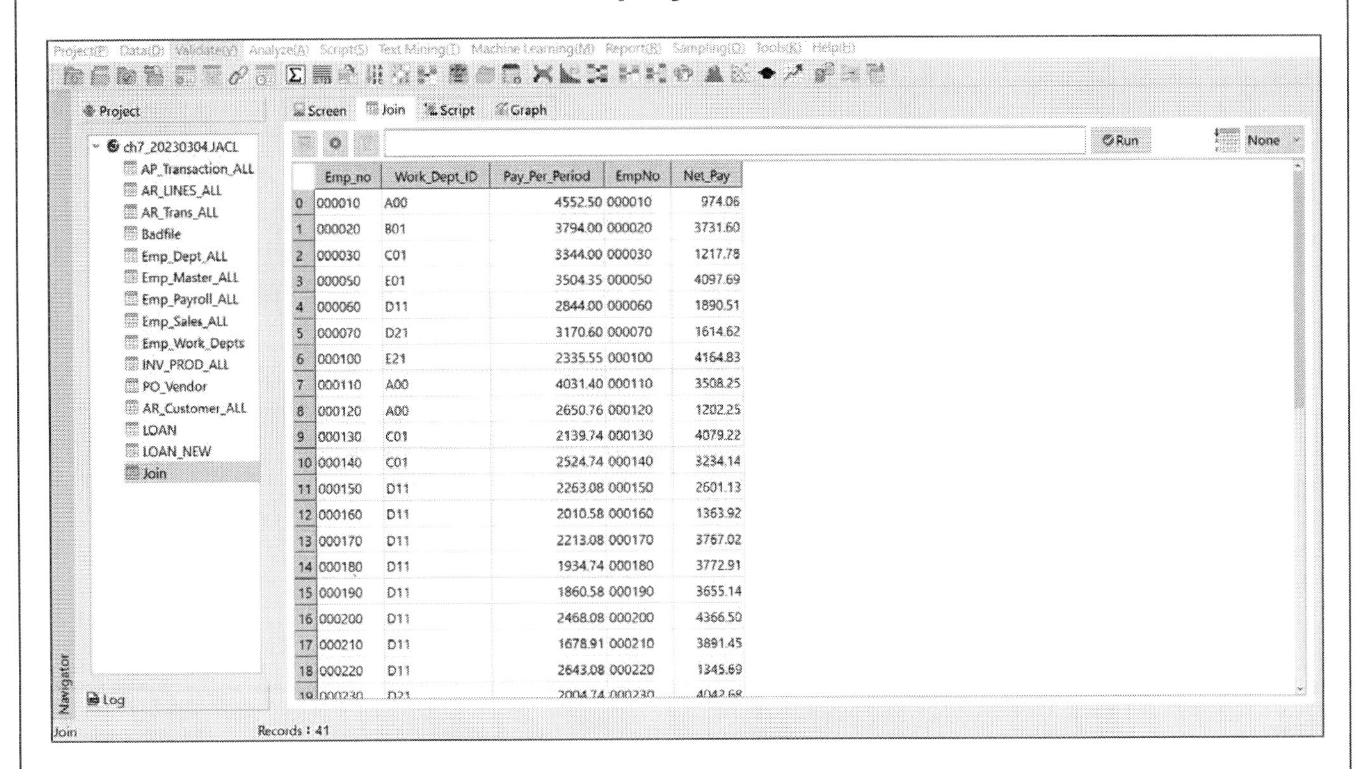

	Emp_no	Work_Dept_ID	Pay_Per_Period	EmpNo	Net_Pay
0	000010	A00	4552.50	000010	974.06
1	000020	B01	3794.00	000020	3731.60
2	000030	C01	3344.00	000030	1217.78
3	000050	E01	3504.35	000050	4097.69
4	000060	D11	2844.00	000060	1890.51
5	000070	D21	3170.60	000070	1614.62
6	000100	E21	2335.55	000100	4164.83
7	000110	A00	4031.40	000110	3508.25
8	000120	A00	2650.76	000120	1202.25
9	000130	C01	2139.74	000130	4079.22
10	000140	C01	2524.74	000140	3234.14
11	000150	D11	2263.08	000150	2601.13
12	000160	D11	2010.58	000160	1363.92
13	000170	D11	2213.08	000170	3767.02
14	000180	D11	1934.74	000180	3772.91
15	000190	D11	1860.58	000190	3655.14
16	000200	D11	2468.08	000200	4366.50
17	000210	D11	1678.91	000210	3891.45
18	000220	D11	2643.08	000220	1345.69
19	000230	D21	2004.74	000230	4042.68

Exercise 7.12

Please use the Join command. Use the employee data table (Emp_Master_ALL) as the primary table and the employee payroll table (Emp_Payroll_ALL) as the secondary table.

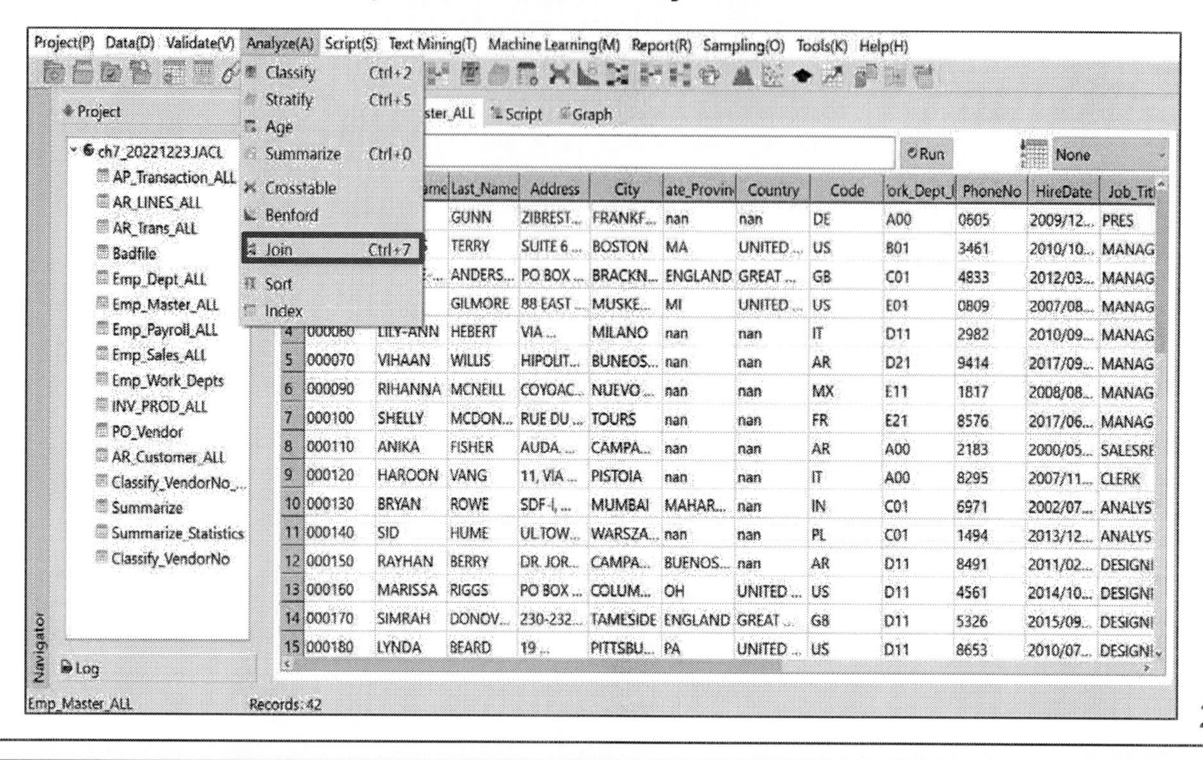

Exercise 7.12

Select Emp_no as primary keys, select all columns from the primary table and list all employees who have not been paid in the payroll.

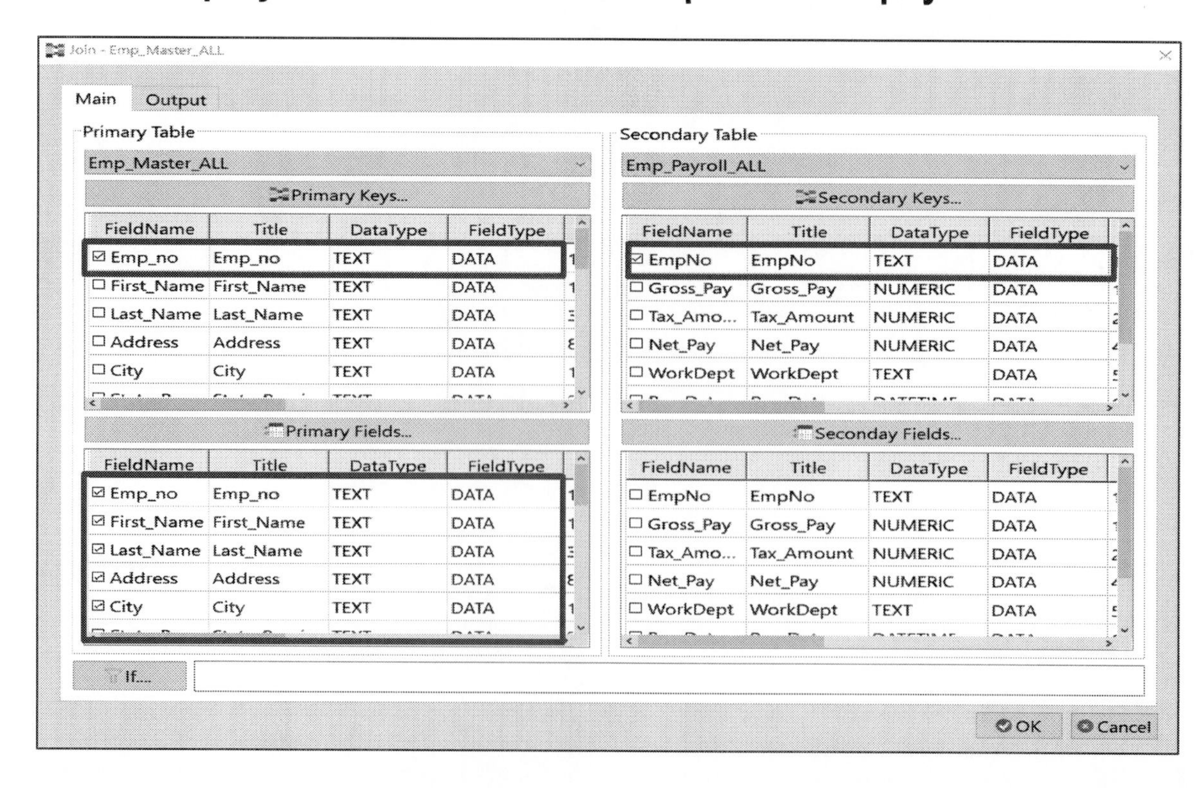

Exercise 7.12

Select the Join Type as "Unmatched Primary"

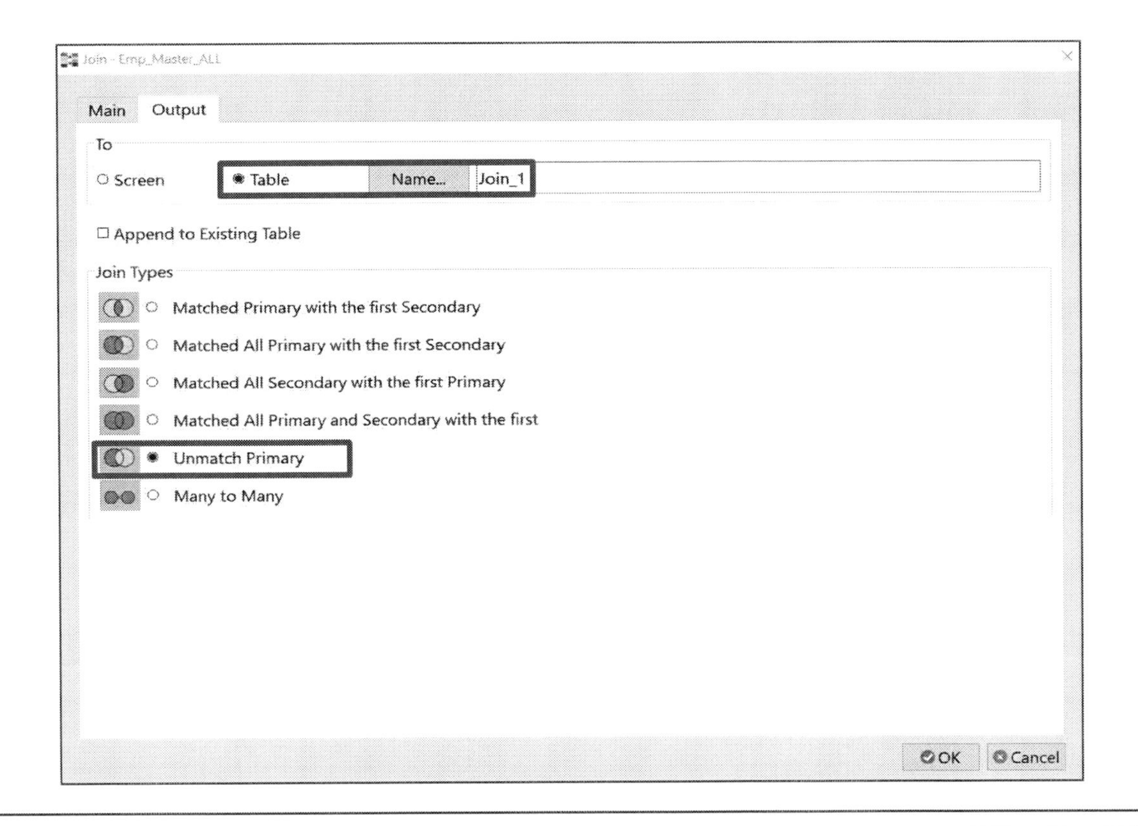

Exercise 7.12

We can view the results in the display area.

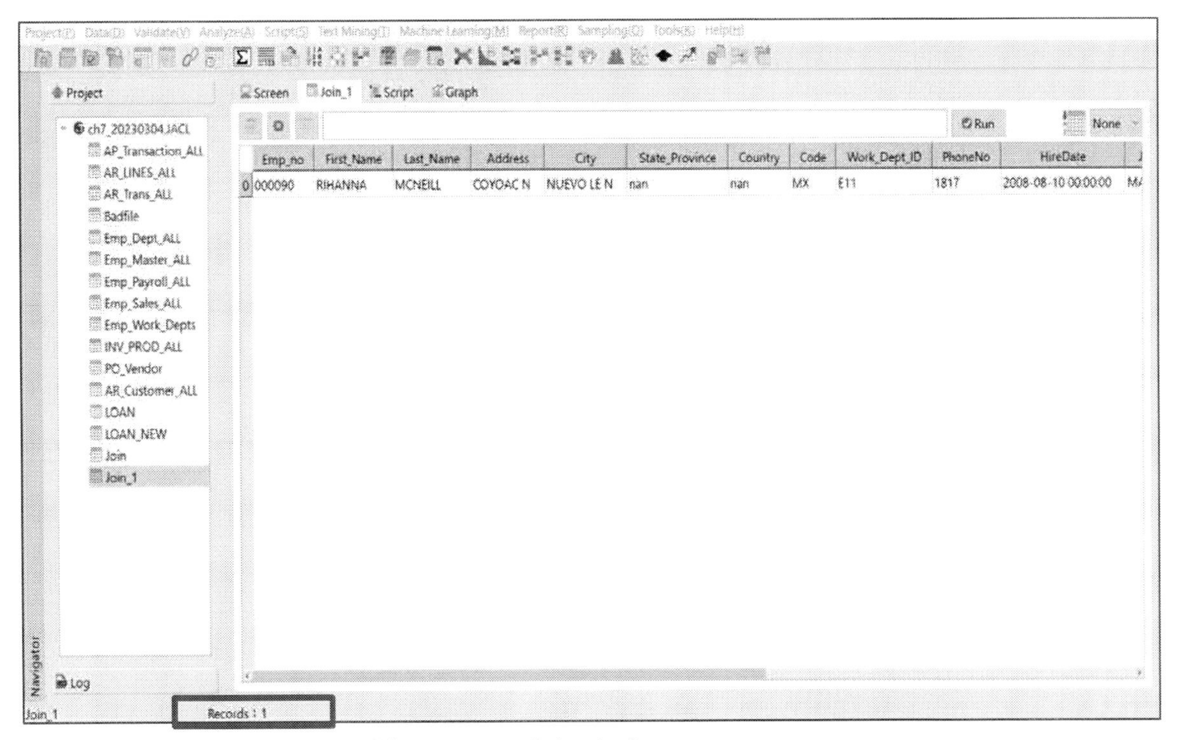

1 instances detected

JCAATs Learning Note:

Chapter 7 - Exercise

- Practice 7.13: Test whether payroll has been issued to any non-employee.

- Practice 7.14: Please compare the balance of accounts receivable (AR_Lines_ALL) of customers and check for abnormal situations where the balance exceeds the approved credit limit in the customer main file (AR_Customer_ALL). Please list the customers with exceeding credit limits and calculate the amount of excess credit limit.

JCAATs - AI Audit Software

Exercise 7.13

Exercise 7.13

Verify whether there has been payroll issued to non-employee master file employees

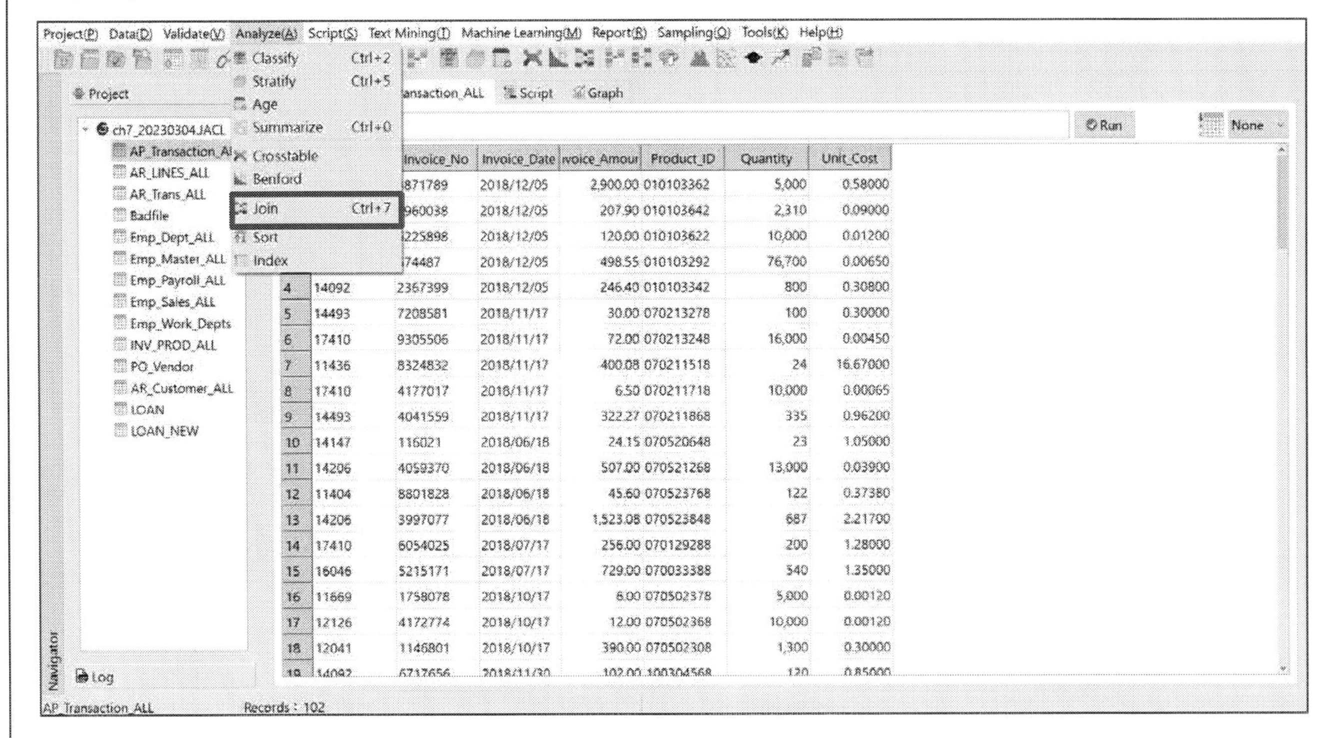

		Invoice_No	Invoice_Date	Invoice_Amount	Product_ID	Quantity	Unit_Cost
		871789	2018/12/05	2,900.00	010103362	5,000	0.58000
		960038	2018/12/05	207.90	010103642	2,310	0.09000
		225898	2018/12/05	120.00	010103622	10,000	0.01200
		74487	2018/12/05	498.55	010103292	76,700	0.00650
4	14092	2367399	2018/12/05	246.40	010103342	800	0.30800
5	14493	7208581	2018/11/17	30.00	070213278	100	0.30000
6	17410	9305506	2018/11/17	72.00	070213248	16,000	0.00450
7	11436	8324832	2018/11/17	400.08	070211518	24	16.67000
8	17410	4177017	2018/11/17	6.50	070211718	10,000	0.00065
9	14493	4041559	2018/11/17	322.27	070211868	335	0.96200
10	14147	116021	2018/06/18	24.15	070520648	23	1.05000
11	14206	4059370	2018/06/18	507.00	070521268	13,000	0.03900
12	11404	8801828	2018/06/18	45.60	070523768	122	0.37380
13	14206	3997077	2018/06/18	1,523.08	070523848	687	2.21700
14	17410	6054025	2018/07/17	256.00	070129288	200	1.28000
15	16046	5215171	2018/07/17	729.00	070033388	540	1.35000
16	11669	1758078	2018/10/17	6.00	070502378	5,000	0.00120
17	12126	4172774	2018/10/17	12.00	070502368	10,000	0.00120
18	12041	1146801	2018/10/17	390.00	070502308	1,300	0.30000

Exercise 7.13

Verify whether there has been payroll issued to non-employee master file employees

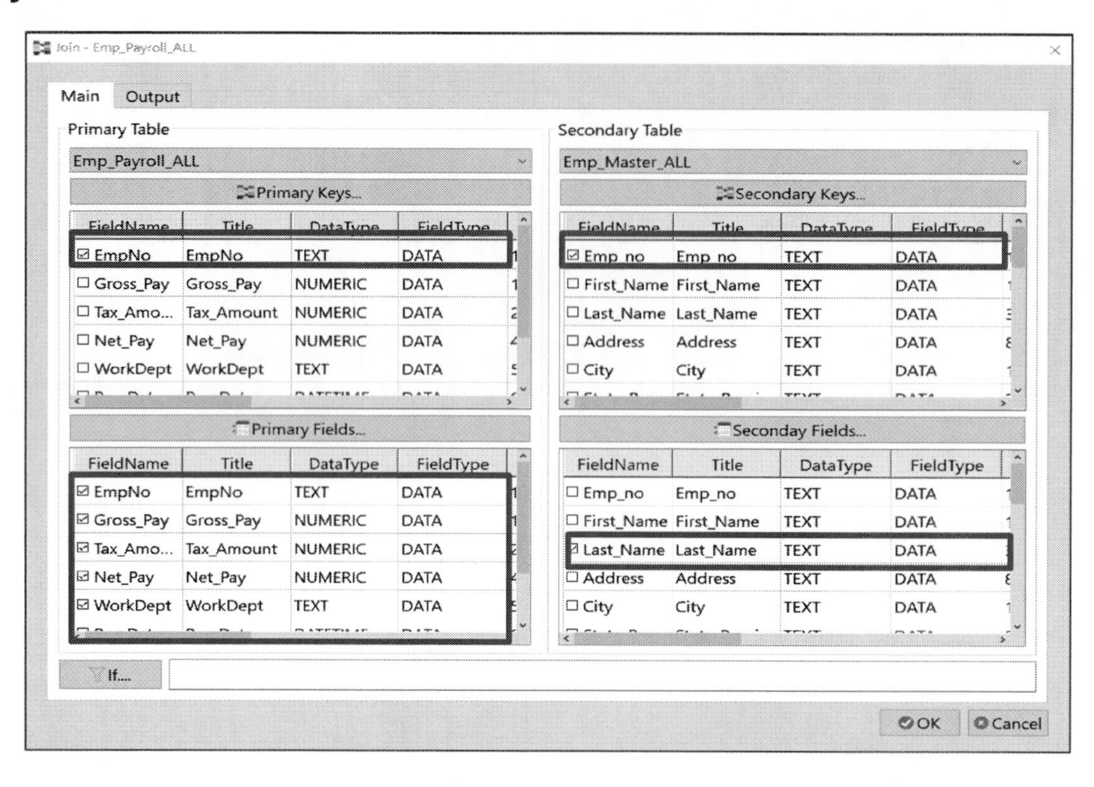

JCAATs-AI Audit Software
Copyright © 2023 JACKSOFT.

Exercise 7.13

Verify whether there has been payroll issued to non-employee master file employees

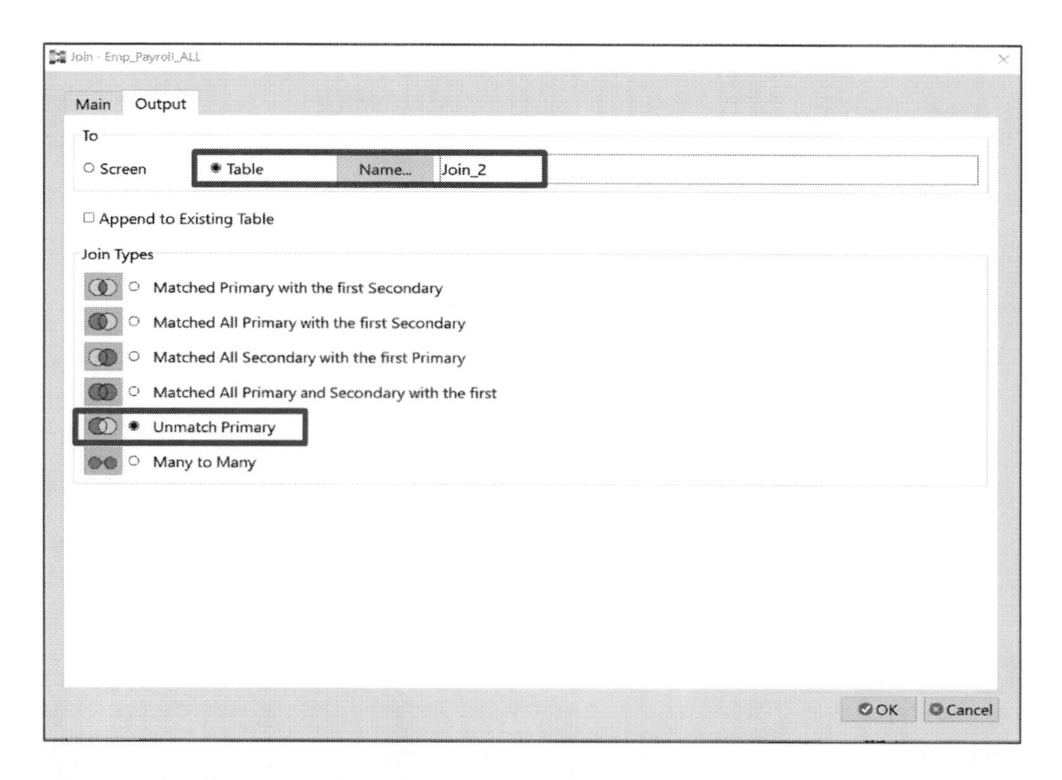

Exercise 7.13

Verify whether there has been payroll issued to non-employee master file employees

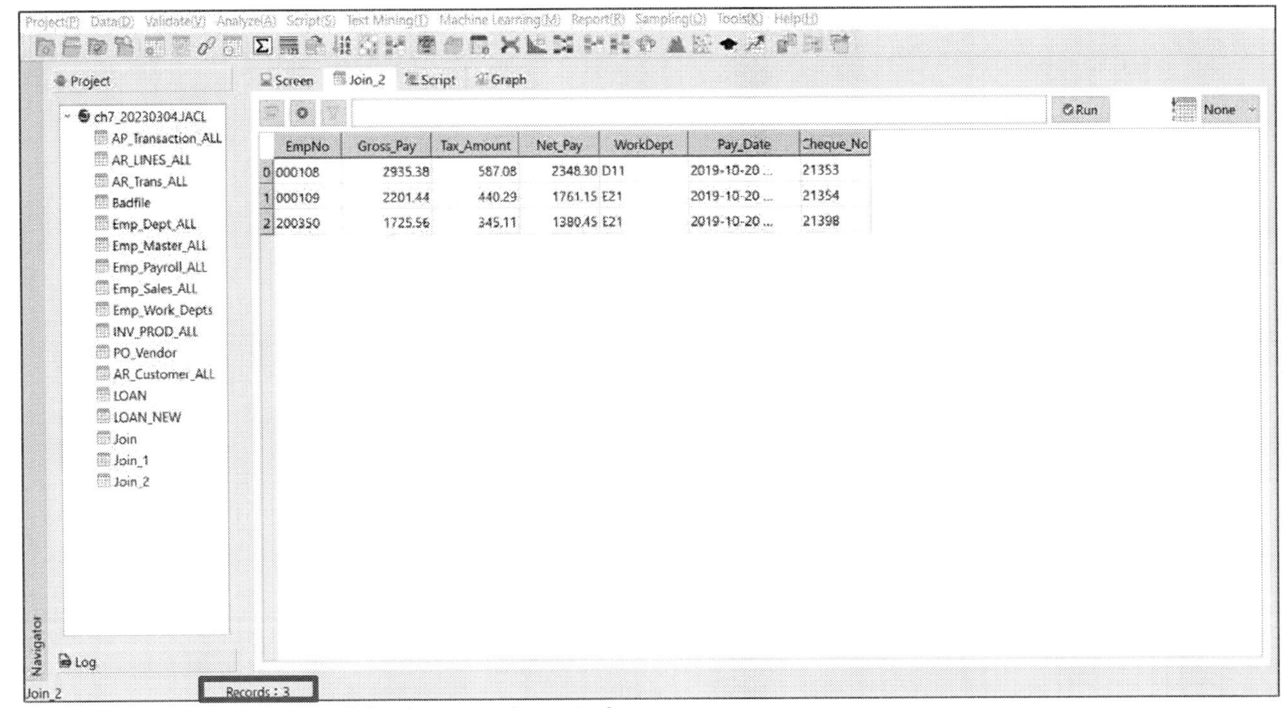

	EmpNo	Gross_Pay	Tax_Amount	Net_Pay	WorkDept	Pay_Date	Cheque_No
0	000108	2935.38	587.08	2348.30	D11	2019-10-20 ...	21353
1	000109	2201.44	440.29	1761.15	E21	2019-10-20 ...	21354
2	200350	1725.56	345.11	1380.45	E21	2019-10-20 ...	21398

3 instances detected

JCAATs - AI Audit Software

Exercise 7.14

Please compare the customer's AR balance (AR_Lines_ALL) with the approved credit limit in the customer master file (AR_Customer_ALL) to identify any abnormal situations. Please provide a list of customers whose AR balance exceeds their credit limit and calculate the amount of excess AR balance.

First, summarize(Classify) the AR customer balances.

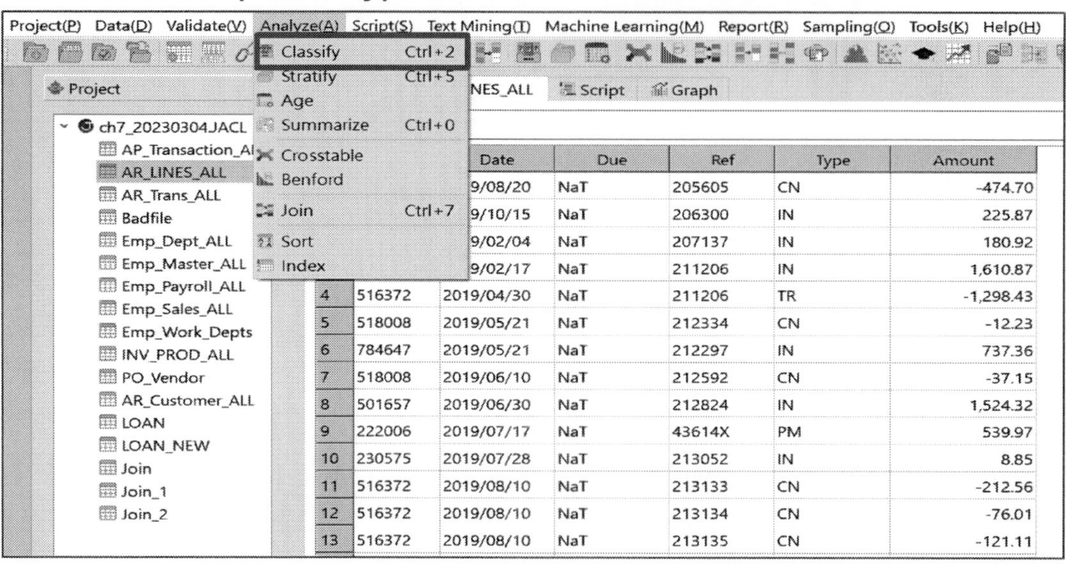

JCAATs-AI Audit Software
Copyright © 2023 JACKSOFT.

Exercise 7.14

Please compare the customer's AR balance (AR_Lines_ALL) with the approved credit limit in the customer master file (AR_Customer_ALL) to identify any abnormal situations. Please provide a list of customers whose AR balance exceeds their credit limit and calculate the amount of excess AR balance.

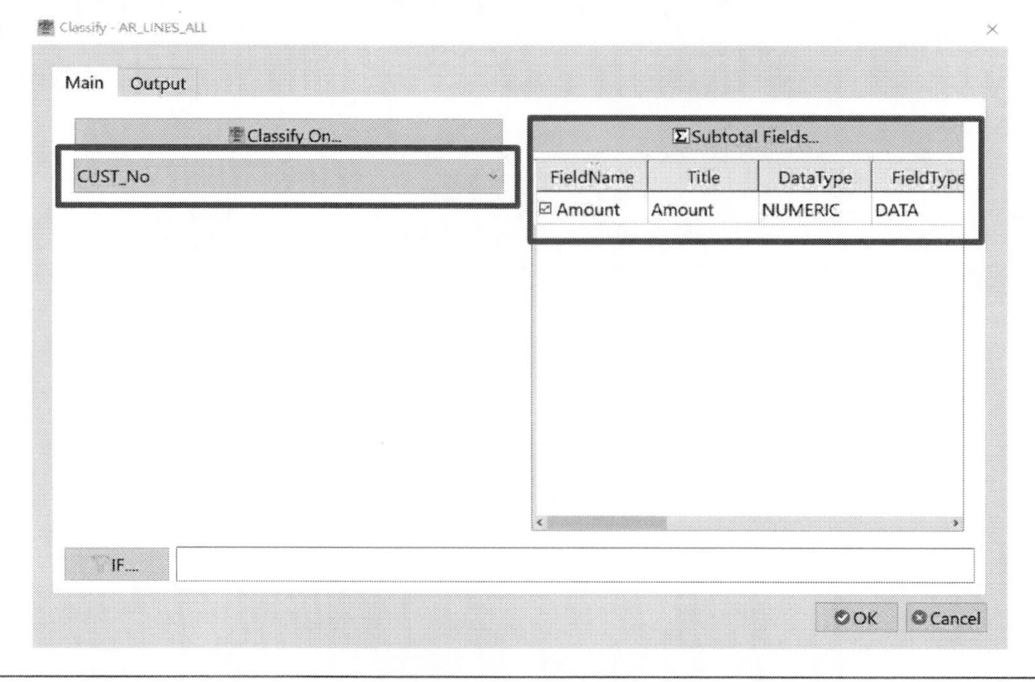

Exercise 7.14

Please compare the customer's AR balance (AR_Lines_ALL) with the approved credit limit in the customer master file (AR_Customer_ALL) to identify any abnormal situations. Please provide a list of customers whose AR balance exceeds their credit limit and calculate the amount of excess AR balance.

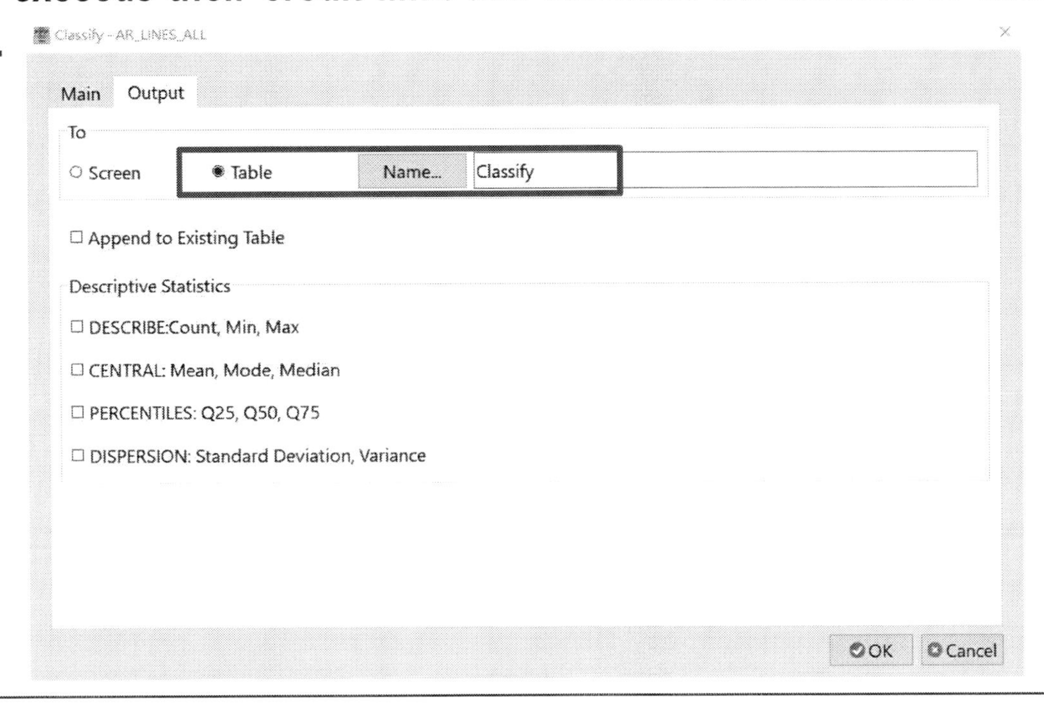

JCAATs-AI Audit Software
Copyright © 2023 JACKSOFT.

Exercise 7.14

Classify command result - Get the total AR amount for each customer.

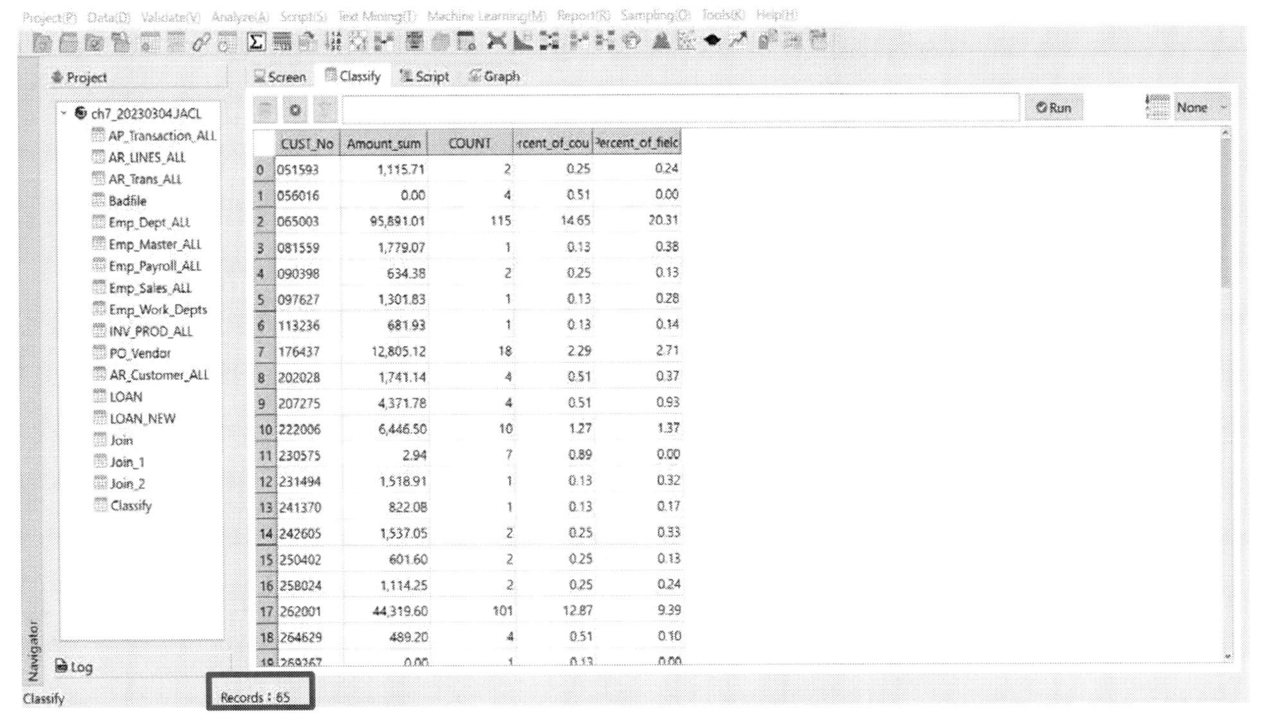

	CUST_No	Amount_sum	COUNT	rcent_of_cou	Percent_of_fielc
0	051593	1,115.71	2	0.25	0.24
1	056016	0.00	4	0.51	0.00
2	065003	95,891.01	115	14.65	20.31
3	081559	1,779.07	1	0.13	0.38
4	090398	634.38	2	0.25	0.13
5	097627	1,301.83	1	0.13	0.28
6	113236	681.93	1	0.13	0.14
7	176437	12,805.12	18	2.29	2.71
8	202028	1,741.14	4	0.51	0.37
9	207275	4,371.78	4	0.51	0.93
10	222006	6,446.50	10	1.27	1.37
11	230575	2.94	7	0.89	0.00
12	231494	1,518.91	1	0.13	0.32
13	241370	822.08	1	0.13	0.17
14	242605	1,537.05	2	0.25	0.33
15	250402	601.60	2	0.25	0.13
16	258024	1,114.25	2	0.25	0.24
17	262001	44,319.60	101	12.87	9.39
18	264629	489.20	4	0.51	0.10
19	269367	0.00	1	0.13	0.00

65 instances detected

Exercise 7.14

Join the customer master file (AR_Customer_ALL) based on the Classify table as the primary, and bringing in the Limit field.

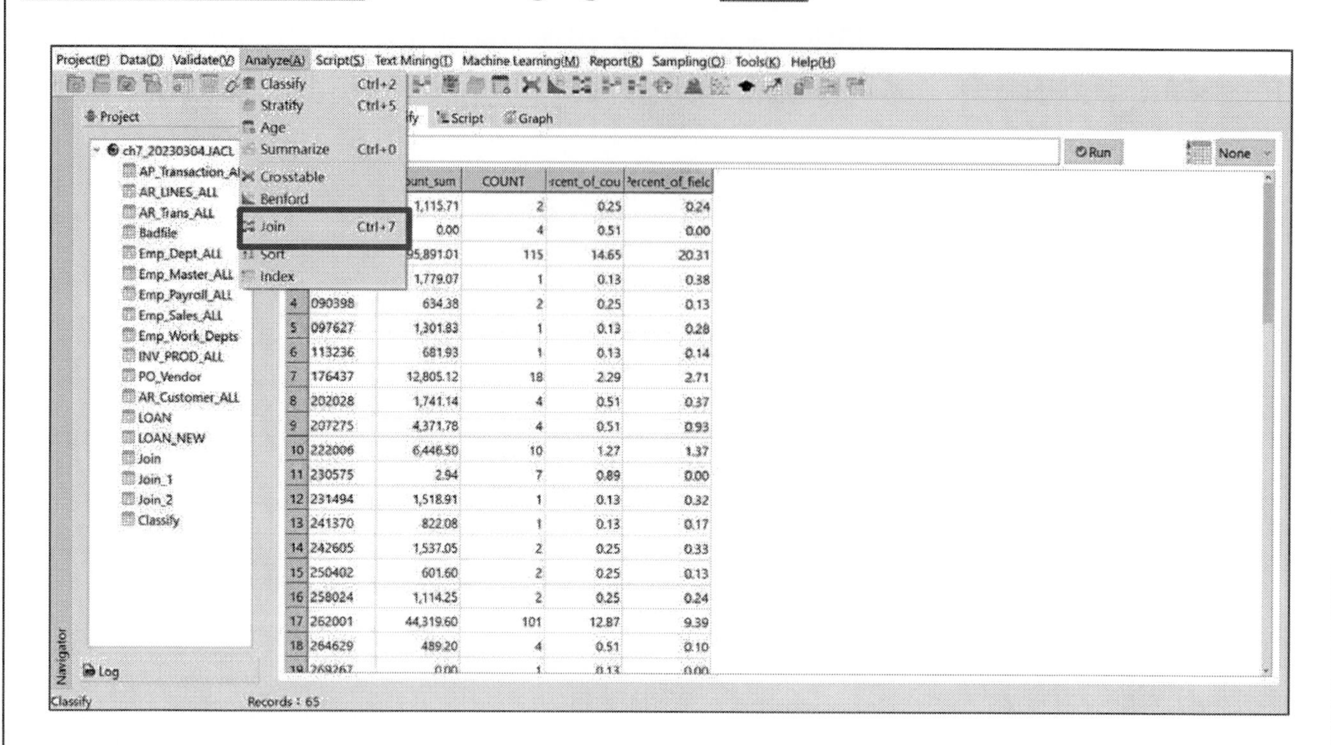

JCAATs-AI Audit Software

Copyright © 2023 JACKSOFT.

Exercise 7.14

Join the customer master file (AR_Customer_ALL) based on the Classify table as the primary, and bringing in the Limit field.

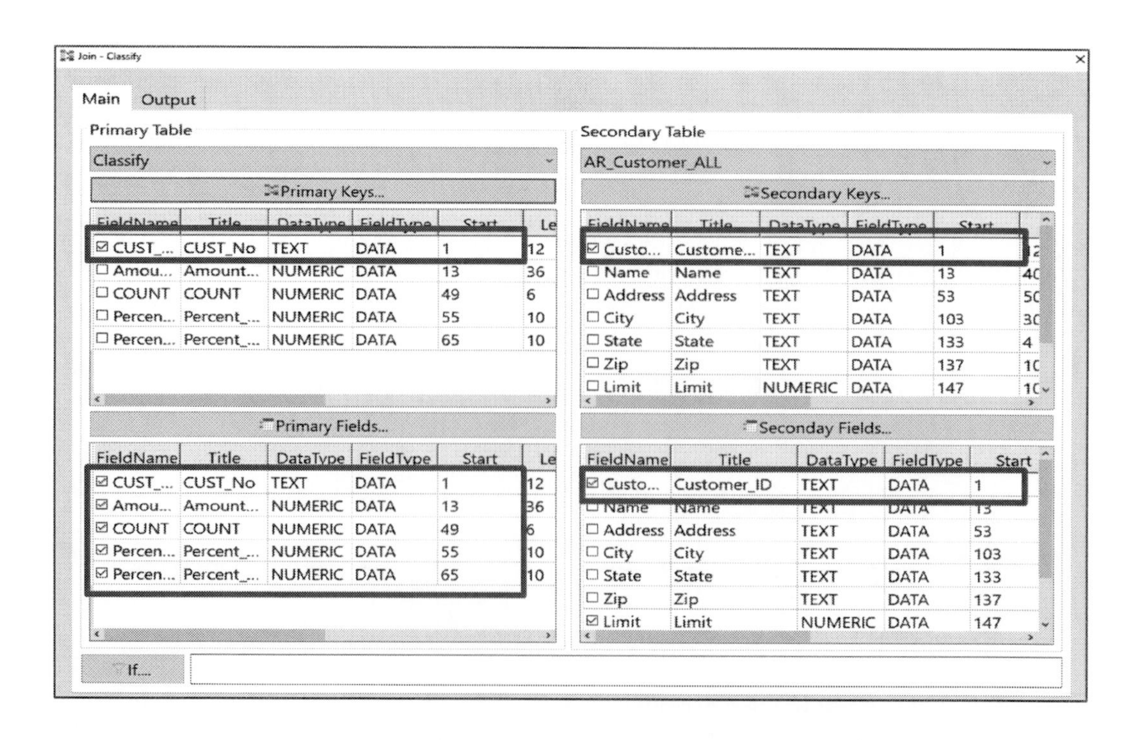

Exercise 7.14

In the output option Join Types select "Matched All Primary With the first Secondary", and enter join_3 for the table name

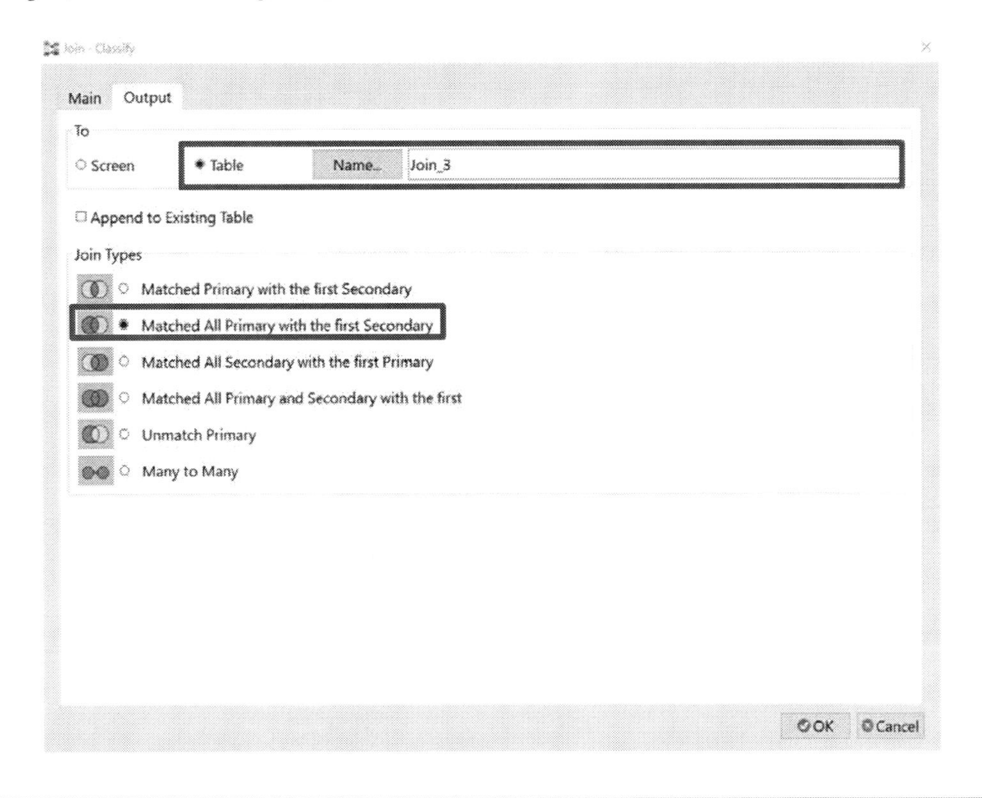

Exercise 7.14

The Join command result

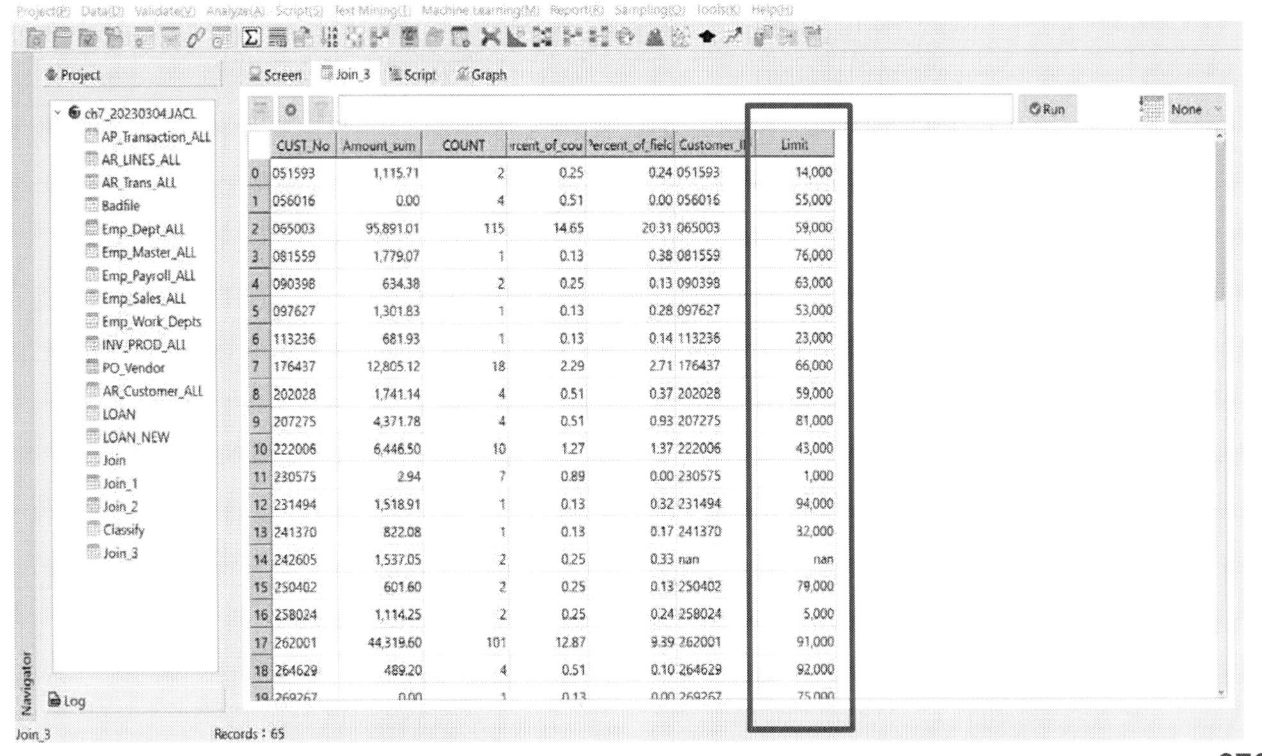

	CUST_No	Amount_sum	COUNT	rcent_of_cou	Percent_of_fielc	Customer_I	Limit
0	051593	1,115.71	2	0.25	0.24	051593	14,000
1	056016	0.00	4	0.51	0.00	056016	55,000
2	065003	95,891.01	115	14.65	20.31	065003	59,000
3	081559	1,779.07	1	0.13	0.38	081559	76,000
4	090398	634.38	2	0.25	0.13	090398	63,000
5	097627	1,301.83	1	0.13	0.28	097627	53,000
6	113236	681.93	1	0.13	0.14	113236	23,000
7	176437	12,805.12	18	2.29	2.71	176437	66,000
8	202028	1,741.14	4	0.51	0.37	202028	59,000
9	207275	4,371.78	4	0.51	0.93	207275	81,000
10	222006	6,446.50	10	1.27	1.37	222006	43,000
11	230575	2.94	7	0.89	0.00	230575	1,000
12	231494	1,518.91	1	0.13	0.32	231494	94,000
13	241370	822.08	1	0.13	0.17	241370	32,000
14	242605	1,537.05	2	0.25	0.33	nan	nan
15	250402	601.60	2	0.25	0.13	250402	79,000
16	258024	1,114.25	2	0.25	0.24	258024	5,000
17	262001	44,319.60	101	12.87	9.39	262001	91,000
18	264629	489.20	4	0.51	0.10	264629	92,000
19	269267	0.00	1	0.13	0.00	269267	75,000

Exercise 7.14

Filter out the customer list that exceeds the credit limit in the Join_3 table.

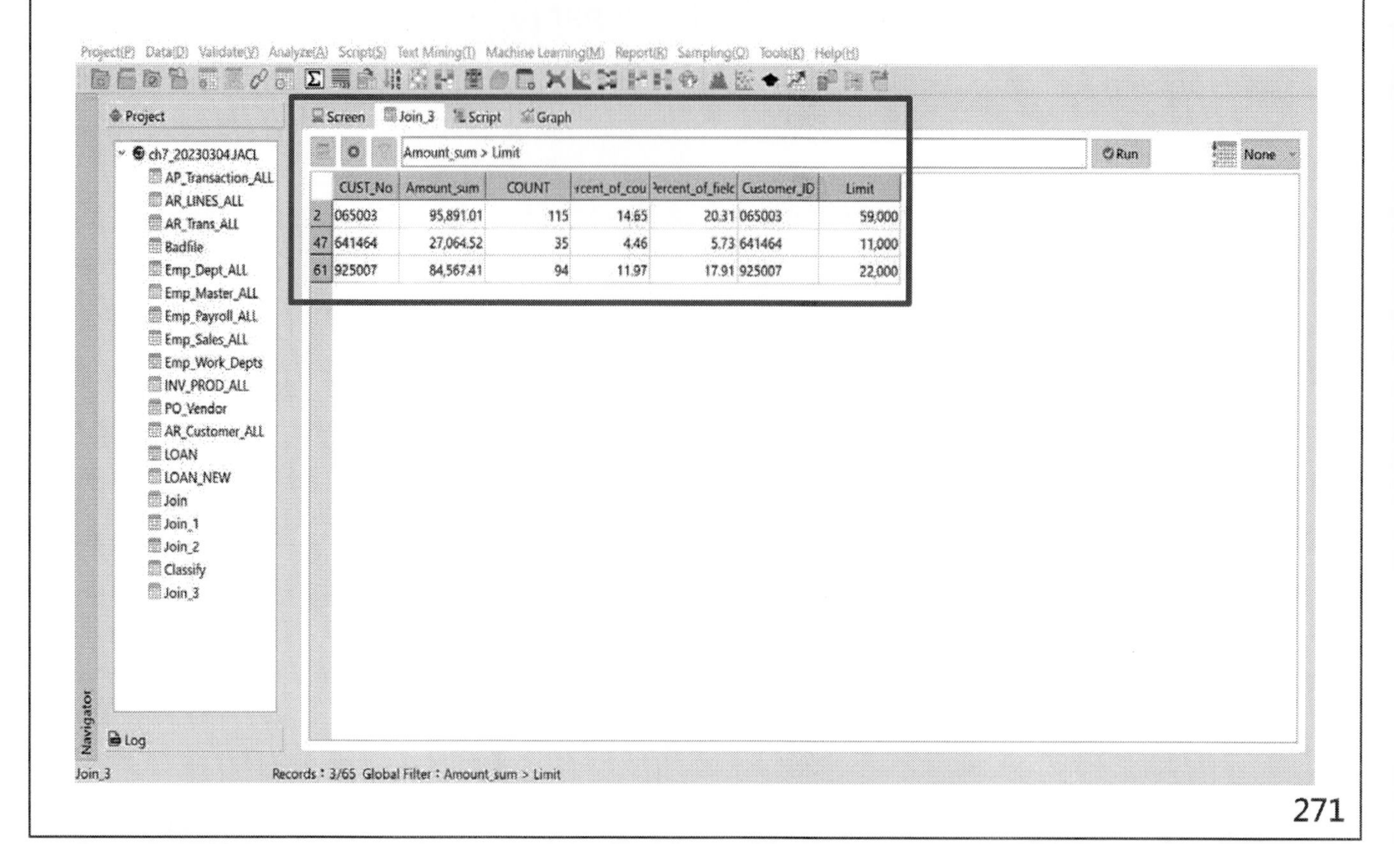

Exercise 7.14

Extract the abnormal details into a result list.

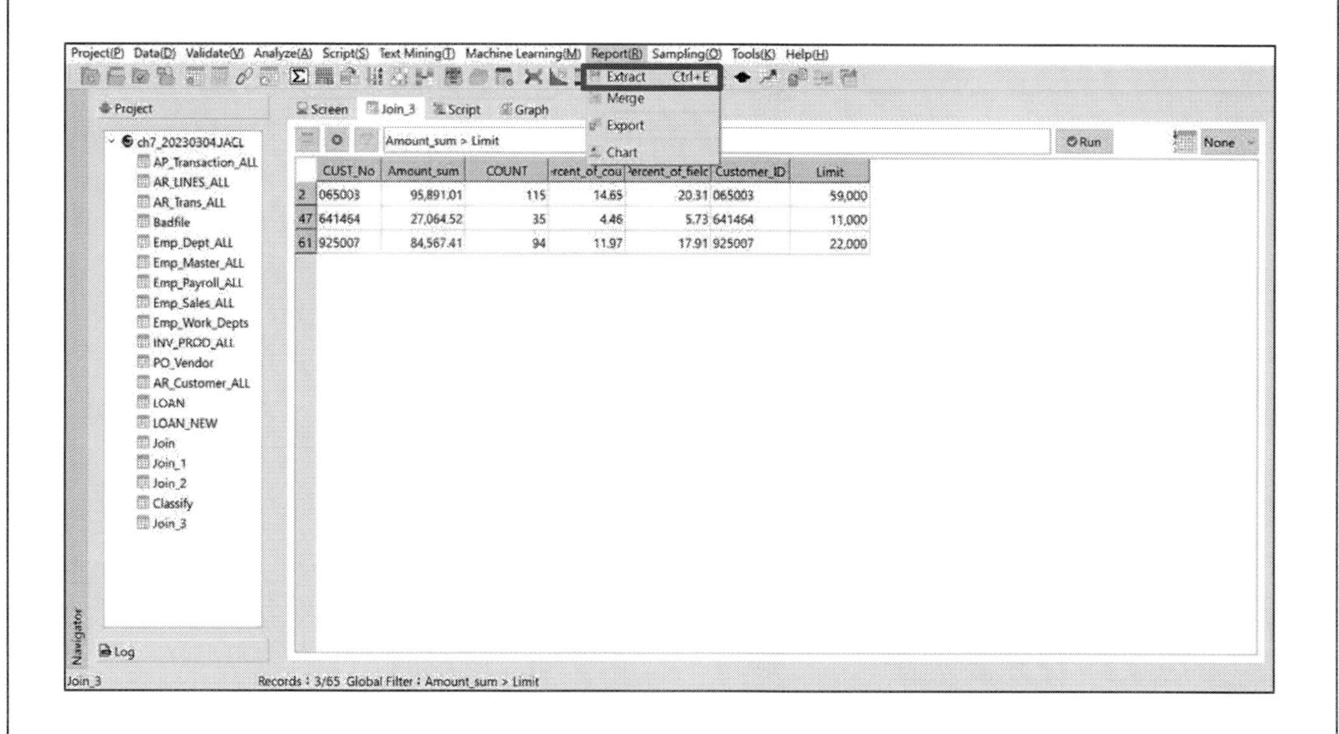

Exercise 7.14

Select all fields and enter the table name "extract" in the output menu.

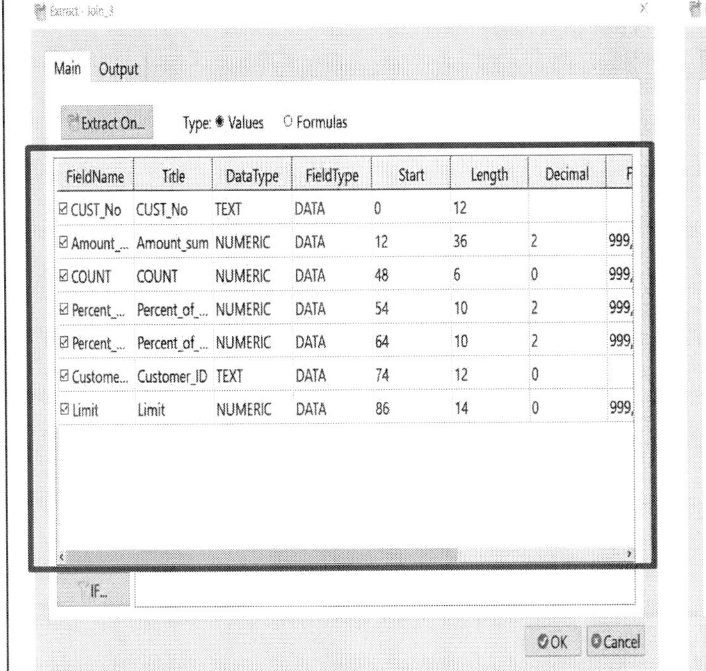

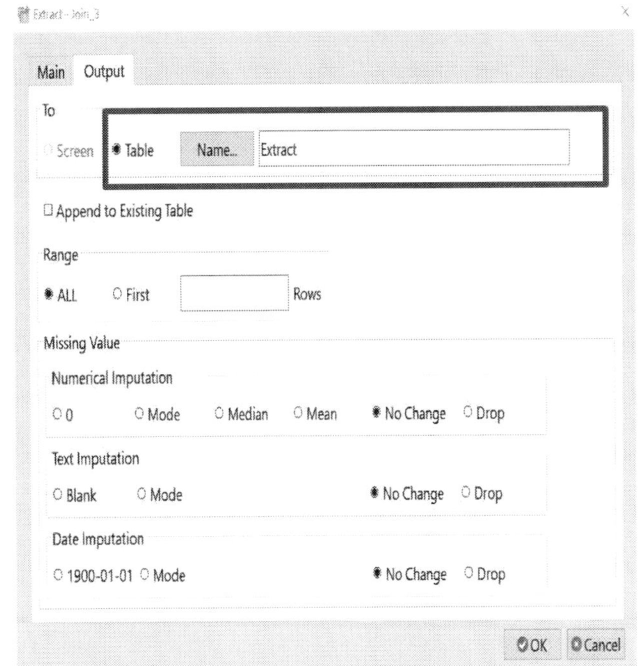

Exercise 7.14

The Extract command result, total of 3 abnormal data.

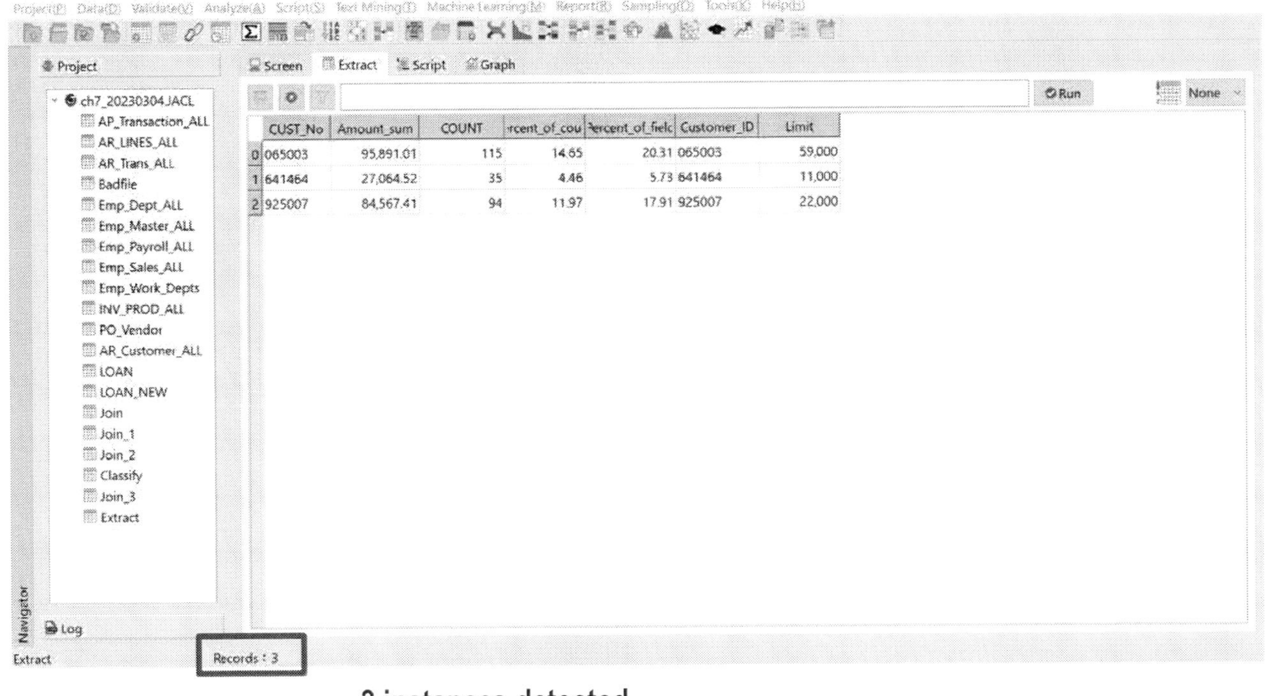

3 instances detected

Chapter 7 - Exercise

- Practice 7.15: Using the **Benford** command, analyze the amount field in the sales detail data table (AR_LINE_ALL) with 1 digital. The numerical starting position is the second digit. Use chi-square test to confirm whether the data conforms to the trend of Benford for subsequent verification.

- Practice 7.16: Using the **Crosstable** command, compare the highest salary amount of department and country codes in the employee payroll table (Emp_Payroll_ALL) and view the result chart.

JCAATs - AI Audit Software

Exercise 7.15

Exercise 7.15

Please use the Benford command to perform a <u>one-digit</u> data analysis on the <u>Amount column</u> of the table (AR_LINE_ALL), with the number starting position being the <u>second digit</u>. Use the chi-square test to confirm whether the data conforms to Benford's trend for subsequent verification.

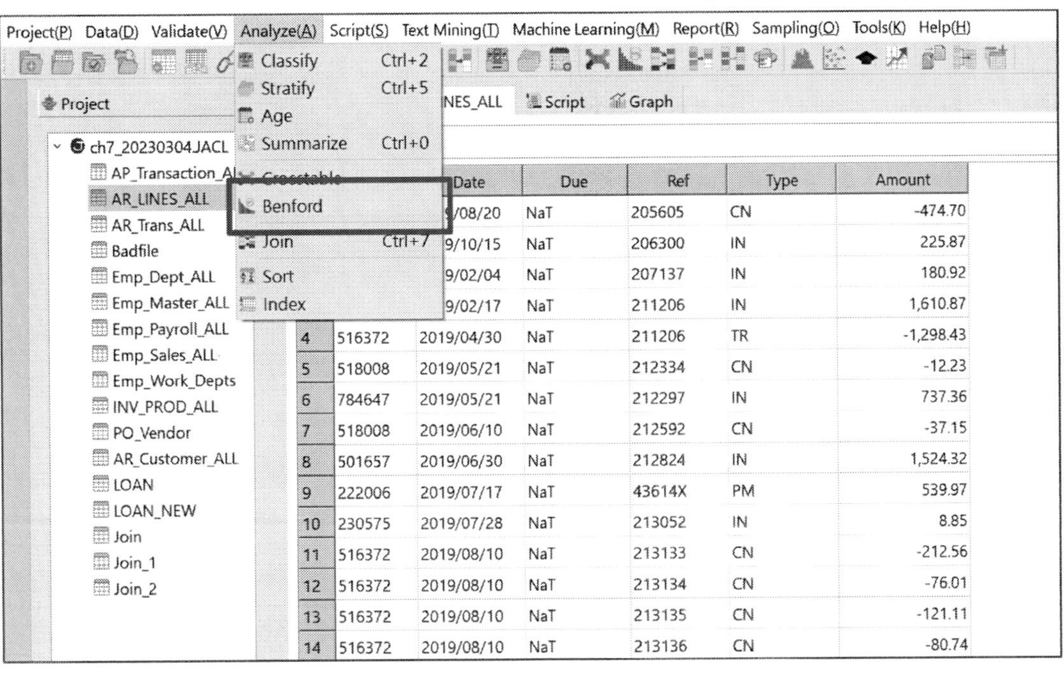

JCAATs-AI Audit Software
Copyright © 2023 JACKSOFT.

Exercise 7.15

Please use the Benford command to perform a <u>one-digit</u> data analysis on the <u>Amount column</u> of the table (AR_LINE_ALL), with the number starting position being the <u>second digit</u>. Use the chi-square test to confirm whether the data conforms to Benford's trend for subsequent verification.

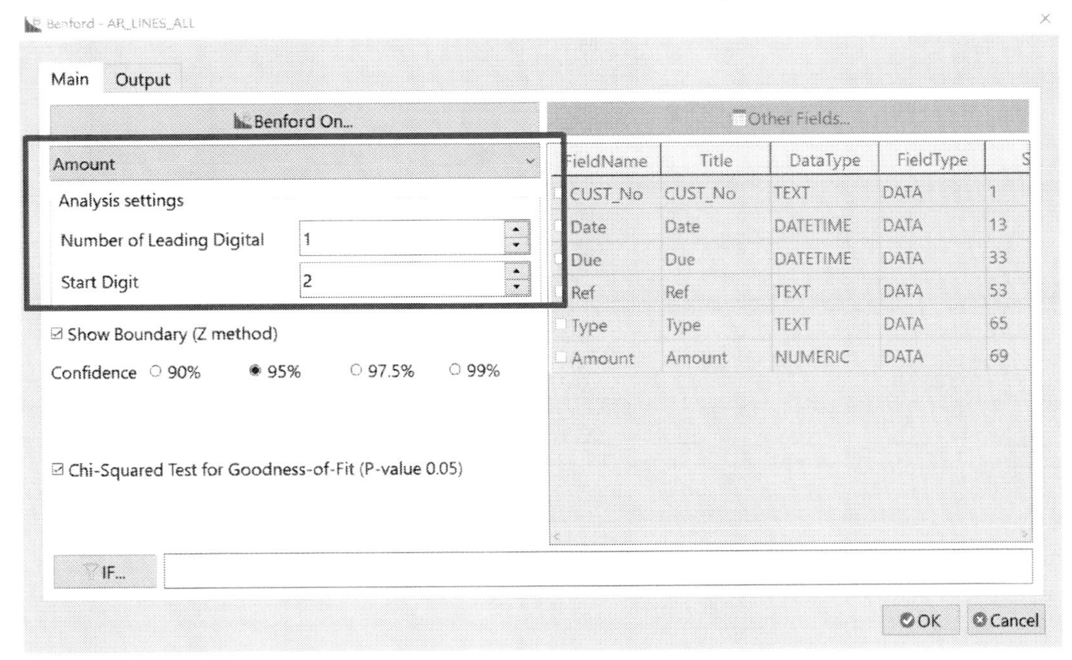

Exercise 7.15

Please use the Benford command to perform a one-digit data analysis on the Amount column of the table (AR_LINE_ALL), with the number starting position being the second digit. Use the chi-square test to confirm whether the data conforms to Benford's trend for subsequent verification.

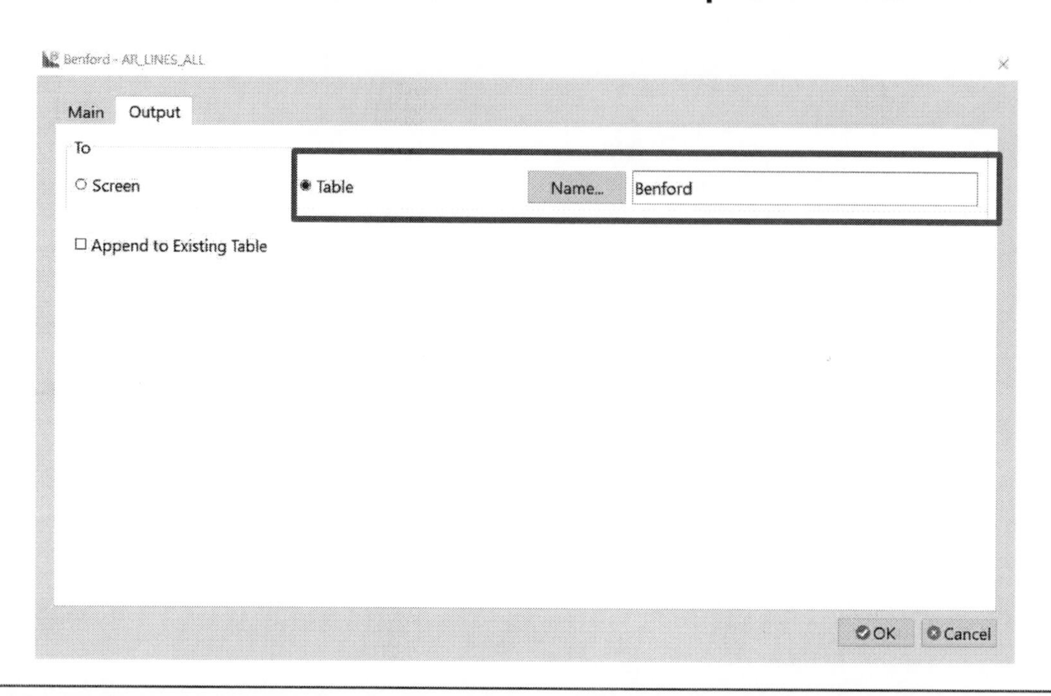

JCAATs-AI Audit Software

Exercise 7.15

Benford command Result

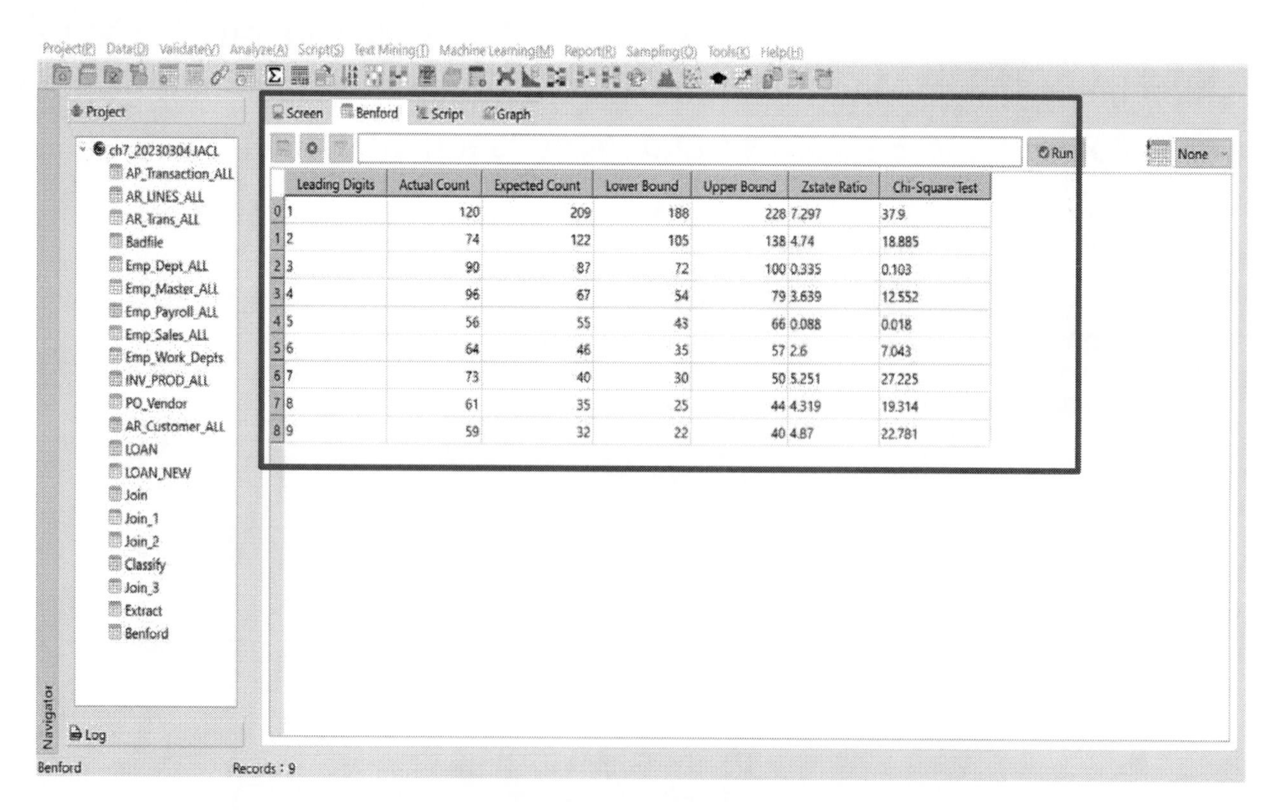

	Leading Digits	Actual Count	Expected Count	Lower Bound	Upper Bound	Zstate Ratio	Chi-Square Test
0	1	120	209	188	228	7.297	37.9
1	2	74	122	105	138	4.74	18.885
2	3	90	87	72	100	0.335	0.103
3	4	96	67	54	79	3.639	12.552
4	5	56	55	43	66	0.088	0.018
5	6	64	46	35	57	2.6	7.043
6	7	73	40	30	50	5.251	27.225
7	8	61	35	25	44	4.319	19.314
8	9	59	32	22	40	4.87	22.781

Exercise 7.15

Benford command Result graph.

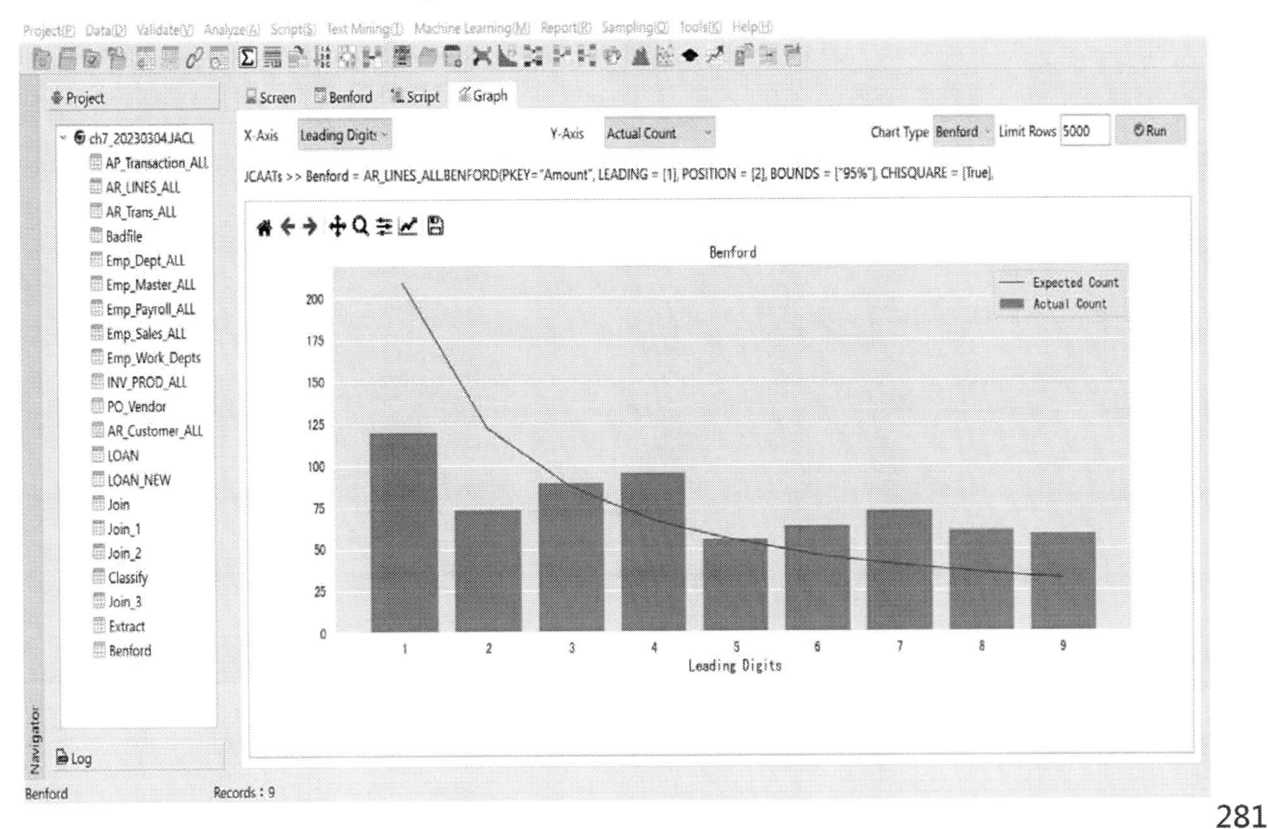

Exercise 7.16

Please use the Crosstable command to cross-reference the highest salary amount by department and country code in the employee salary table (Emp_Master_ALL), and view the result chart.

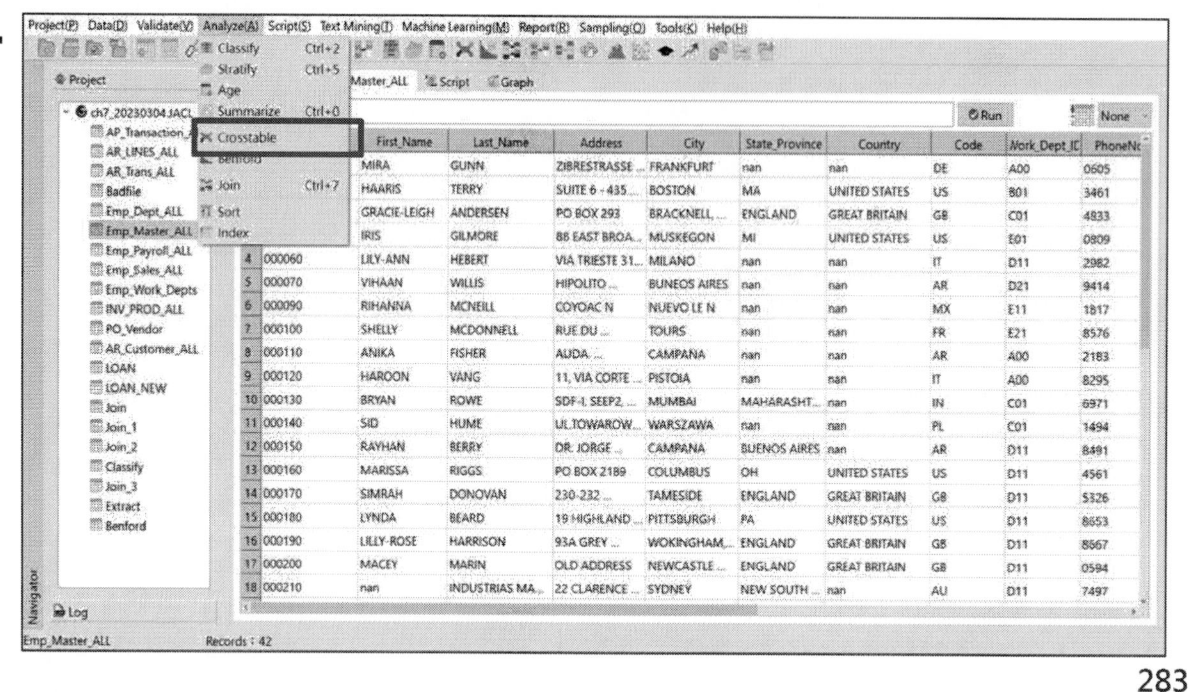

Exercise 7.16

Please use the Crosstable command to cross-reference the highest salary amount by department and country code in the employee salary table (Emp_Master_ALL), and view the result chart.

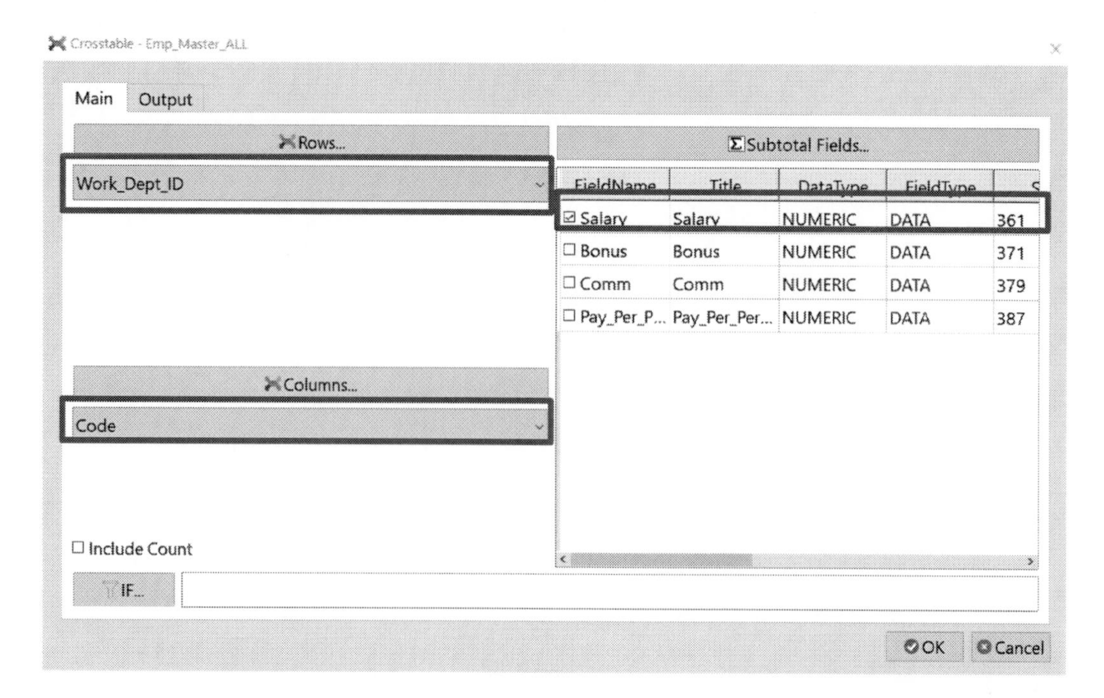

Exercise 7.16

Please use Crosstable command to compare the highest salary amount for each department and country code in the employee salary table (Emp_Master_ALL), and view the result graph.

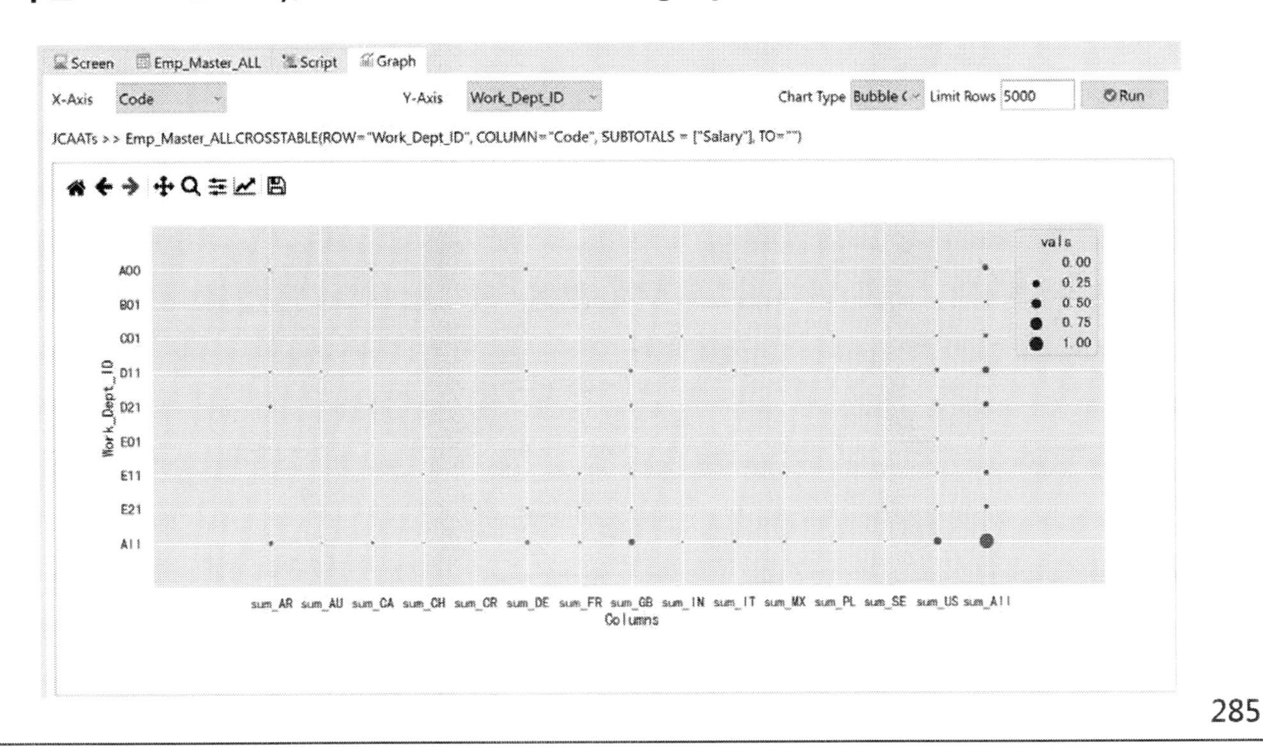

JCAATs-AI Audit Software
Copyright © 2023 JACKSOFT.

JCAATs Learning Note:

Exercise 8.1: A Case Study

- Please design an automated audit script that can auto-download the list of rejected vendors from the government open data and perform the audit tasks in case 7.19. Each of the audit tasks will become one script.

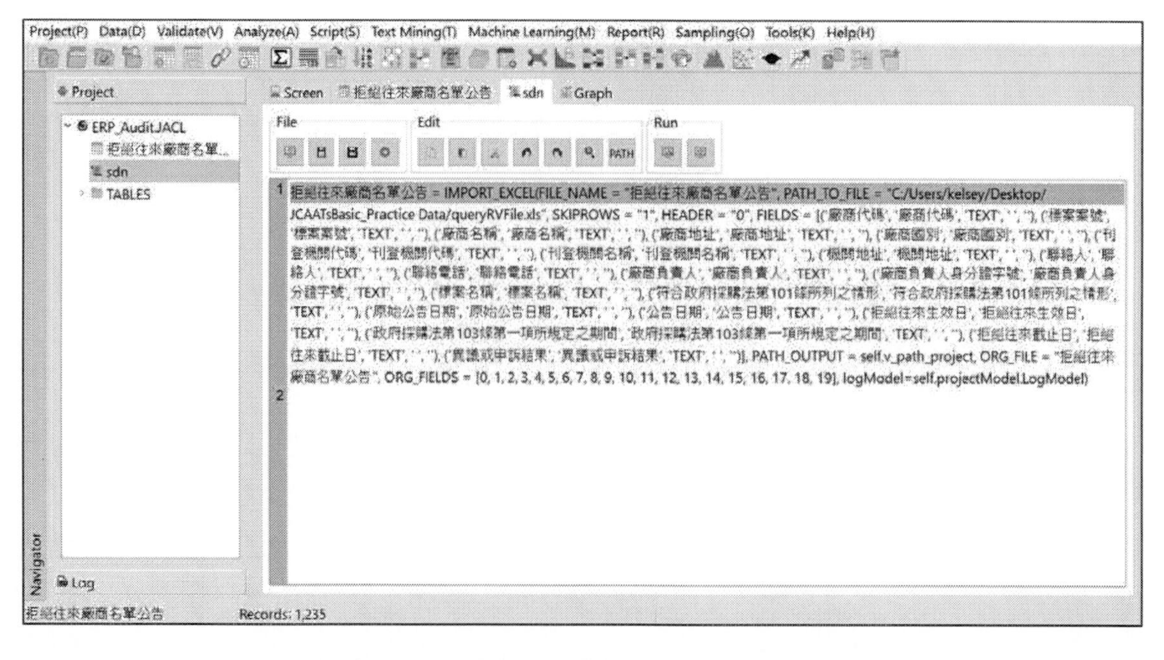

JCAATs - AI Audit Software

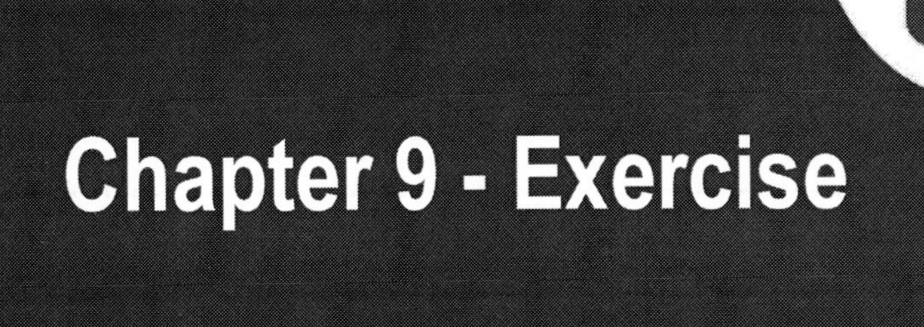

Chapter 9 - Exercise

Chapter 9 - Exercise

- Exercise 9.1: Perform text mining analysis on the FSC's internal control penalty cases, select cases within a certain period, use the word cloud command to examine high-risk keywords that should be noted for subsequent strengthening of audits.

- Exercise 9.2: Using the keyword command, perform text mining on penalty cases to identify high-risk keywords.

- Exercise 9.3: Using sentiment analysis commands, conduct anomaly checks on loan cases and identify high-risk cases that require further investigation.

Exercise 9.1

Conduct text mining analysis on the internal control penalty cases of the FSC, focusing on cases from a specific time frame. Use the word cloud function to identify high-risk keywords.

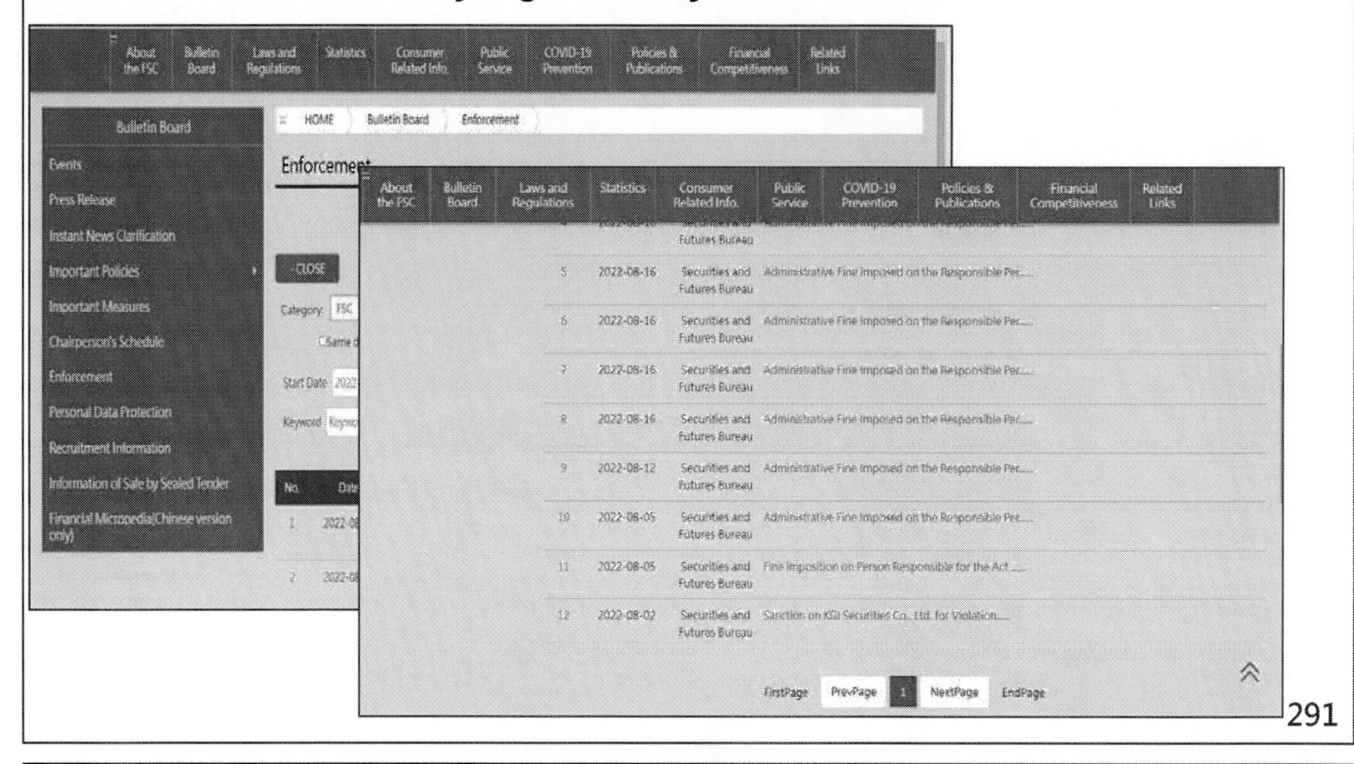

JCAATs-AI Audit Software
Copyright © 2023 JACKSOFT.

Exercise 9.1

Conduct text mining analysis on the internal control penalty cases of the FSC, focusing on cases from a specific time frame. Use the word cloud function to identify high-risk keywords.

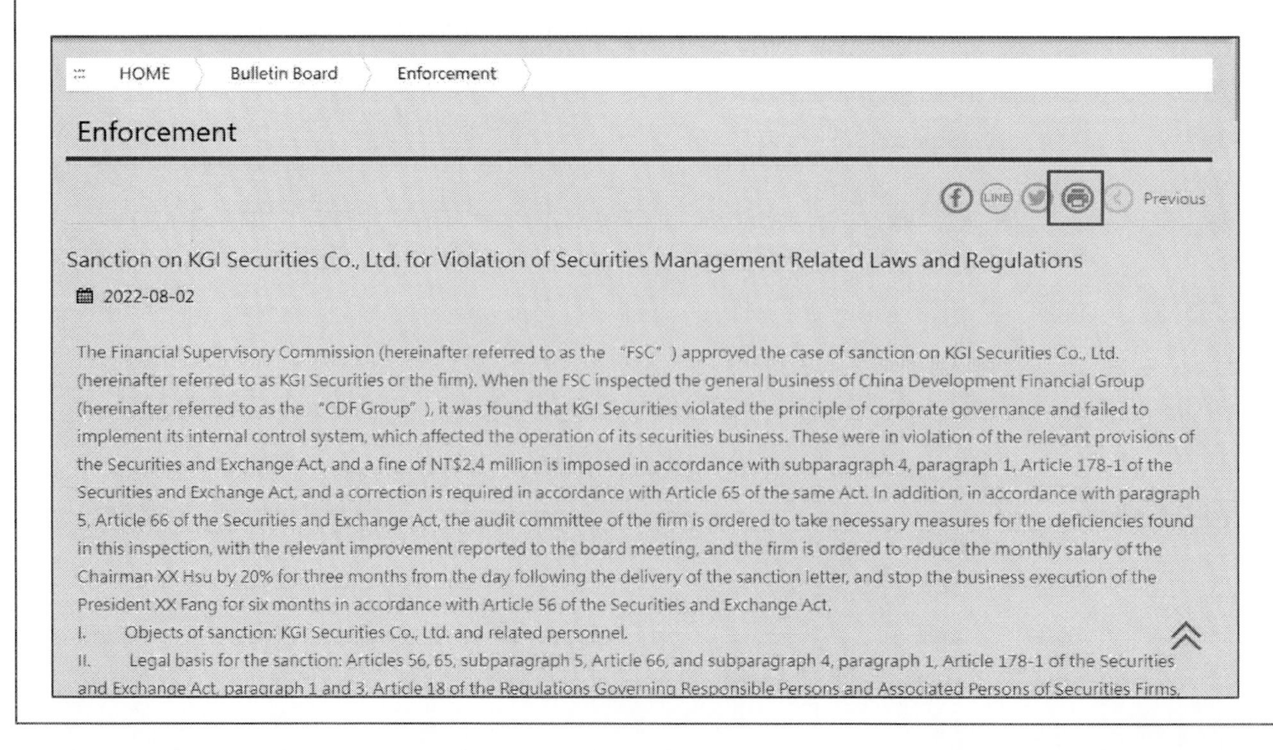

::: HOME > Bulletin Board > Enforcement

Enforcement

Previous

Sanction on KGI Securities Co., Ltd. for Violation of Securities Management Related Laws and Regulations

2022-08-02

The Financial Supervisory Commission (hereinafter referred to as the "FSC") approved the case of sanction on KGI Securities Co., Ltd. (hereinafter referred to as KGI Securities or the firm). When the FSC inspected the general business of China Development Financial Group (hereinafter referred to as the "CDF Group"), it was found that KGI Securities violated the principle of corporate governance and failed to implement its internal control system, which affected the operation of its securities business. These were in violation of the relevant provisions of the Securities and Exchange Act, and a fine of NT$2.4 million is imposed in accordance with subparagraph 4, paragraph 1, Article 178-1 of the Securities and Exchange Act, and a correction is required in accordance with Article 65 of the same Act. In addition, in accordance with paragraph 5, Article 66 of the Securities and Exchange Act, the audit committee of the firm is ordered to take necessary measures for the deficiencies found in this inspection, with the relevant improvement reported to the board meeting, and the firm is ordered to reduce the monthly salary of the Chairman XX Hsu by 20% for three months from the day following the delivery of the sanction letter, and stop the business execution of the President XX Fang for six months in accordance with Article 56 of the Securities and Exchange Act.

I. Objects of sanction: KGI Securities Co., Ltd. and related personnel.

II. Legal basis for the sanction: Articles 56, 65, subparagraph 5, Article 66, and subparagraph 4, paragraph 1, Article 178-1 of the Securities and Exchange Act, paragraph 1 and 3, Article 18 of the Regulations Governing Responsible Persons and Associated Persons of Securities Firms,

Exercise 9.1

Conduct text mining analysis on the internal control penalty cases of the FSC, focusing on cases from a specific time frame. Use the word cloud function to identify high-risk keywords.

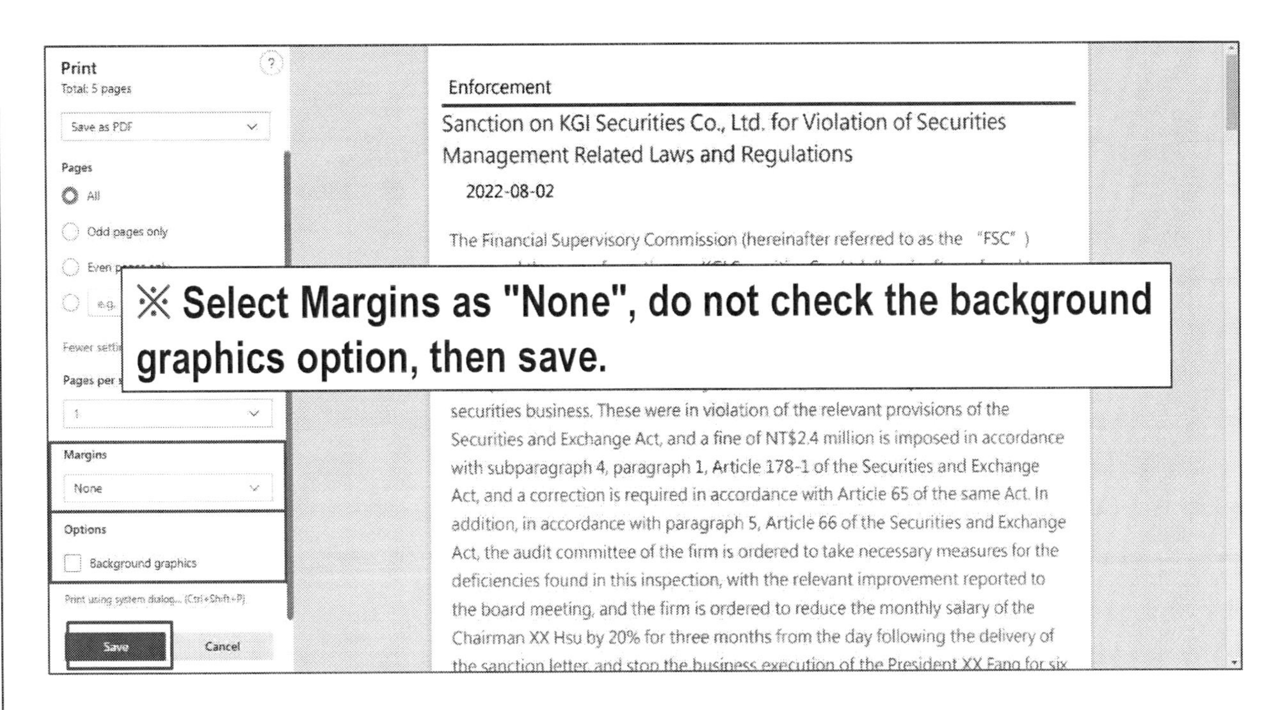

Exercise 9.1

Conduct text mining analysis on the internal control penalty cases of the FSC, focusing on cases from a specific time frame. Use the word cloud function to identify high-risk keywords.

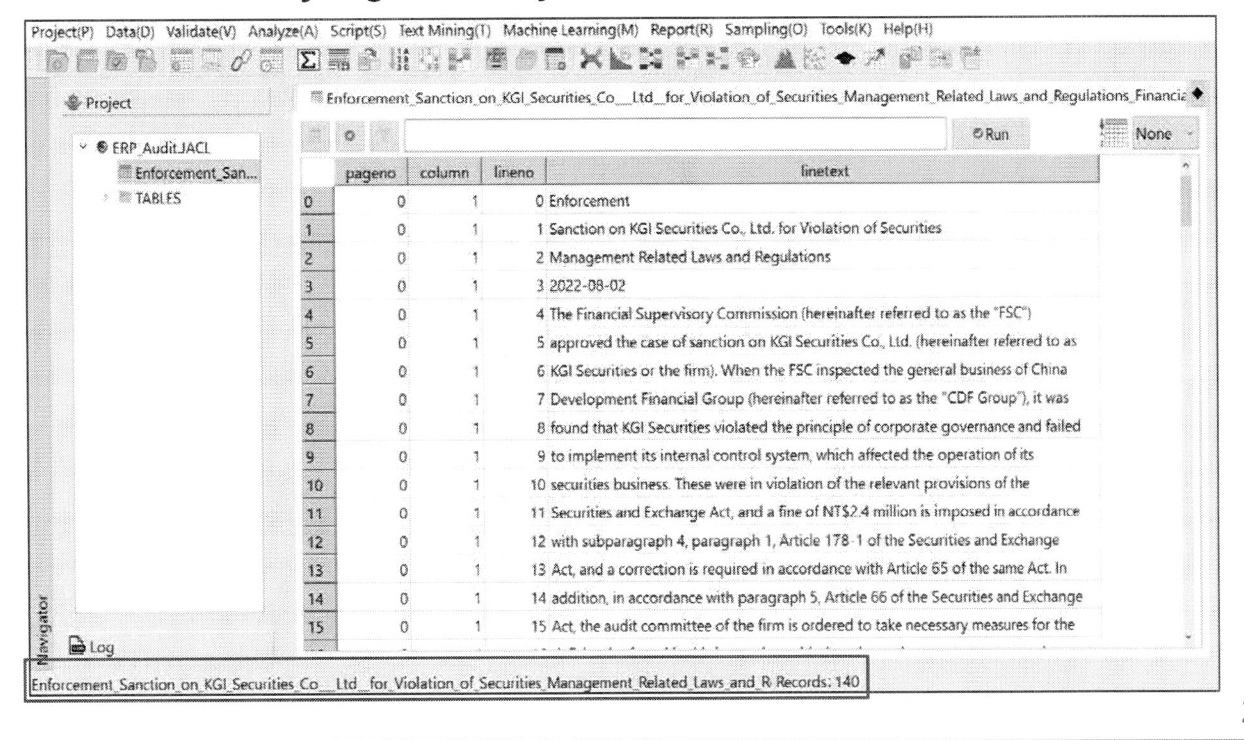

Exercise 9.1

Conduct text mining analysis on the internal control penalty cases of the FSC, focusing on cases from a specific time frame. Use the word cloud function to identify high-risk keywords.

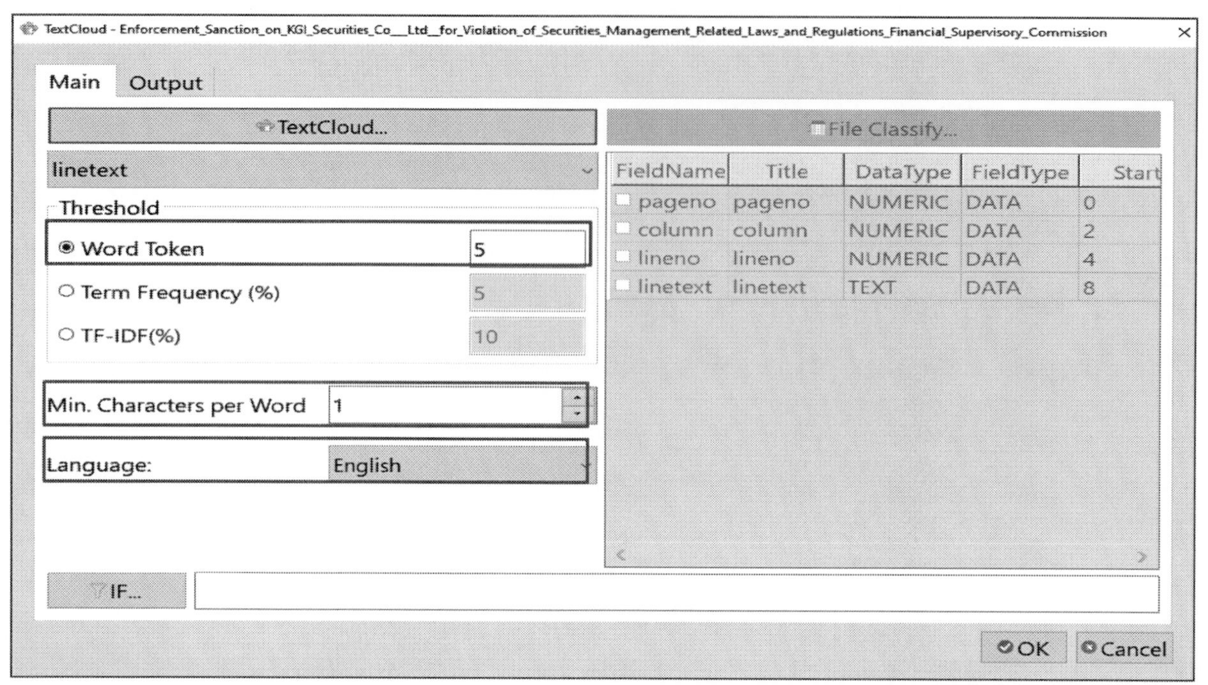

JCAATs-AI Audit Software
Copyright © 2023 JACKSOFT.

Exercise 9.1

Conduct text mining analysis on the internal control penalty cases of the FSC, focusing on cases from a specific time frame. Use the word cloud function to identify high-risk keywords.

Exercise 9.2

Using the keyword command, perform text mining on penalty cases to identify high-risk keywords.

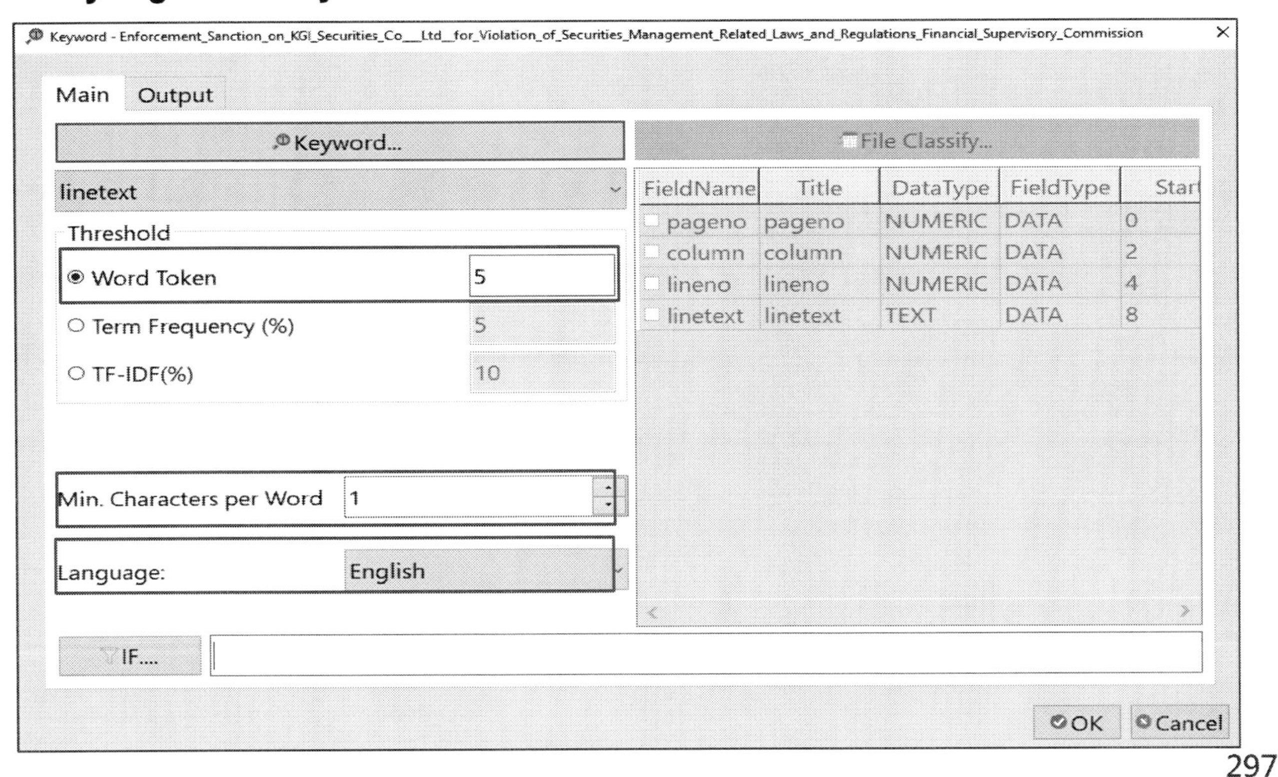

JCAATs-AI Audit Software

Exercise 9.2

Using the keyword command, perform text mining on penalty cases to identify high-risk keywords.

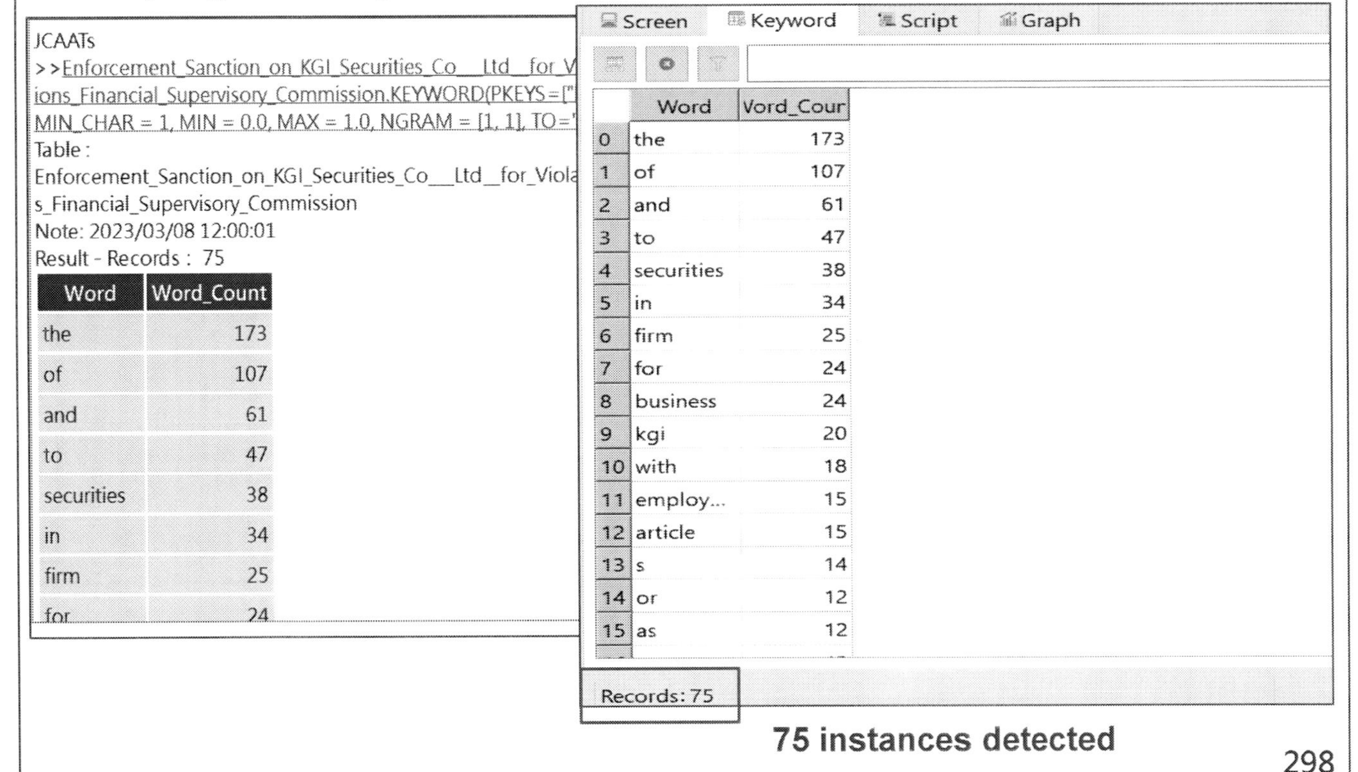

75 instances detected

Exercise 9.3

Using sentiment analysis commands, conduct anomaly checks on loan cases and identify high-risk cases that require further investigation.

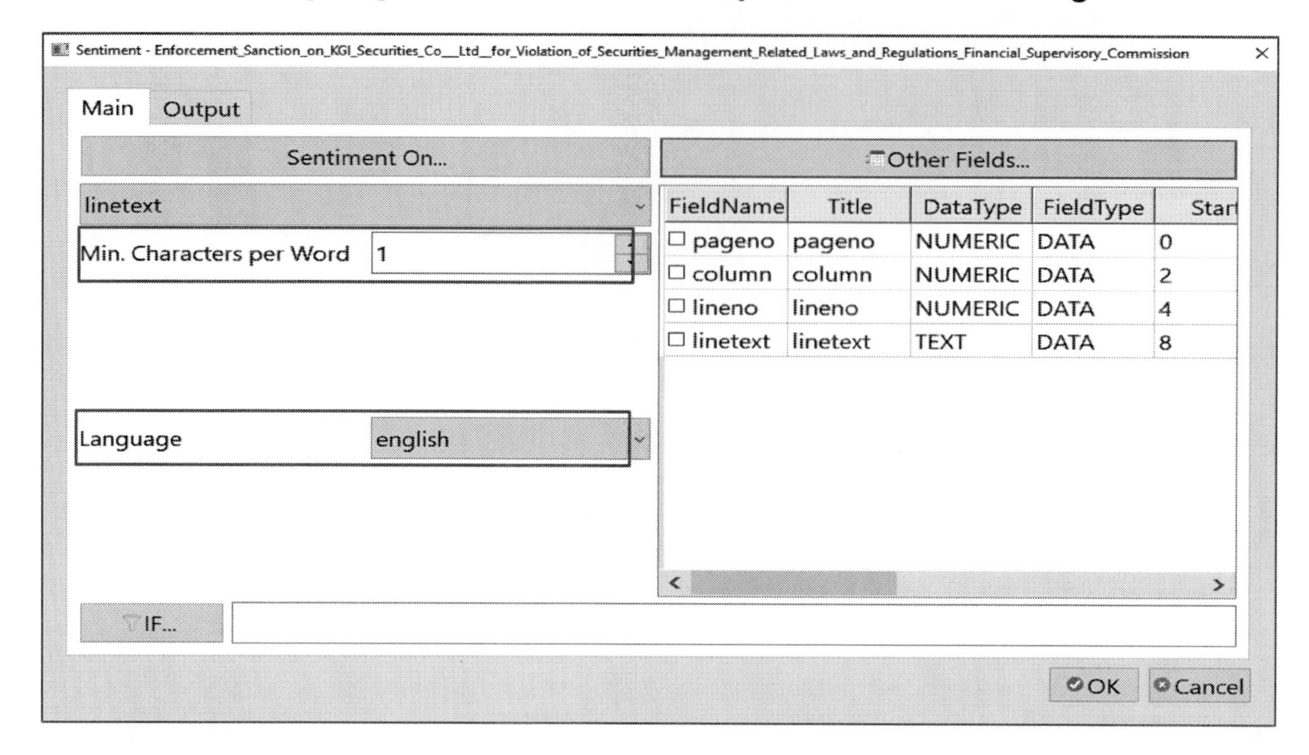

Exercise 9.3

Using sentiment analysis commands, conduct anomaly checks on loan cases and identify high-risk cases that require further investigation.

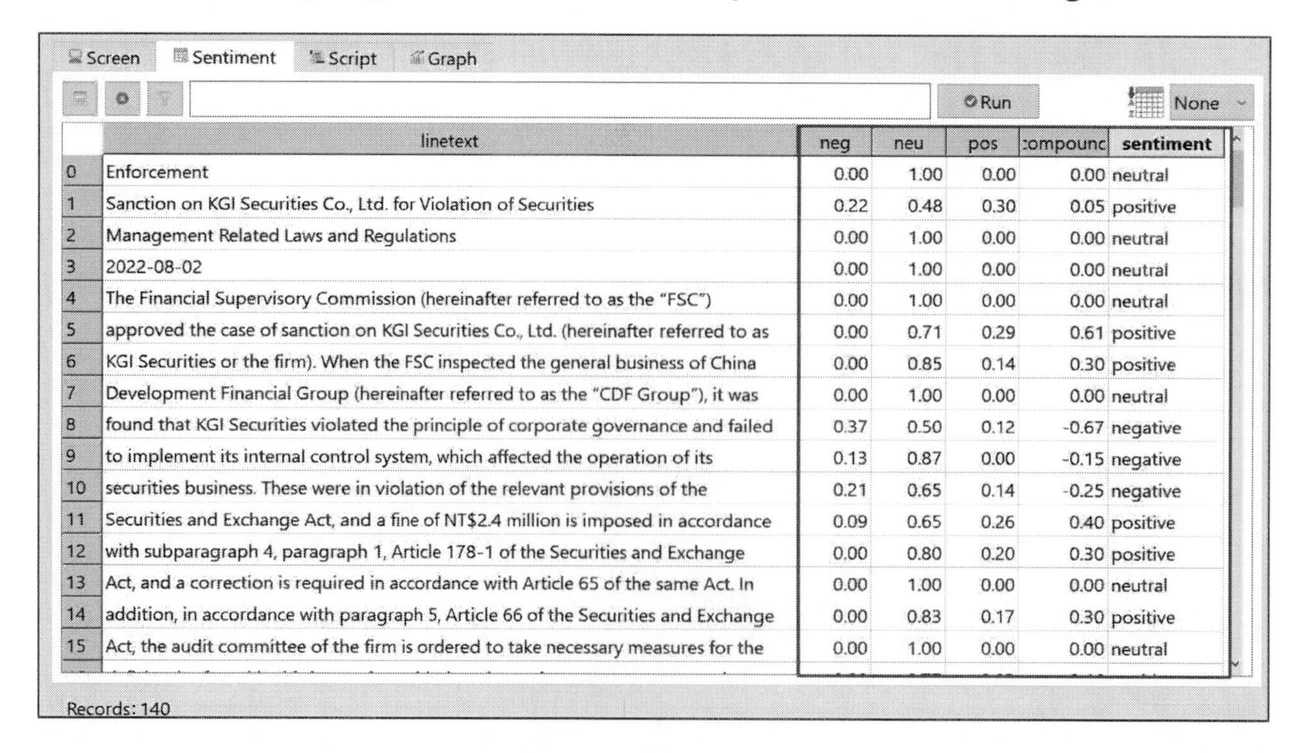

	linetext	neg	neu	pos	compound	sentiment
0	Enforcement	0.00	1.00	0.00	0.00	neutral
1	Sanction on KGI Securities Co., Ltd. for Violation of Securities	0.22	0.48	0.30	0.05	positive
2	Management Related Laws and Regulations	0.00	1.00	0.00	0.00	neutral
3	2022-08-02	0.00	1.00	0.00	0.00	neutral
4	The Financial Supervisory Commission (hereinafter referred to as the "FSC")	0.00	1.00	0.00	0.00	neutral
5	approved the case of sanction on KGI Securities Co., Ltd. (hereinafter referred to as	0.00	0.71	0.29	0.61	positive
6	KGI Securities or the firm). When the FSC inspected the general business of China	0.00	0.85	0.14	0.30	positive
7	Development Financial Group (hereinafter referred to as the "CDF Group"), it was	0.00	1.00	0.00	0.00	neutral
8	found that KGI Securities violated the principle of corporate governance and failed	0.37	0.50	0.12	-0.67	negative
9	to implement its internal control system, which affected the operation of its	0.13	0.87	0.00	-0.15	negative
10	securities business. These were in violation of the relevant provisions of the	0.21	0.65	0.14	-0.25	negative
11	Securities and Exchange Act, and a fine of NT$2.4 million is imposed in accordance	0.09	0.65	0.26	0.40	positive
12	with subparagraph 4, paragraph 1, Article 178-1 of the Securities and Exchange	0.00	0.80	0.20	0.30	positive
13	Act, and a correction is required in accordance with Article 65 of the same Act. In	0.00	1.00	0.00	0.00	neutral
14	addition, in accordance with paragraph 5, Article 66 of the Securities and Exchange	0.00	0.83	0.17	0.30	positive
15	Act, the audit committee of the firm is ordered to take necessary measures for the	0.00	1.00	0.00	0.00	neutral

Records: 140

Exercise 9.3

Using sentiment analysis commands, conduct anomaly checks on loan cases and identify high-risk cases that require further investigation.

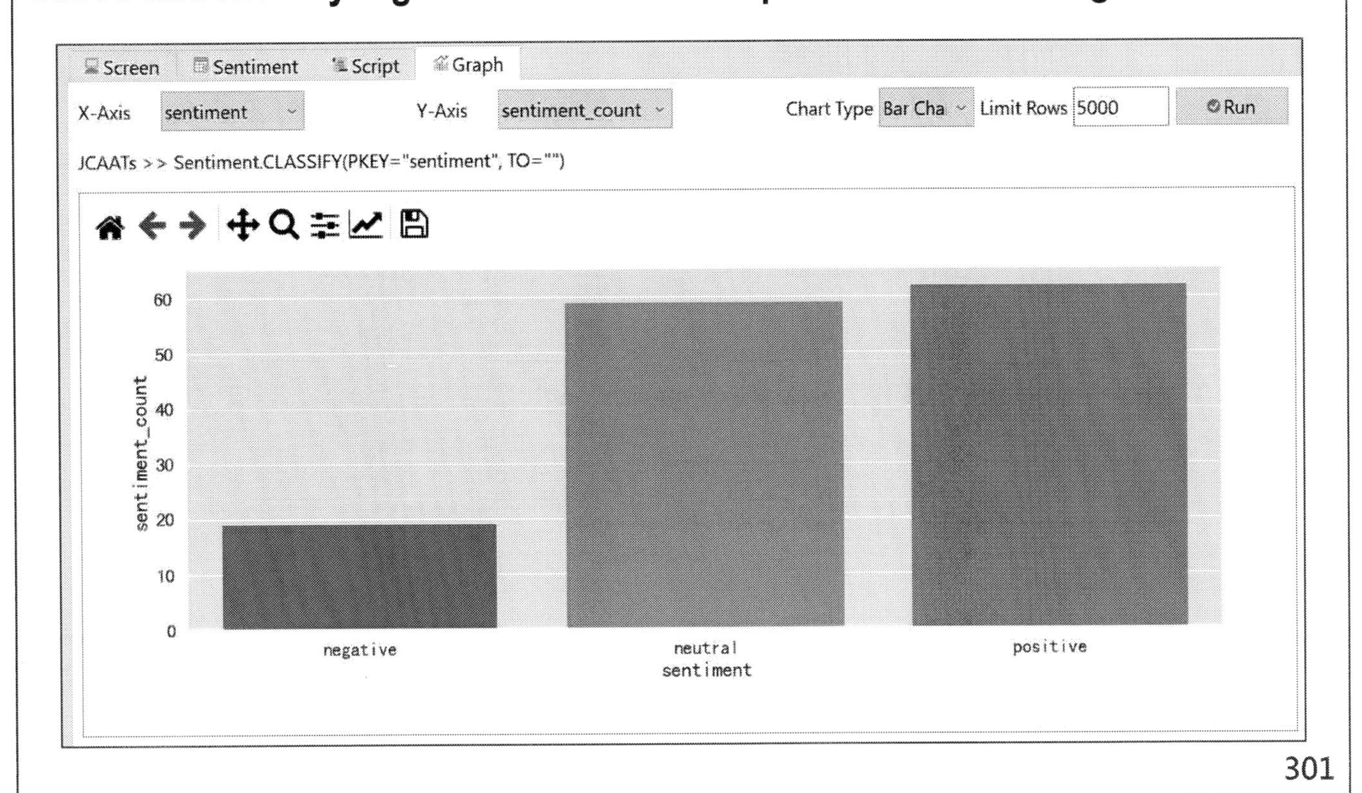

JCAATs Learning Note:

JCAATs - AI Audit Software

Chapter 10 - Exercise

- Exercise 10.1: Please perform supervised machine learning on the loan data file (LOAN). The target field to be learned (LOAN_DECISION) is whether the loan is approved, and the learning object field is other feature fields (you can choose other fields at will). The machine learning model used is logistic regression, and the learning path uses the initial settings to create the knowledge model: Loan_ML. Please list the **Confusion Matrix** of this knowledge model and explain it.
- Exercise 10.2: List the relevant learning performance indicators of the knowledge model: **Loan_ML** in Exercise 10.1, including accuracy, precision, recall, F1, etc., and discuss the uses of these indicators and the degree of superiority or inferiority of this learning effect. Should other prediction models be used for further learning?
- Exercise 10.3: List the indicators in the PerformanceMetrics of the knowledge model: Loan_ML in Exercise 10.1, and explain the use and meaning of the Intercept value. Explain the use and meaning of the Importance value of the indicator.

Chapter 10 - Exercise

Exercise 10.4: Sherry wants to conduct a pre-audit on loan defaults after the loan is granted to predict the existing risk of current loan data. The steps are as follows:

STEP 1: Use the Extract command to extract the data in the LOAN data table where LOAN_DECISION == "Yes" and create a new data table: Approved _Loan Table.

STEP 2: Supervised machine learning is applied to the Approved_Loan Table, using the decision tree model. The target field for learning (PS_IS_VIOLATE) is whether there is a breach of contract, and the learning object fields are other feature fields. The learning pipeline uses the initial settings to create the knowledge model: **Breach_ML**.

Please use the above steps to practice and list the confusion matrix and learning performance indicators of your knowledge model.

Chapter 10 - Exercise

- Exercise 10.5: Please use the knowledge model "**Loan_ML**" to predict which new loan applications in the LOAN_NEW file will be approved. How many records will be predicted to approve?

- Exercise 10.6: Please use the knowledge model "**Breach_ML**" to predict which loan applications in the LOAN_NEW file may breach contract? How many records will be predicted to breach contract?

- Exercise 10.7: Please use a Join command to list the number of cases in the LOAN_NEW file that are predicted to be approved for a loan and predicted to breach contract.

Exercise 10.1

Please perform supervised machine learning on the loan data file (LOAN). The target field to be learned (LOAN_DECISION) is whether the loan is approved, and the learning object field is other feature fields (you can choose other fields at will). The machine learning model used is logistic regression, and the learning path uses the initial settings to create the knowledge model: Loan_ML. Please list the Confusion Matrix of this knowledge model and explain it.

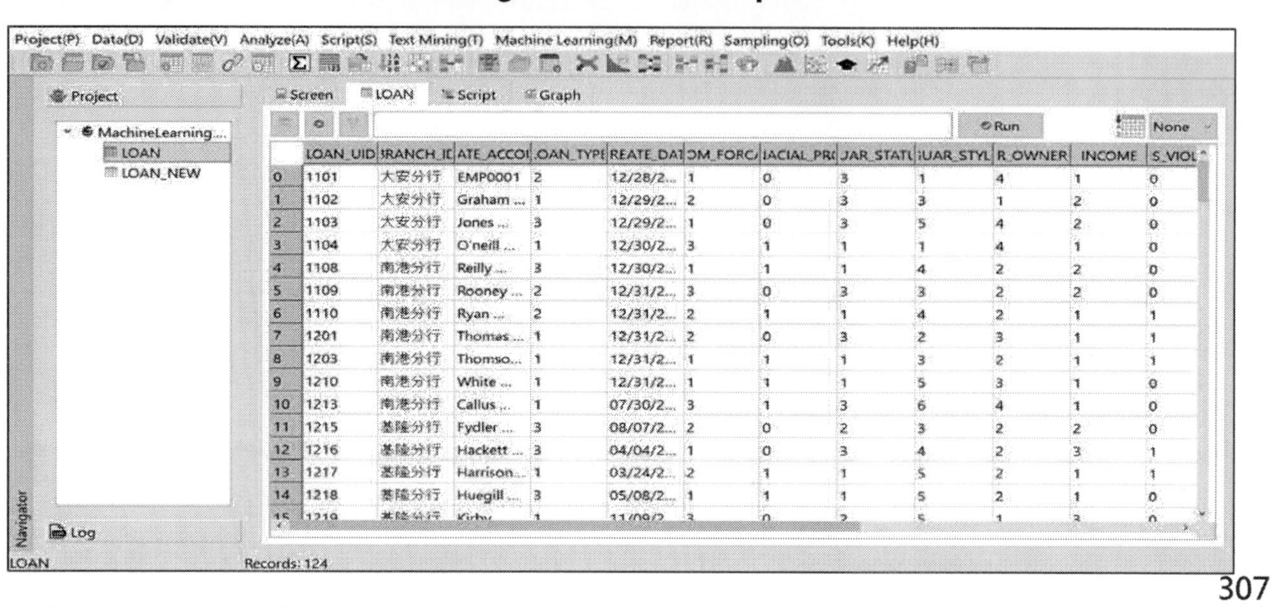

Exercise 10.1

Please perform supervised machine learning on the loan data file (LOAN). The target field to be learned (LOAN_DECISION) is whether the loan is approved, and the learning object field is other feature fields (you can choose other fields at will). The machine learning model used is logistic regression, and the learning path uses the initial settings to create the knowledge model: Loan_ML. Please list the Confusion Matrix of this knowledge model and explain it.

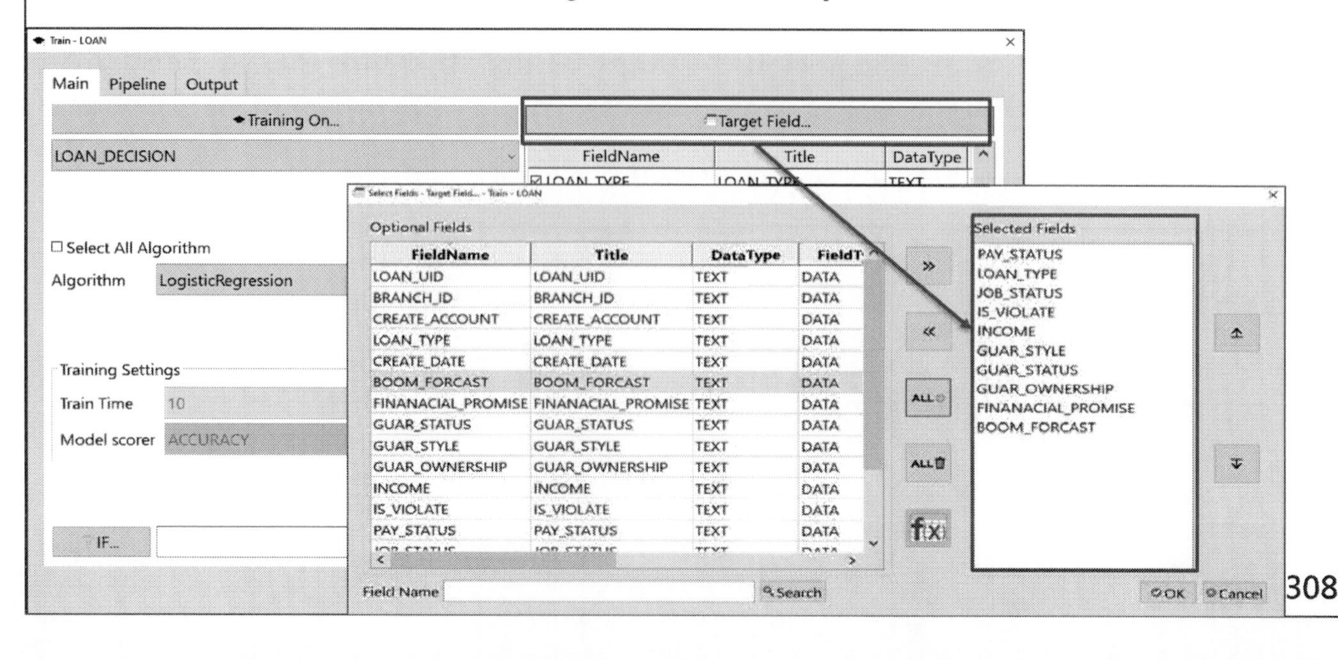

Exercise 10.1

Please perform supervised machine learning on the loan data file (LOAN). The target field to be learned (LOAN_DECISION) is whether the loan is approved, and the learning object field is other feature fields (you can choose other fields at will). The machine learning model used is logistic regression, and the learning path uses the initial settings to create the knowledge model: Loan_ML. Please list the Confusion Matrix of this knowledge model and explain it.

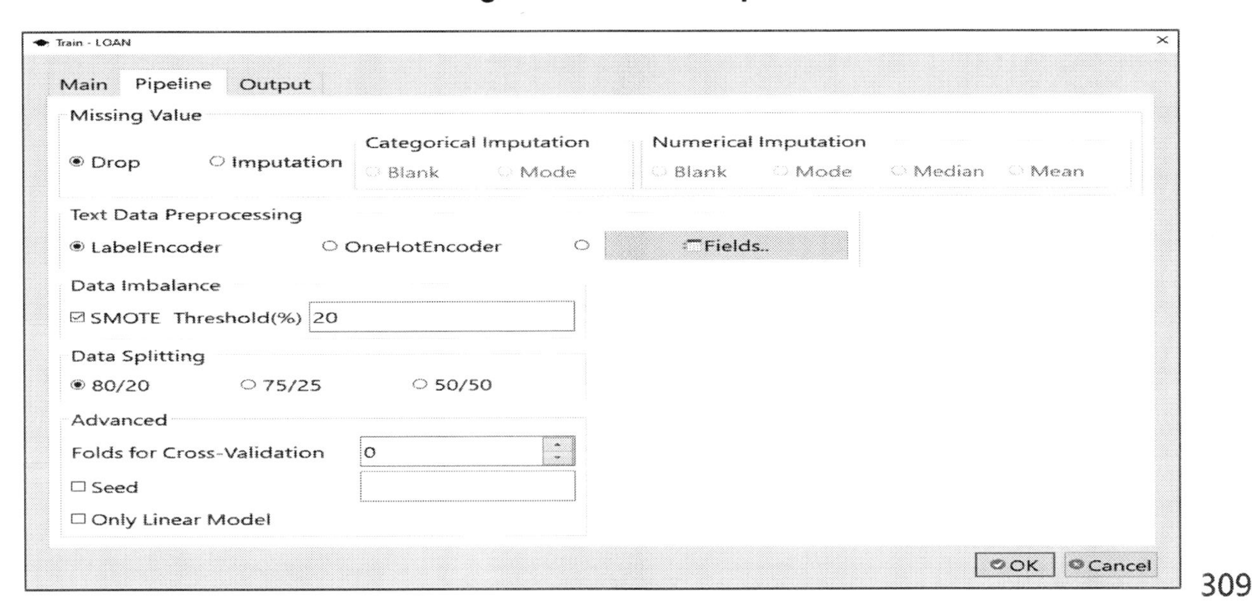

JCAATs-AI Audit Software
Copyright © 2023 JACKSOFT.

Exercise 10.1

Please perform supervised machine learning on the loan data file (LOAN). The target field to be learned (LOAN_DECISION) is whether the loan is approved, and the learning object field is other feature fields (you can choose other fields at will). The machine learning model used is logistic regression, and the learning path uses the initial settings to create the knowledge model: Loan_ML. Please list the Confusion Matrix of this knowledge model and explain it.

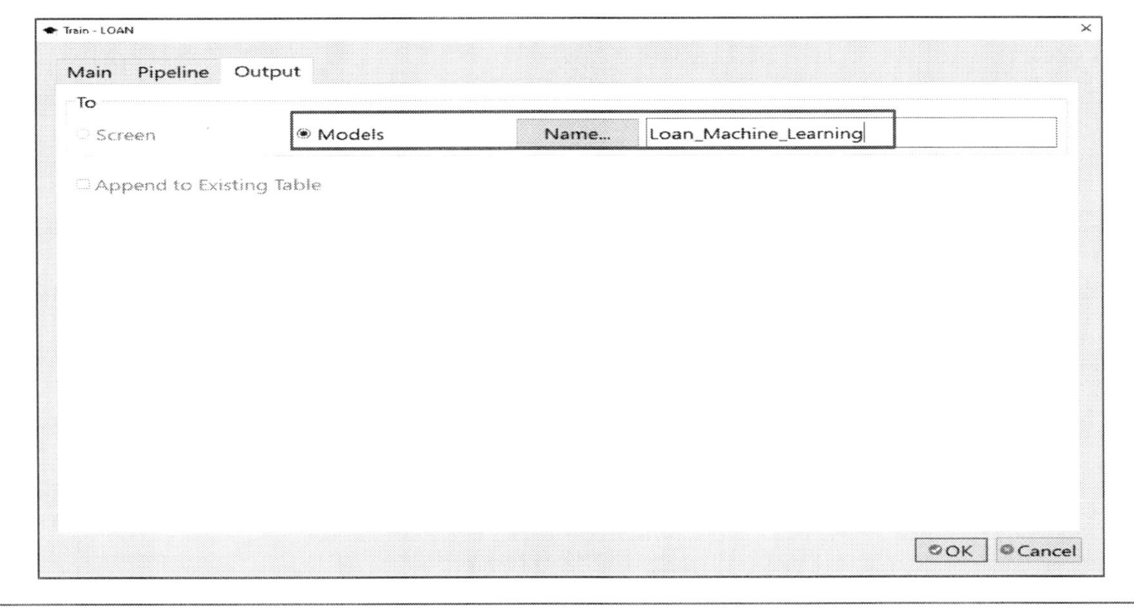

Exercise 10.1

Please perform supervised machine learning on the loan data file (LOAN). The target field to be learned (LOAN_DECISION) is whether the loan is approved, and the learning object field is other feature fields (you can choose other fields at will). The machine learning model used is logistic regression, and the learning path uses the initial settings to create the knowledge model: Loan_ML. Please list the Confusion Matrix of this knowledge model and explain it.

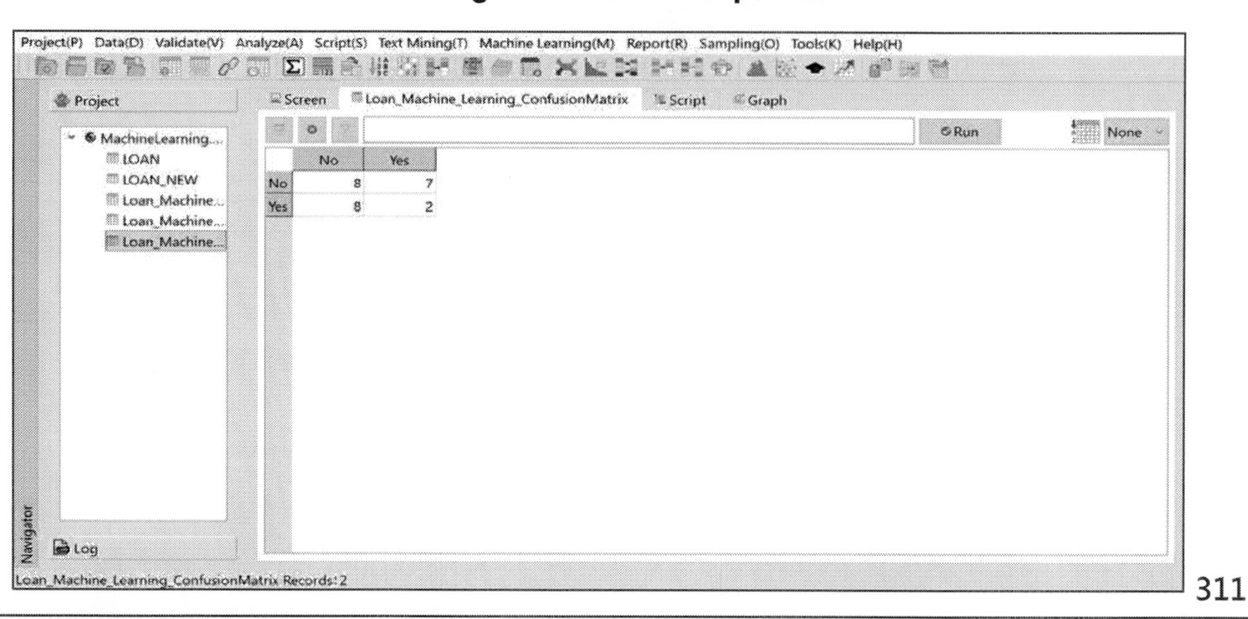

	No	Yes
No	8	7
Yes	8	2

JCAATs-AI Audit Software
Copyright © 2023 JACKSOFT.

Exercise 10.1

Please perform supervised machine learning on the loan data file (LOAN). The target field to be learned (LOAN_DECISION) is whether the loan is approved, and the learning object field is other feature fields (you can choose other fields at will). The machine learning model used is logistic regression, and the learning path uses the initial settings to create the knowledge model: Loan_ML. Please list the Confusion Matrix of this knowledge model and explain it.

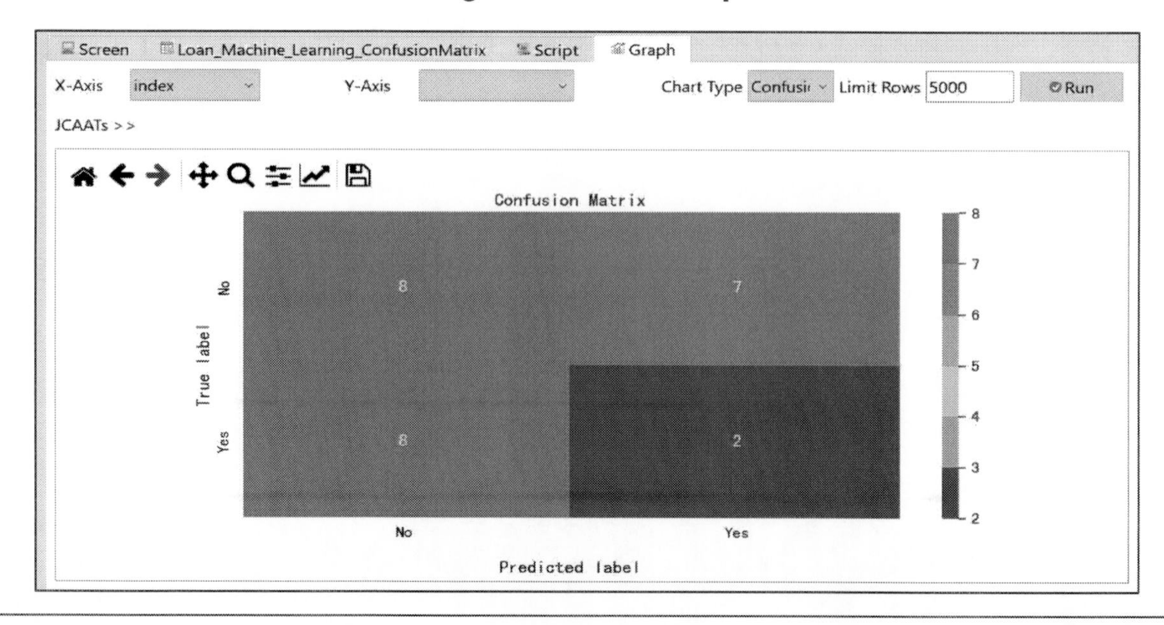

Exercise 10.2

List the relevant learning performance indicators of the knowledge model: Loan_ML in Exercise 10.1, including accuracy, precision, recall, F1, etc., and discuss the uses of these indicators and the degree of superiority or inferiority of this learning effect.

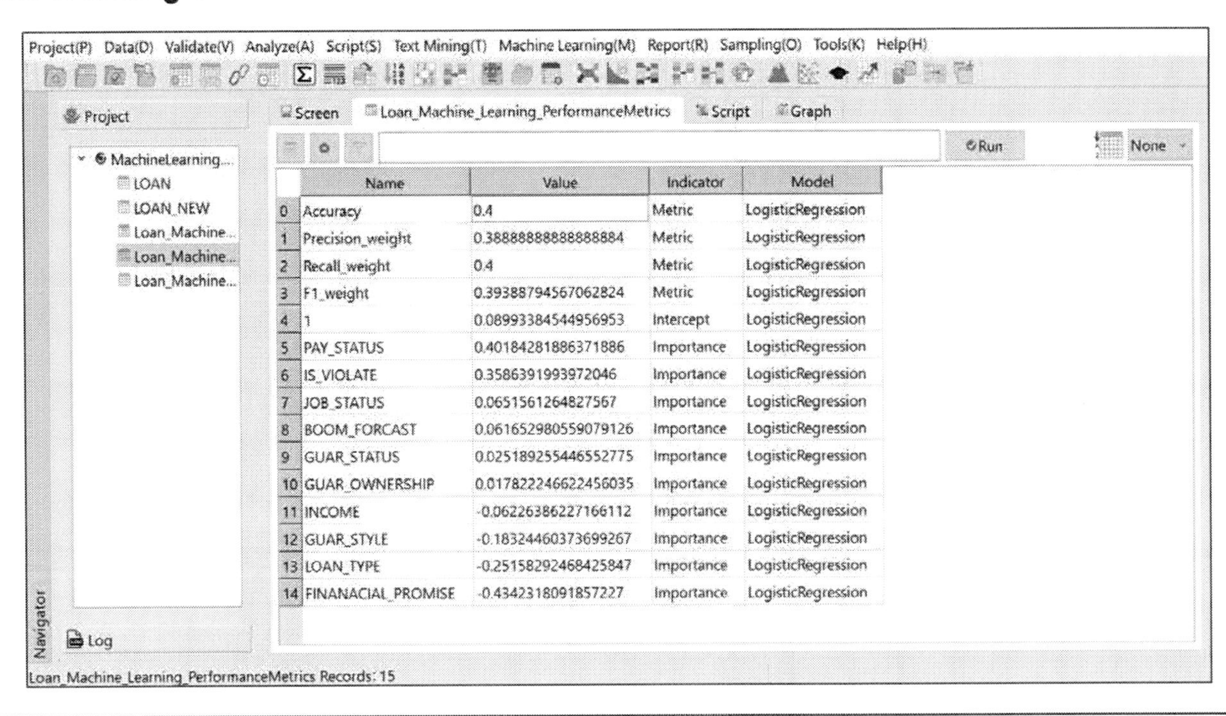

	Name	Value	Indicator	Model
0	Accuracy	0.4	Metric	LogisticRegression
1	Precision_weight	0.3888888888888884	Metric	LogisticRegression
2	Recall_weight	0.4	Metric	LogisticRegression
3	F1_weight	0.39388794567062824	Metric	LogisticRegression
4	1	0.08993384544956953	Intercept	LogisticRegression
5	PAY_STATUS	0.40184281886371886	Importance	LogisticRegression
6	IS_VIOLATE	0.3586391993972046	Importance	LogisticRegression
7	JOB_STATUS	0.0651561264827567	Importance	LogisticRegression
8	BOOM_FORCAST	0.061652980559079126	Importance	LogisticRegression
9	GUAR_STATUS	0.025189255446552775	Importance	LogisticRegression
10	GUAR_OWNERSHIP	0.017822246622456035	Importance	LogisticRegression
11	INCOME	-0.06226386227166112	Importance	LogisticRegression
12	GUAR_STYLE	-0.18324460373699267	Importance	LogisticRegression
13	LOAN_TYPE	-0.25158292468425847	Importance	LogisticRegression
14	FINANACIAL_PROMISE	-0.4342318091857227	Importance	LogisticRegression

Loan_Machine_Learning_PerformanceMetrics Records: 15

JCAATs-AI Audit Software

Exercise 10.2

List the relevant learning performance indicators of the knowledge model: Loan_ML in Exercise 10.1, including accuracy, precision, recall, F1, etc., and discuss the uses of these indicators and the degree of superiority or inferiority of this learning effect.

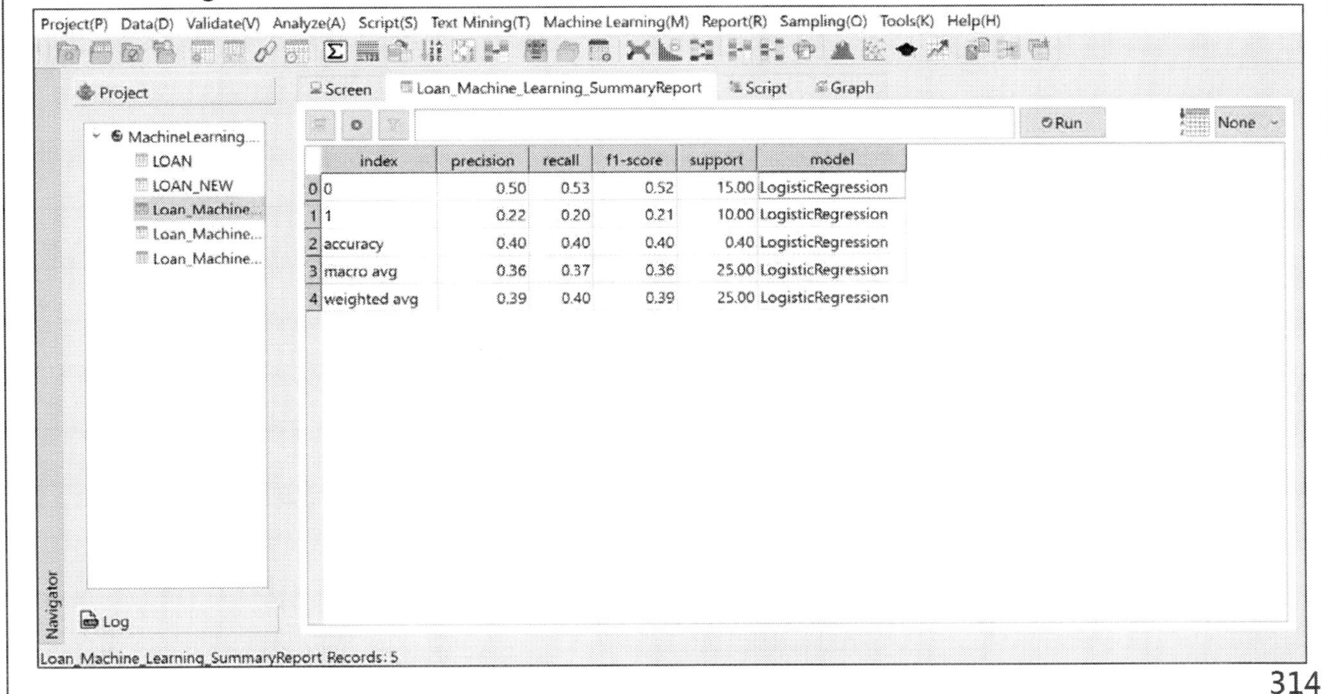

	index	precision	recall	f1-score	support	model
0	0	0.50	0.53	0.52	15.00	LogisticRegression
1	1	0.22	0.20	0.21	10.00	LogisticRegression
2	accuracy	0.40	0.40	0.40	0.40	LogisticRegression
3	macro avg	0.36	0.37	0.36	25.00	LogisticRegression
4	weighted avg	0.39	0.40	0.39	25.00	LogisticRegression

Loan_Machine_Learning_SummaryReport Records: 5

Exercise 10.3

List the indicators in the PerformanceMetrics of the knowledge model: Loan_ML in Exercise 10.1, and explain the use and meaning of the Intercept value. Explain the use and meaning of the Importance value of the indicator.

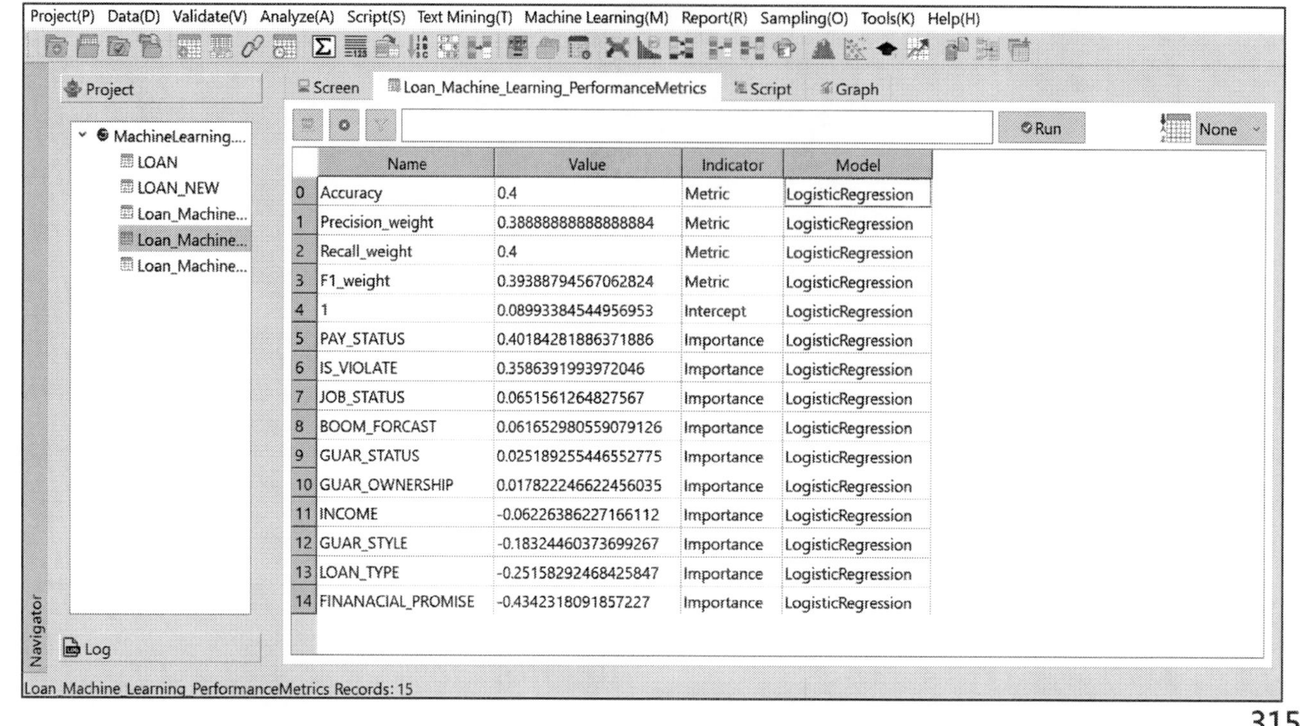

	Name	Value	Indicator	Model
0	Accuracy	0.4	Metric	LogisticRegression
1	Precision_weight	0.38888888888888884	Metric	LogisticRegression
2	Recall_weight	0.4	Metric	LogisticRegression
3	F1_weight	0.39388794567062824	Metric	LogisticRegression
4	1	0.08993384544956953	Intercept	LogisticRegression
5	PAY_STATUS	0.40184281886371886	Importance	LogisticRegression
6	IS_VIOLATE	0.3586391993972046	Importance	LogisticRegression
7	JOB_STATUS	0.0651561264827567	Importance	LogisticRegression
8	BOOM_FORCAST	0.061652980559079126	Importance	LogisticRegression
9	GUAR_STATUS	0.025189255446552775	Importance	LogisticRegression
10	GUAR_OWNERSHIP	0.017822246622456035	Importance	LogisticRegression
11	INCOME	-0.06226386227166112	Importance	LogisticRegression
12	GUAR_STYLE	-0.18324460373699267	Importance	LogisticRegression
13	LOAN_TYPE	-0.25158292468425847	Importance	LogisticRegression
14	FINANACIAL_PROMISE	-0.4342318091857227	Importance	LogisticRegression

Loan_Machine_Learning_PerformanceMetrics Records: 15

Exercise 10.4

Sherry wants to conduct a pre-audit on loan defaults after the loan is granted to predict the existing risk of current loan data.

STEP 1: Use the Extract command to extract the data in the LOAN data table where LOAN_DECISION == "Yes" and create a new data table: Approved _Loan Table.

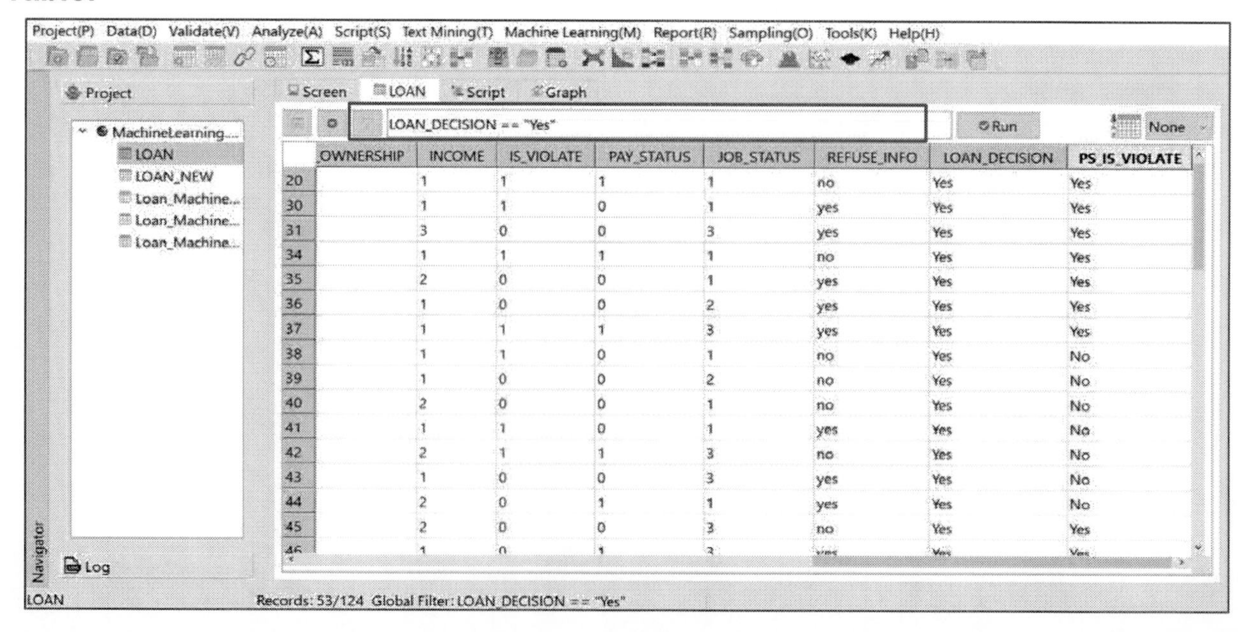

LOAN_DECISION == "Yes"

	OWNERSHIP	INCOME	IS_VIOLATE	PAY_STATUS	JOB_STATUS	REFUSE_INFO	LOAN_DECISION	PS_IS_VIOLATE
20		1	1	1	1	no	Yes	Yes
30		1	1	0	1	yes	Yes	Yes
31		3	0	0	3	yes	Yes	Yes
34		1	1	1	1	no	Yes	Yes
35		2	0	0	1	yes	Yes	Yes
36		1	0	0	2	yes	Yes	Yes
37		1	1	1	3	yes	Yes	Yes
38		1	1	0	1	no	Yes	No
39		1	0	0	2	no	Yes	No
40		2	0	0	1	no	Yes	No
41		1	1	0	1	yes	Yes	No
42		2	1	1	3	no	Yes	No
43		1	0	0	3	yes	Yes	No
44		2	0	1	1	yes	Yes	No
45		2	0	0	3	no	Yes	Yes

Records: 53/124 Global Filter: LOAN_DECISION == "Yes"

Exercise 10.4

Sherry wants to conduct a pre-audit on loan defaults after the loan is granted to predict the existing risk of current loan data.

STEP 1: Use the Extract command to extract the data in the LOAN data table where LOAN_DECISION == "Yes" and create a new data table: Approved _Loan Table.

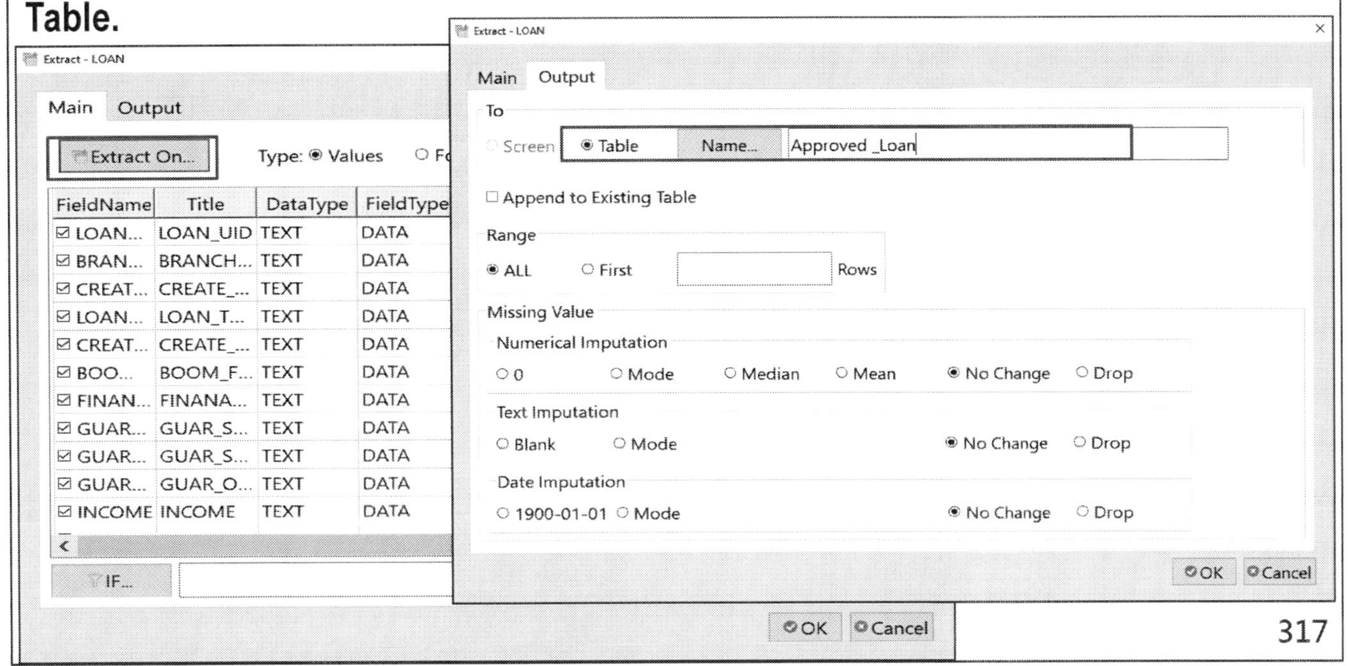

Exercise 10.4

Sherry wants to conduct a pre-audit on loan defaults after the loan is granted to predict the existing risk of current loan data.

STEP 1: Use the Extract command to extract the data in the LOAN data table where LOAN_DECISION == "Yes" and create a new data table: Approved _Loan Table.

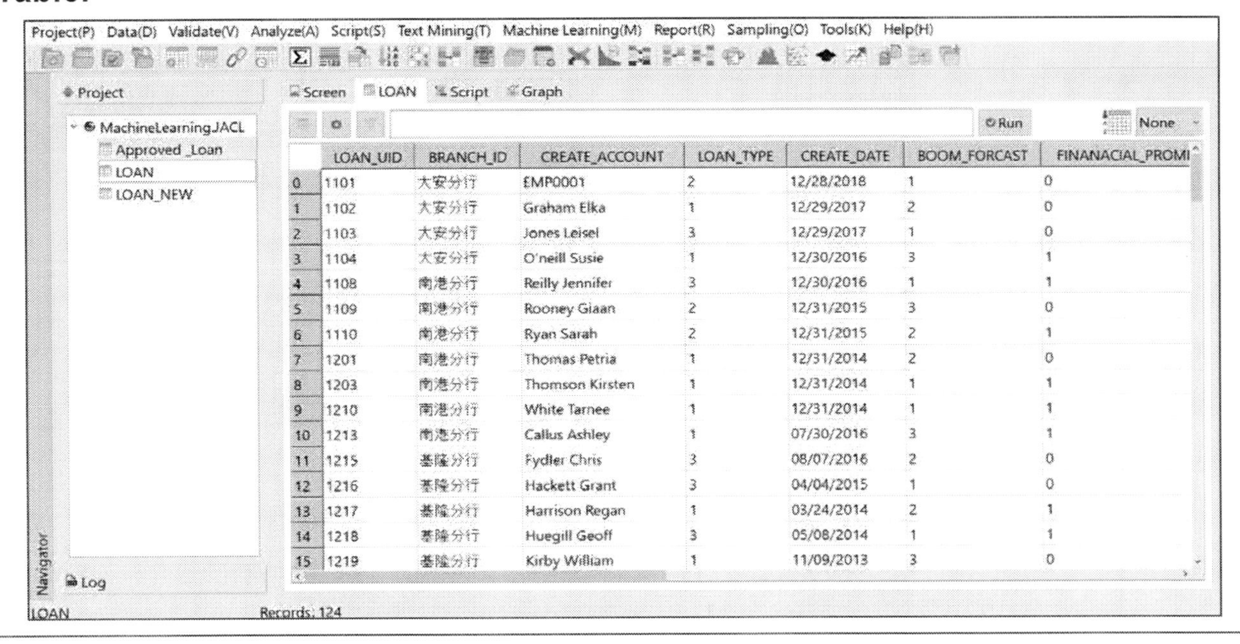

	LOAN_UID	BRANCH_ID	CREATE_ACCOUNT	LOAN_TYPE	CREATE_DATE	BOOM_FORCAST	FINANACIAL_PROMI
0	1101	大安分行	EMP0001	2	12/28/2018	1	0
1	1102	大安分行	Graham Elka	1	12/29/2017	2	0
2	1103	大安分行	Jones Leisel	3	12/29/2017	1	0
3	1104	大安分行	O'neill Susie	1	12/30/2016	3	1
4	1108	南港分行	Reilly Jennifer	3	12/30/2016	1	1
5	1109	南港分行	Rooney Giaan	2	12/31/2015	3	0
6	1110	南港分行	Ryan Sarah	2	12/31/2015	2	1
7	1201	南港分行	Thomas Petria	1	12/31/2014	2	0
8	1203	南港分行	Thomson Kirsten	1	12/31/2014	1	1
9	1210	南港分行	White Tarnee	1	12/31/2014	1	1
10	1213	南港分行	Callus Ashley	1	07/30/2016	3	1
11	1215	基隆分行	Fydler Chris	3	08/07/2016	2	0
12	1216	基隆分行	Hackett Grant	3	04/04/2015	1	0
13	1217	基隆分行	Harrison Regan	1	03/24/2014	2	1
14	1218	基隆分行	Huegill Geoff	3	05/08/2014	1	1
15	1219	基隆分行	Kirby William	1	11/09/2013	3	0

Exercise 10.4

STEP 2: Supervised machine learning is applied to the Approved_Loan Table, using the decision tree model. The target field for learning (PS_IS_VIOLATE) is whether there is a breach of contract, and the learning object fields are other feature fields. The learning pipeline uses the initial settings to create the knowledge model: Breach_ML.

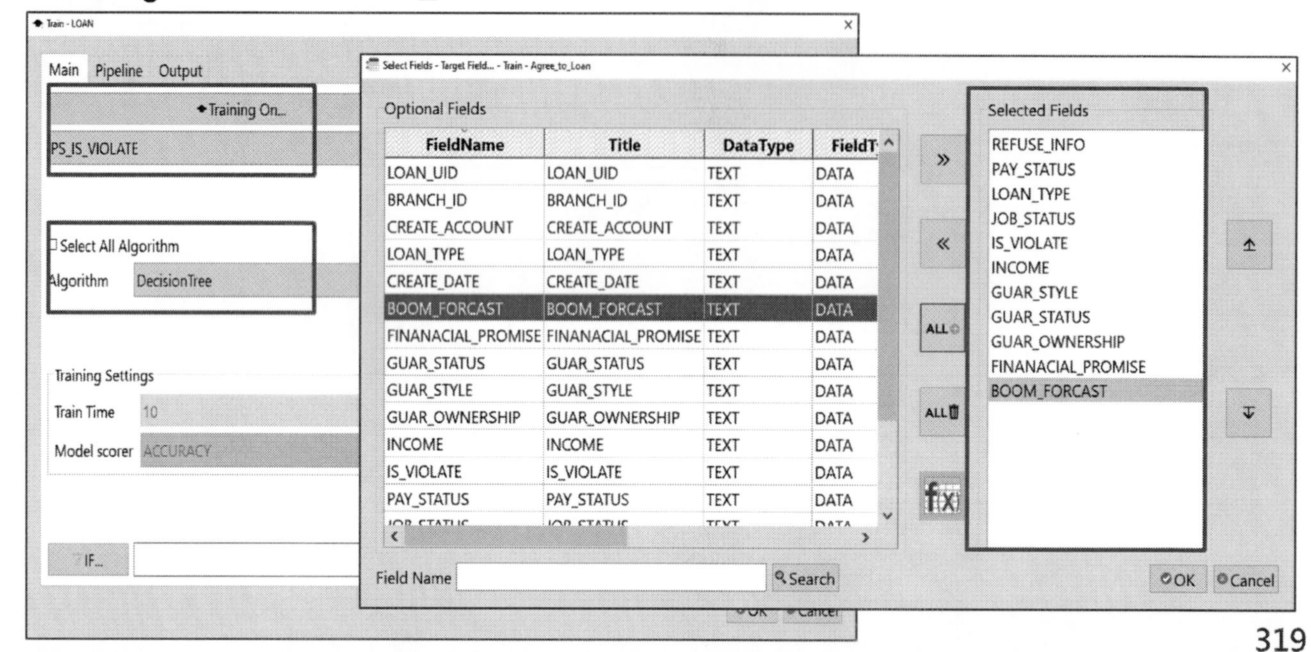

Exercise 10.4

STEP 2: Supervised machine learning is applied to the Approved_Loan Table, using the decision tree model. The target field for learning (PS_IS_VIOLATE) is whether there is a breach of contract, and the learning object fields are other feature fields. The learning pipeline uses the initial settings to create the knowledge model: Breach_ML.

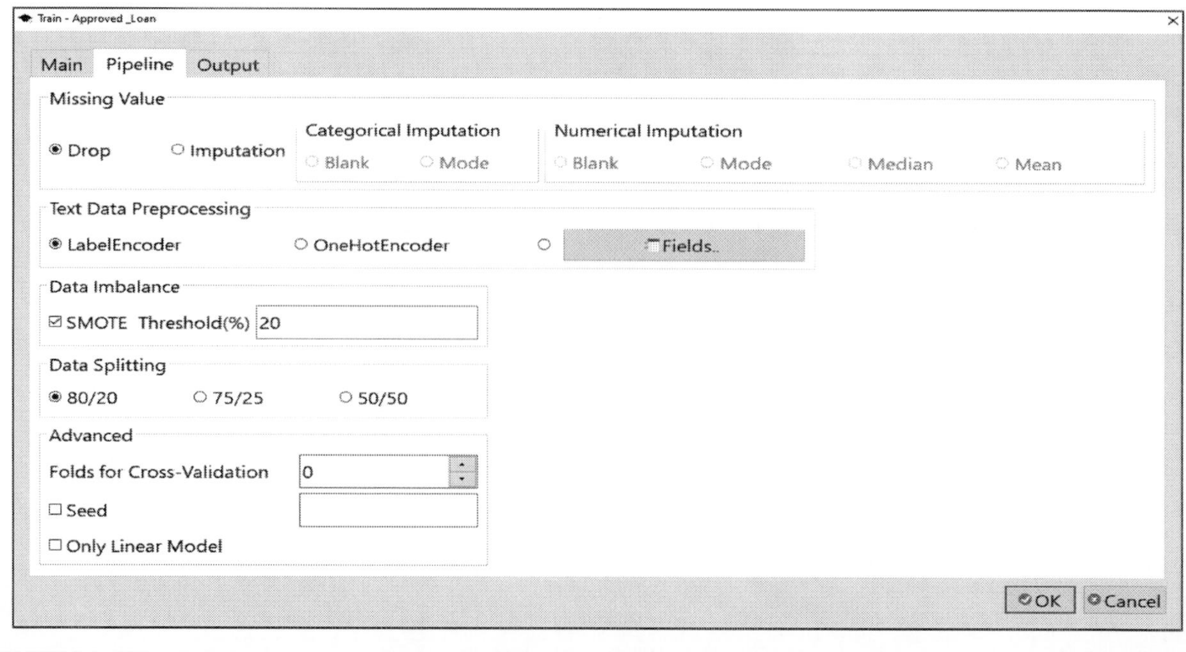

Exercise 10.4

STEP 2: Supervised machine learning is applied to the Approved_Loan Table, using the decision tree model. The target field for learning (PS_IS_VIOLATE) is whether there is a breach of contract, and the learning object fields are other feature fields. The learning pipeline uses the initial settings to create the knowledge model: Breach_ML.

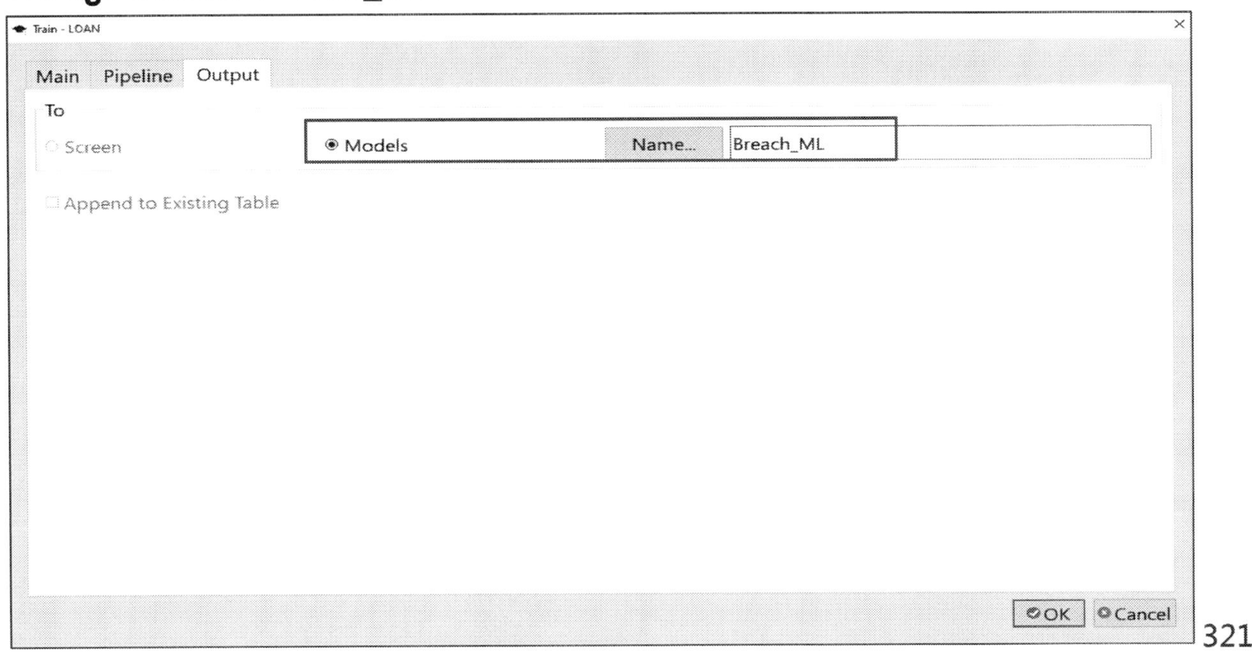

Exercise 10.4

Please use the above steps to practice and list the confusion matrix and learning performance indicators of your knowledge model.

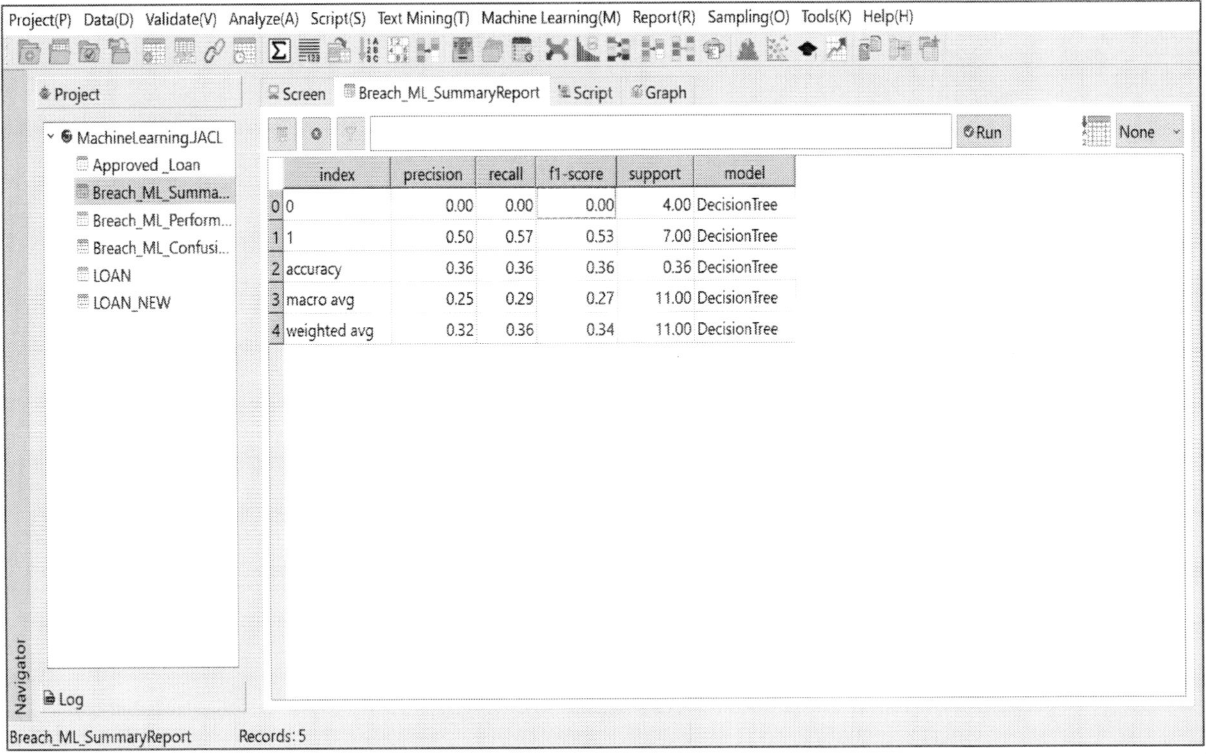

	index	precision	recall	f1-score	support	model
0	0	0.00	0.00	0.00	4.00	DecisionTree
1	1	0.50	0.57	0.53	7.00	DecisionTree
2	accuracy	0.36	0.36	0.36	0.36	DecisionTree
3	macro avg	0.25	0.29	0.27	11.00	DecisionTree
4	weighted avg	0.32	0.36	0.34	11.00	DecisionTree

Exercise 10.4

Please use the above steps to practice and list the confusion matrix and learning performance indicators of your knowledge model.

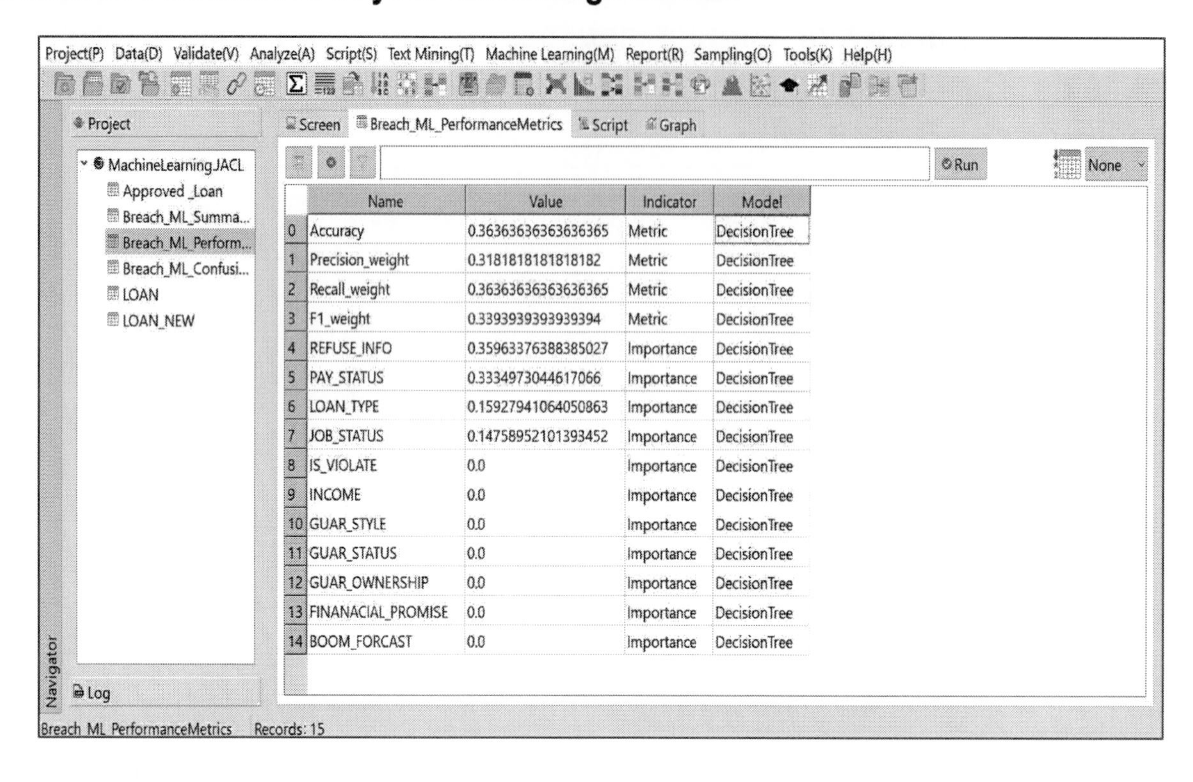

	Name	Value	Indicator	Model
0	Accuracy	0.36363636363636365	Metric	DecisionTree
1	Precision_weight	0.3181818181818182	Metric	DecisionTree
2	Recall_weight	0.36363636363636365	Metric	DecisionTree
3	F1_weight	0.3393939393939394	Metric	DecisionTree
4	REFUSE_INFO	0.35963376388385027	Importance	DecisionTree
5	PAY_STATUS	0.3334973044617066	Importance	DecisionTree
6	LOAN_TYPE	0.15927941064050863	Importance	DecisionTree
7	JOB_STATUS	0.14758952101393452	Importance	DecisionTree
8	IS_VIOLATE	0.0	Importance	DecisionTree
9	INCOME	0.0	Importance	DecisionTree
10	GUAR_STYLE	0.0	Importance	DecisionTree
11	GUAR_STATUS	0.0	Importance	DecisionTree
12	GUAR_OWNERSHIP	0.0	Importance	DecisionTree
13	FINANACIAL_PROMISE	0.0	Importance	DecisionTree
14	BOOM_FORCAST	0.0	Importance	DecisionTree

JCAATs-AI Audit Software

Exercise 10.4

Please use the above steps to practice and list the confusion matrix and learning performance indicators of your knowledge model.

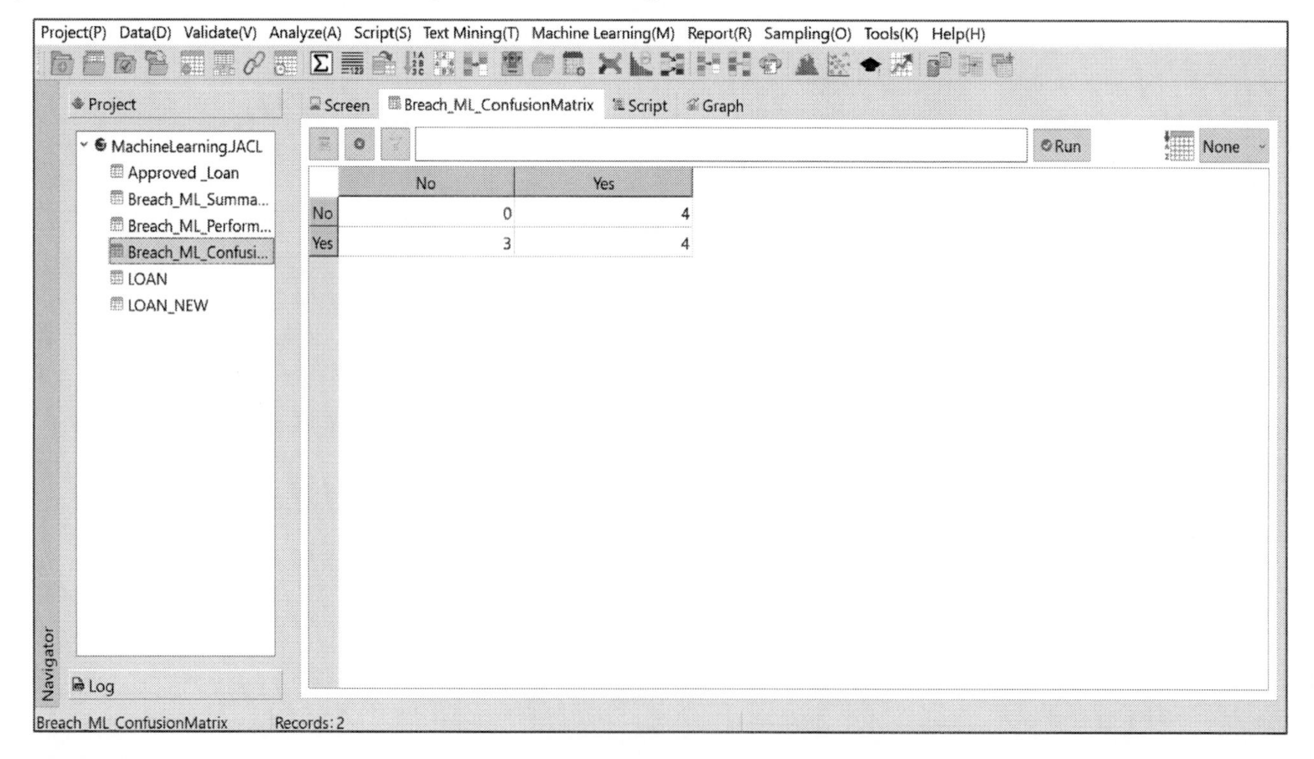

	No	Yes
No	0	4
Yes	3	4

Exercise 10.4

Please use the above steps to practice and list the confusion matrix and learning performance indicators of your knowledge model.

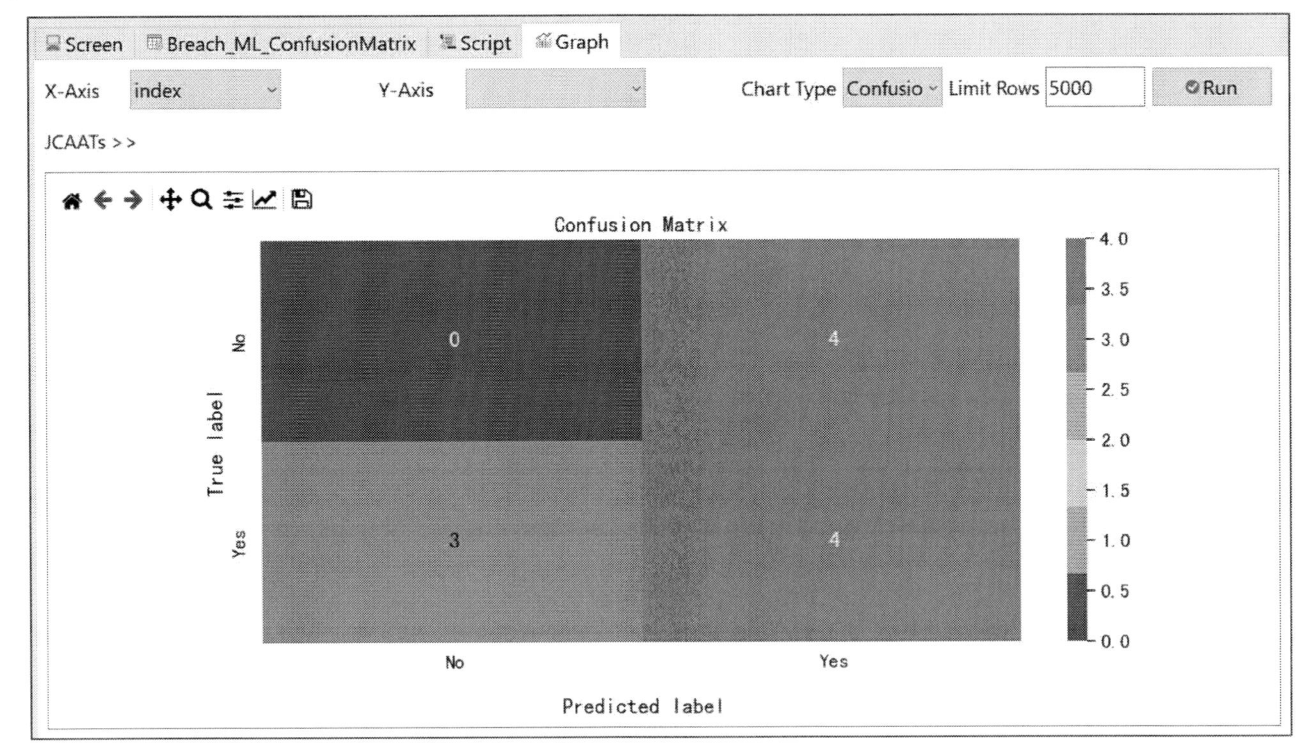

Exercise 10.5

Please use the knowledge model "Loan_ML" to predict which new loan applications in the LOAN_NEW file will be approved. How many records will be predicted to approve?

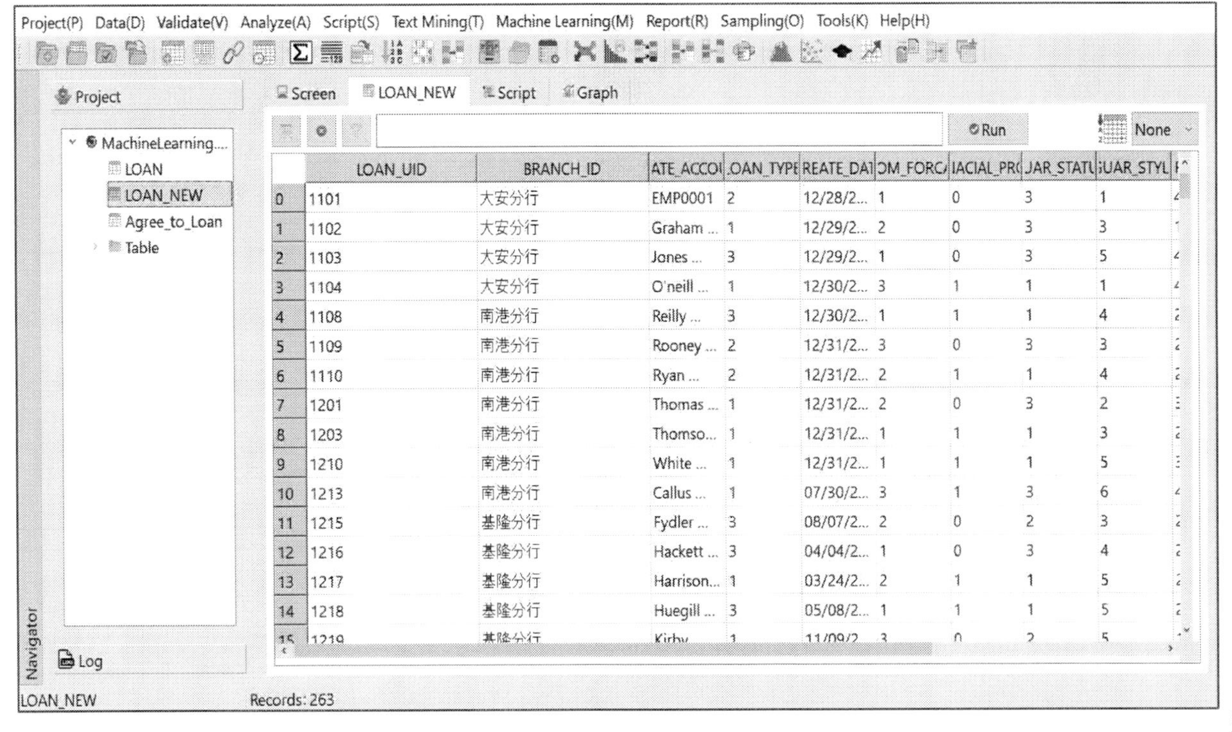

	LOAN_UID	BRANCH_ID	ATE_ACCOI	.OAN_TYPE	REATE_DAT	OM_FORC	IACIAL_PR	JAR_STATU	UAR_STYL
0	1101	大安分行	EMP0001	2	12/28/2...	1	0	3	1
1	1102	大安分行	Graham ...	1	12/29/2...	2	0	3	3
2	1103	大安分行	Jones ...	3	12/29/2...	1	0	3	5
3	1104	大安分行	O'neill ...	1	12/30/2...	3	1	1	1
4	1108	南港分行	Reilly ...	3	12/30/2...	1	1	1	4
5	1109	南港分行	Rooney ...	2	12/31/2...	3	0	3	3
6	1110	南港分行	Ryan ...	2	12/31/2...	2	1	1	4
7	1201	南港分行	Thomas ...	1	12/31/2...	2	0	3	2
8	1203	南港分行	Thomso...	1	12/31/2...	1	1	1	3
9	1210	南港分行	White ...	1	12/31/2...	1	1	1	5
10	1213	南港分行	Callus ...	1	07/30/2...	3	1	3	6
11	1215	基隆分行	Fydler ...	3	08/07/2...	2	0	2	3
12	1216	基隆分行	Hackett ...	3	04/04/2...	1	0	3	4
13	1217	基隆分行	Harrison...	1	03/24/2...	2	1	1	5
14	1218	基隆分行	Huegill ...	3	05/08/2...	1	1	1	5
15	1219	基隆分行	Kirby ...	1	11/09/2...	3	0	2	5

Exercise 10.5

Please use the knowledge model "Loan_ML" to predict which new loan applications in the LOAN_NEW file will be approved. How many records will be predicted to approve?

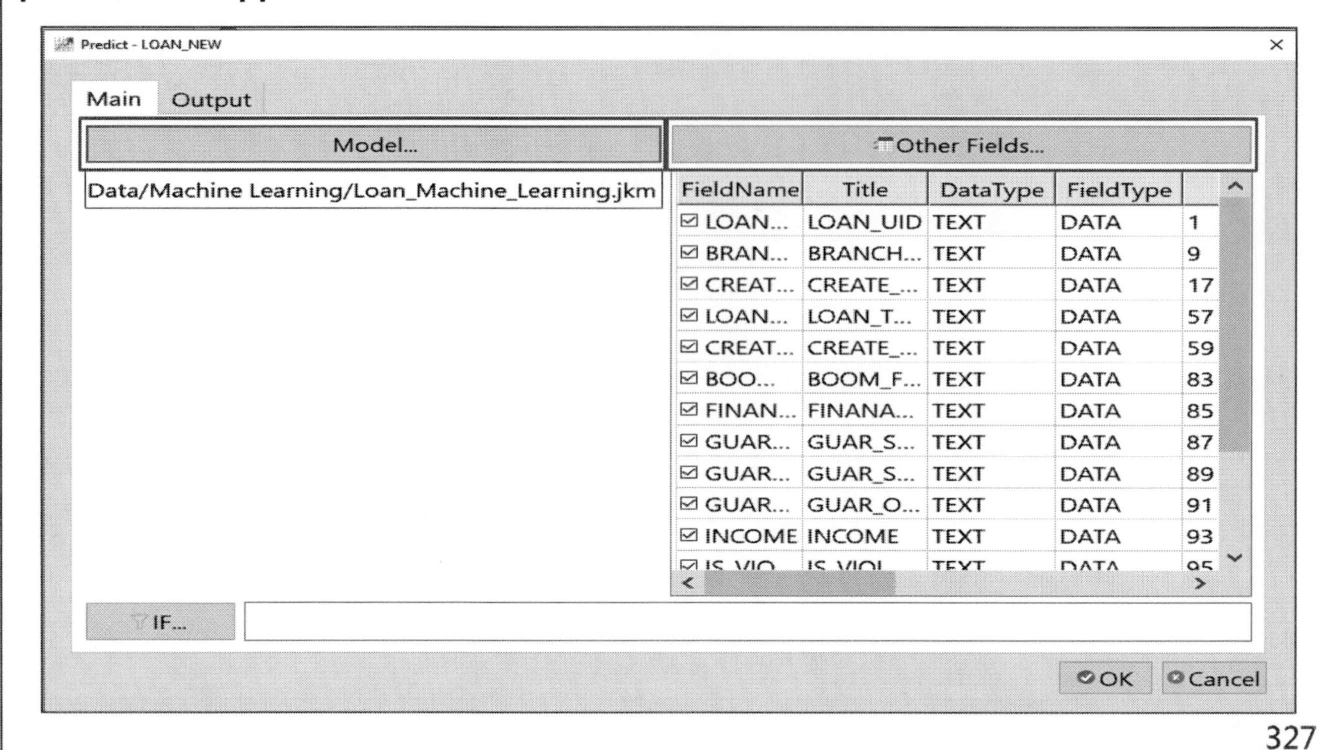

Exercise 10.5

Please use the knowledge model "Loan_ML" to predict which new loan applications in the LOAN_NEW file will be approved. How many records will be predicted to approve?

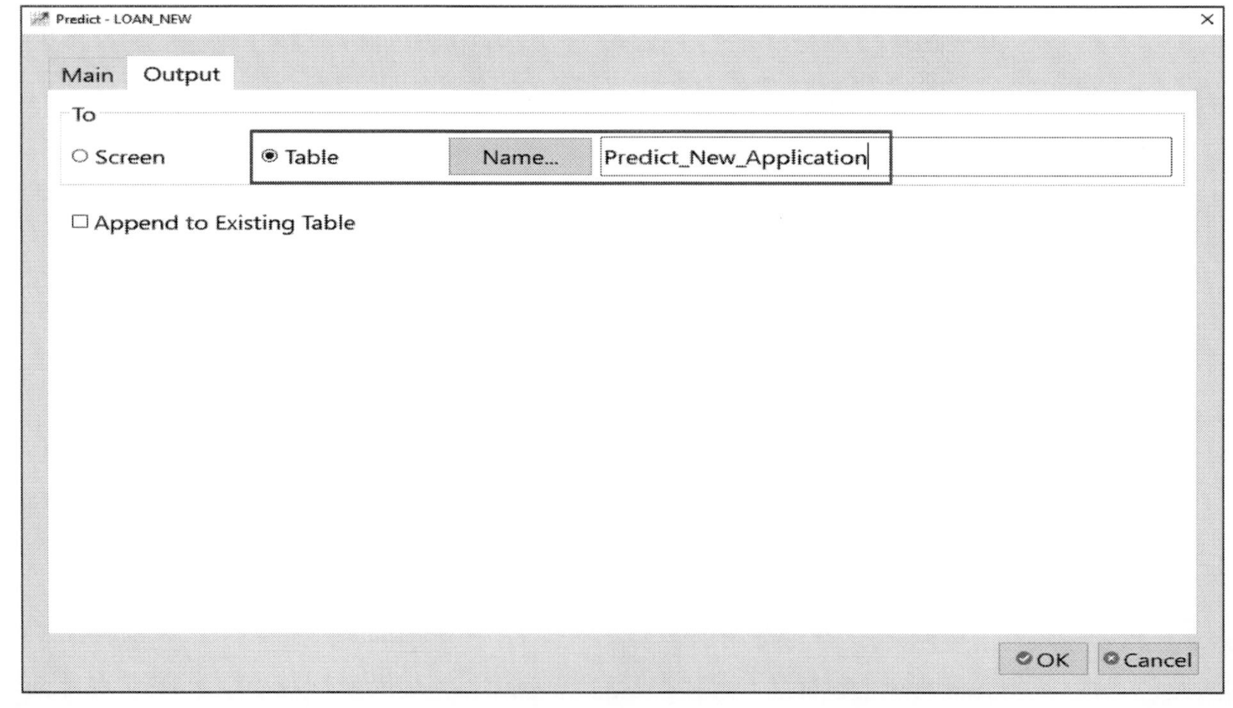

Exercise 10.5

Please use the knowledge model "Loan_ML" to predict which new loan applications in the LOAN_NEW file will be approved. How many records will be predicted to approve?

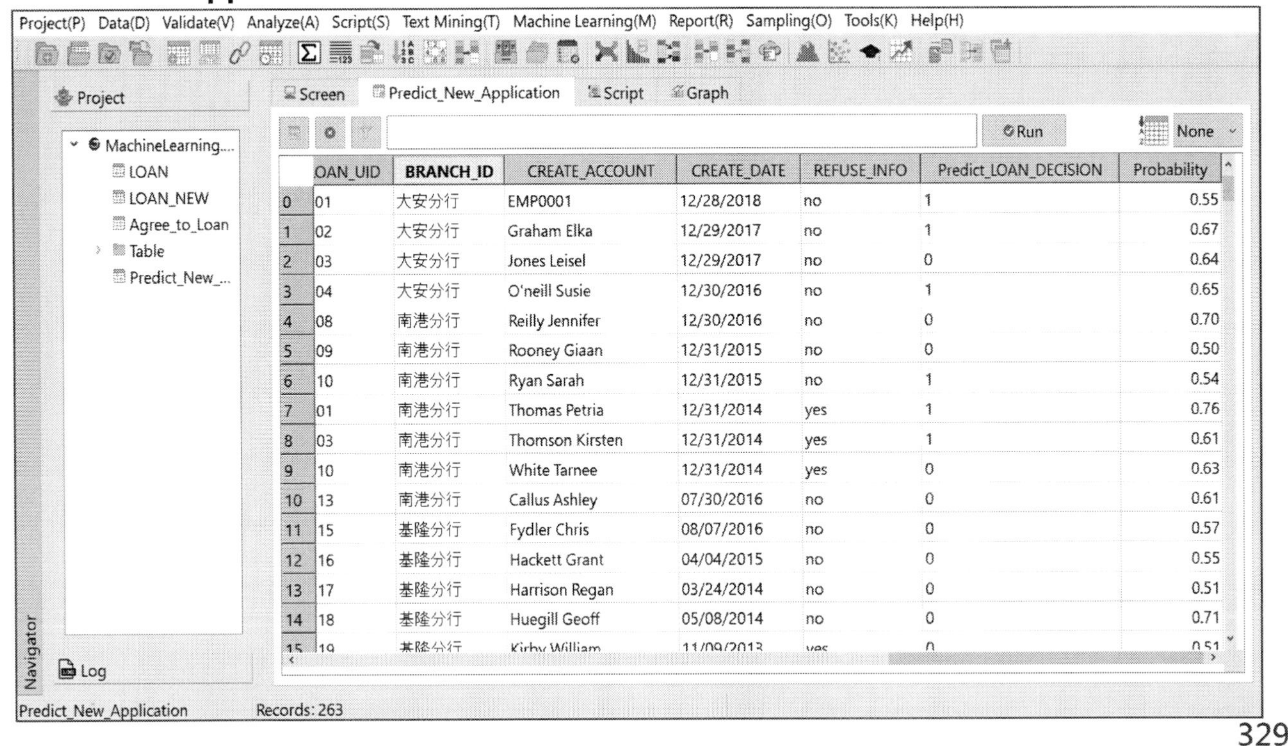

	OAN_UID	BRANCH_ID	CREATE_ACCOUNT	CREATE_DATE	REFUSE_INFO	Predict_LOAN_DECISION	Probability
0	01	大安分行	EMP0001	12/28/2018	no	1	0.55
1	02	大安分行	Graham Elka	12/29/2017	no	1	0.67
2	03	大安分行	Jones Leisel	12/29/2017	no	0	0.64
3	04	大安分行	O'neill Susie	12/30/2016	no	1	0.65
4	08	南港分行	Reilly Jennifer	12/30/2016	no	0	0.70
5	09	南港分行	Rooney Giaan	12/31/2015	no	0	0.50
6	10	南港分行	Ryan Sarah	12/31/2015	no	1	0.54
7	01	南港分行	Thomas Petria	12/31/2014	yes	1	0.76
8	03	南港分行	Thomson Kirsten	12/31/2014	yes	1	0.61
9	10	南港分行	White Tarnee	12/31/2014	yes	0	0.63
10	13	南港分行	Callus Ashley	07/30/2016	no	0	0.61
11	15	基隆分行	Fydler Chris	08/07/2016	no	0	0.57
12	16	基隆分行	Hackett Grant	04/04/2015	no	0	0.55
13	17	基隆分行	Harrison Regan	03/24/2014	no	0	0.51
14	18	基隆分行	Huegill Geoff	05/08/2014	no	0	0.71

JCAATs-AI Audit Software
Copyright © 2023 JACKSOFT.

Exercise 10.5

Please use the knowledge model "Loan_ML" to predict which new loan applications in the LOAN_NEW file will be approved. How many records will be predicted to approve?

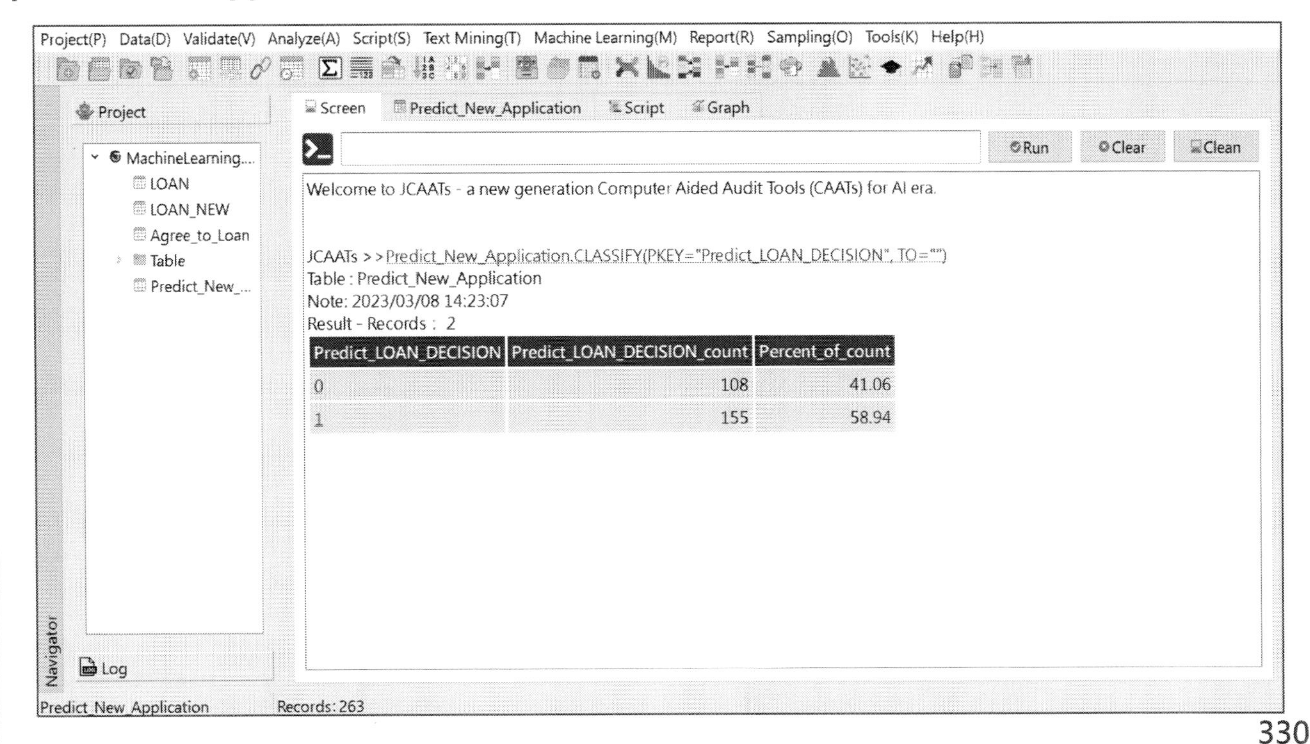

Predict_LOAN_DECISION	Predict_LOAN_DECISION_count	Percent_of_count
0	108	41.06
1	155	58.94

Exercise 10.6

Please use the knowledge model "Breach_ML" to predict which loan applications in the LOAN_NEW file may breach contract? How many records will be predicted to breach contract?

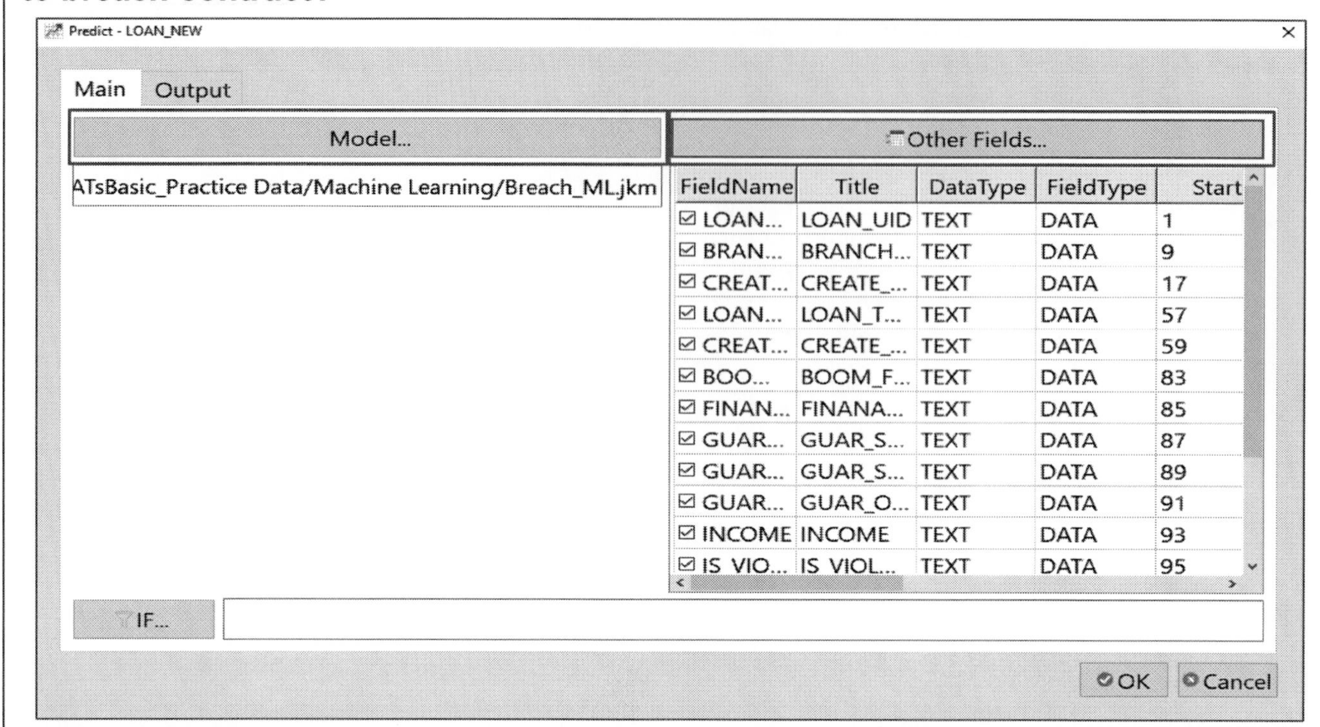

JCAATs-AI Audit Software
Copyright © 2023 JACKSOFT.

Exercise 10.6

Please use the knowledge model "Breach_ML" to predict which loan applications in the LOAN_NEW file may breach contract? How many records will be predicted to breach contract?

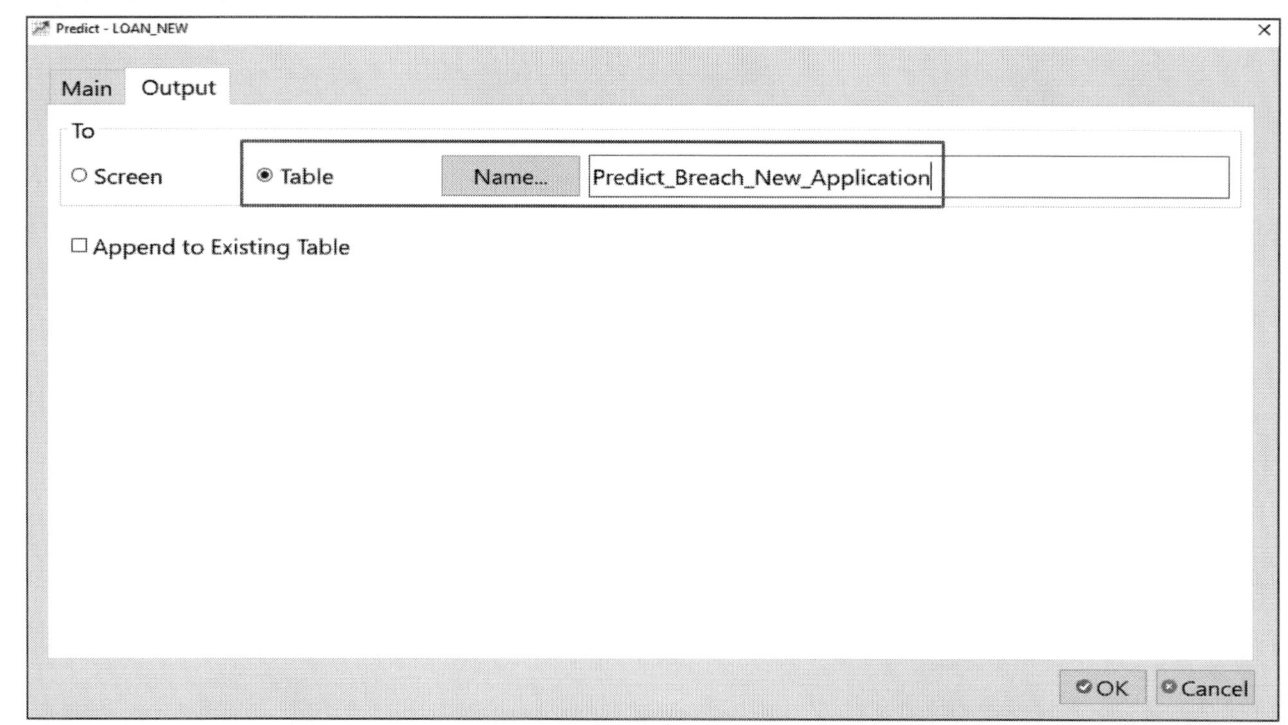

Exercise 10.6

Please use the knowledge model "Breach_ML" to predict which loan applications in the LOAN_NEW file may breach contract? How many records will be predicted to breach contract?

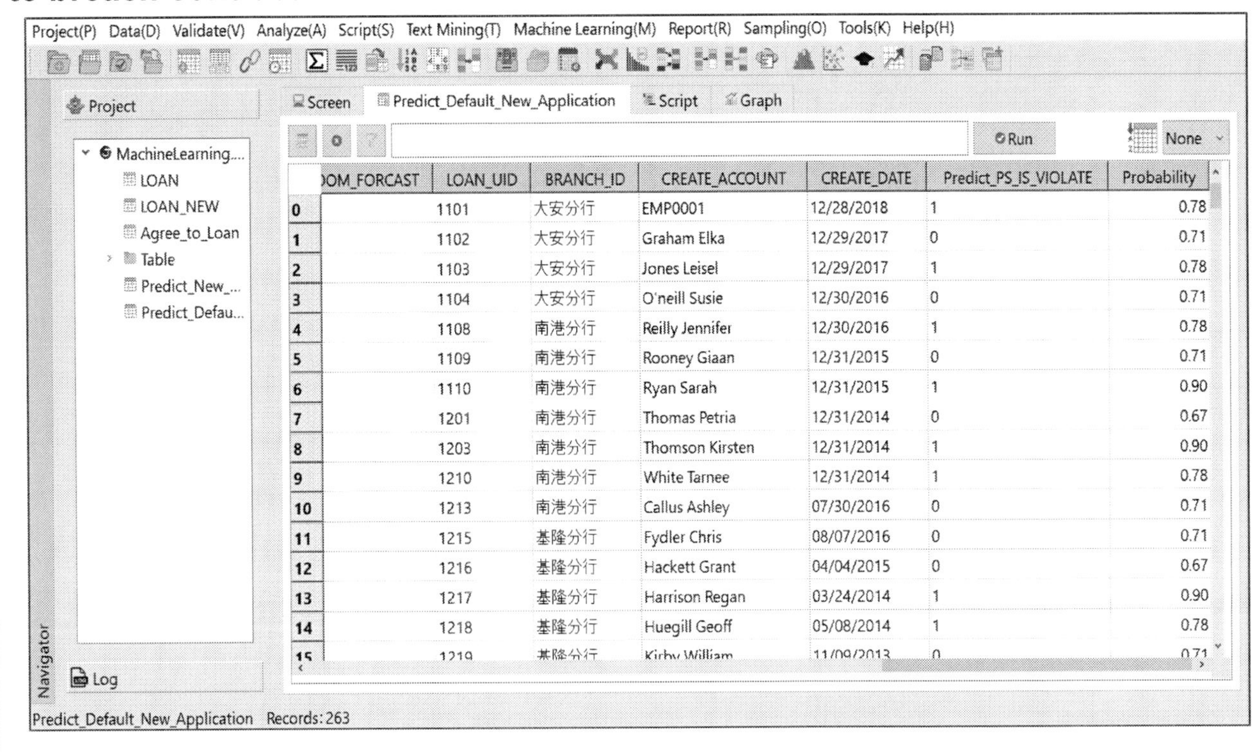

	OM_FORCAST	LOAN_UID	BRANCH_ID	CREATE_ACCOUNT	CREATE_DATE	Predict_PS_IS_VIOLATE	Probability
0		1101	大安分行	EMP0001	12/28/2018	1	0.78
1		1102	大安分行	Graham Elka	12/29/2017	0	0.71
2		1103	大安分行	Jones Leisel	12/29/2017	1	0.78
3		1104	大安分行	O'neill Susie	12/30/2016	0	0.71
4		1108	南港分行	Reilly Jennifer	12/30/2016	1	0.78
5		1109	南港分行	Rooney Giaan	12/31/2015	0	0.71
6		1110	南港分行	Ryan Sarah	12/31/2015	1	0.90
7		1201	南港分行	Thomas Petria	12/31/2014	0	0.67
8		1203	南港分行	Thomson Kirsten	12/31/2014	1	0.90
9		1210	南港分行	White Tarnee	12/31/2014	1	0.78
10		1213	南港分行	Callus Ashley	07/30/2016	0	0.71
11		1215	基隆分行	Fydler Chris	08/07/2016	0	0.71
12		1216	基隆分行	Hackett Grant	04/04/2015	0	0.67
13		1217	基隆分行	Harrison Regan	03/24/2014	1	0.90
14		1218	基隆分行	Huegill Geoff	05/08/2014	1	0.78
15		1219	基隆分行	Kirby William	11/09/2013	0	0.71

Exercise 10.6

Please use the knowledge model "Breach_ML" to predict which loan applications in the LOAN_NEW file may breach contract? How many records will be predicted to breach contract?

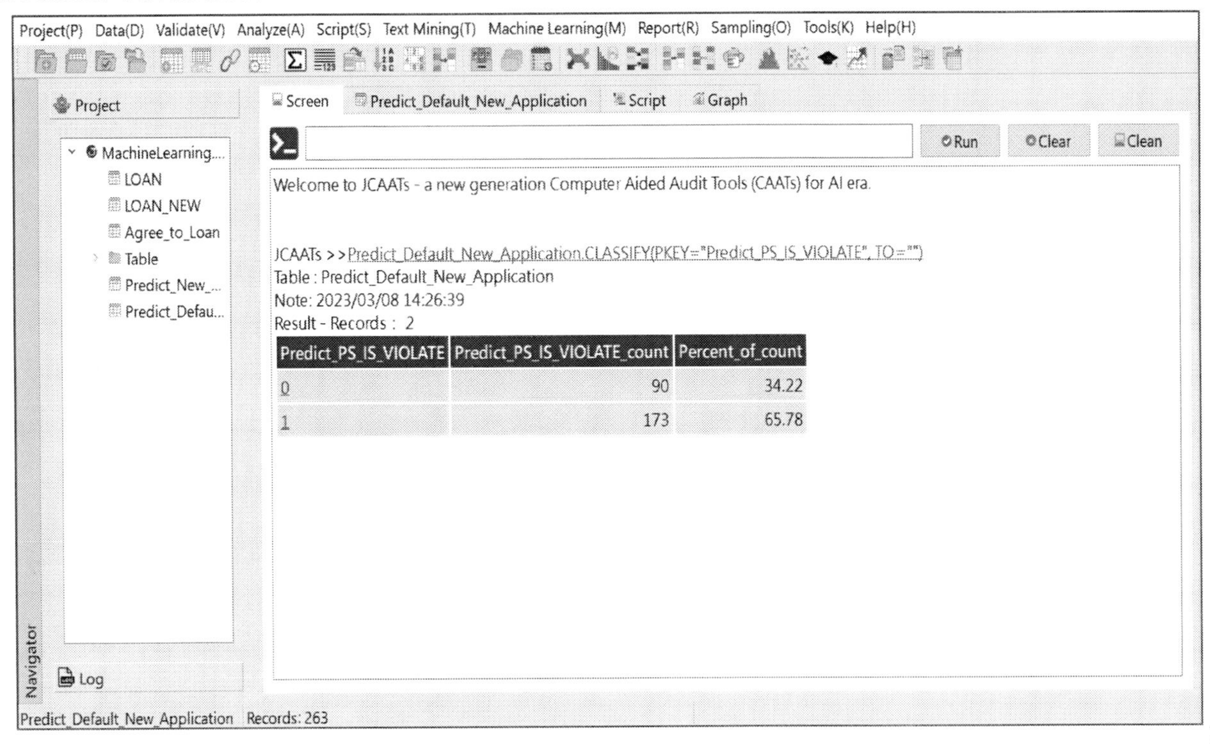

Predict_PS_IS_VIOLATE	Predict_PS_IS_VIOLATE_count	Percent_of_count
0	90	34.22
1	173	65.78

Exercise 10.7

Please use a Join command to list the number of cases in the LOAN_NEW file that are predicted to be approved for a loan and predicted to breach contract.

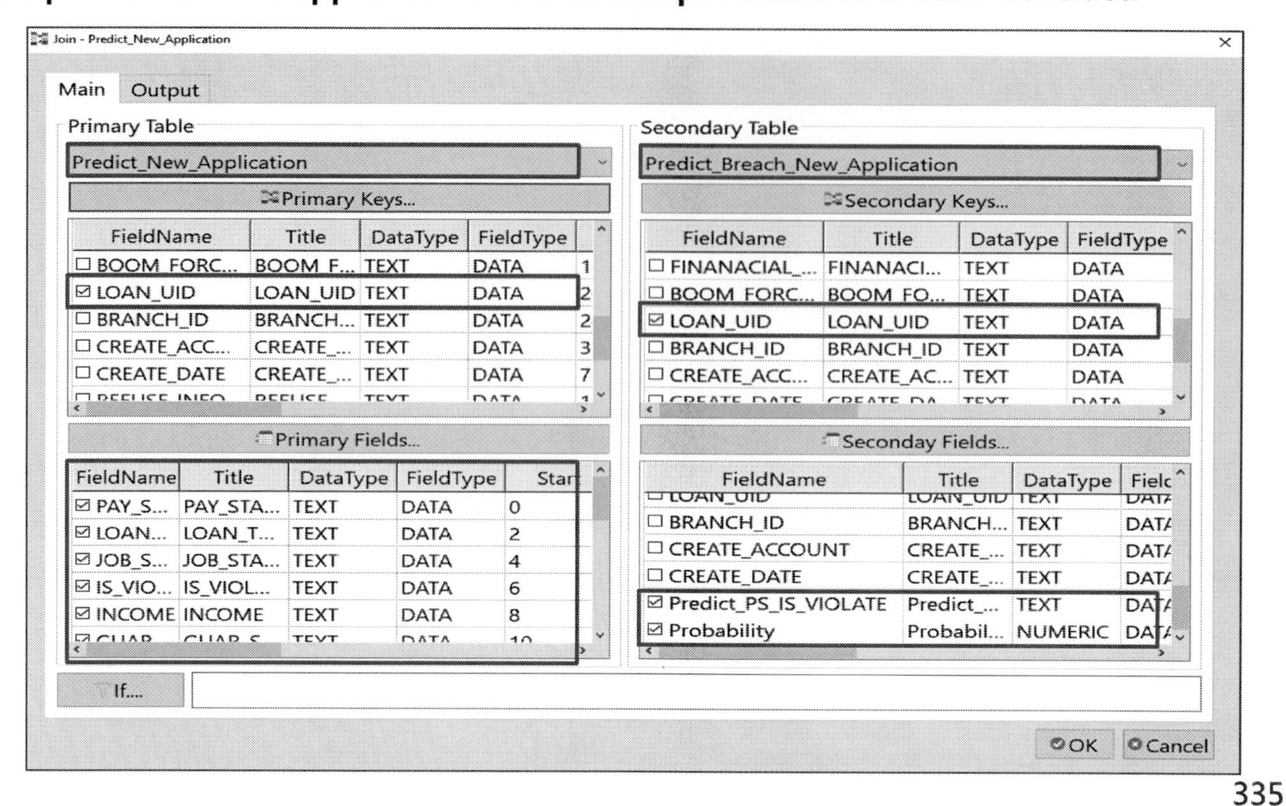

Exercise 10.7

Please use a Join command to list the number of cases in the LOAN_NEW file that are predicted to be approved for a loan and predicted to breach contract.

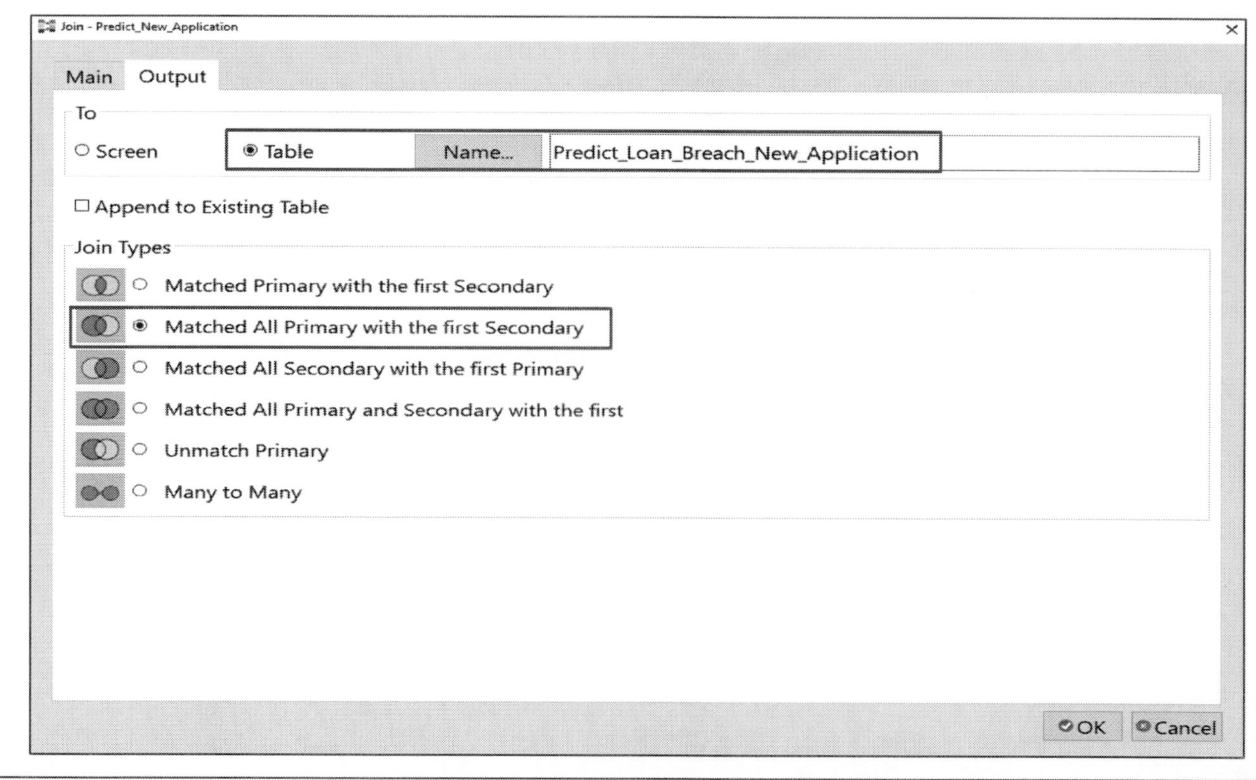

Exercise 10.7

Please use a Join command to list the number of cases in the LOAN_NEW file that are predicted to be approved for a loan and predicted to breach contract.

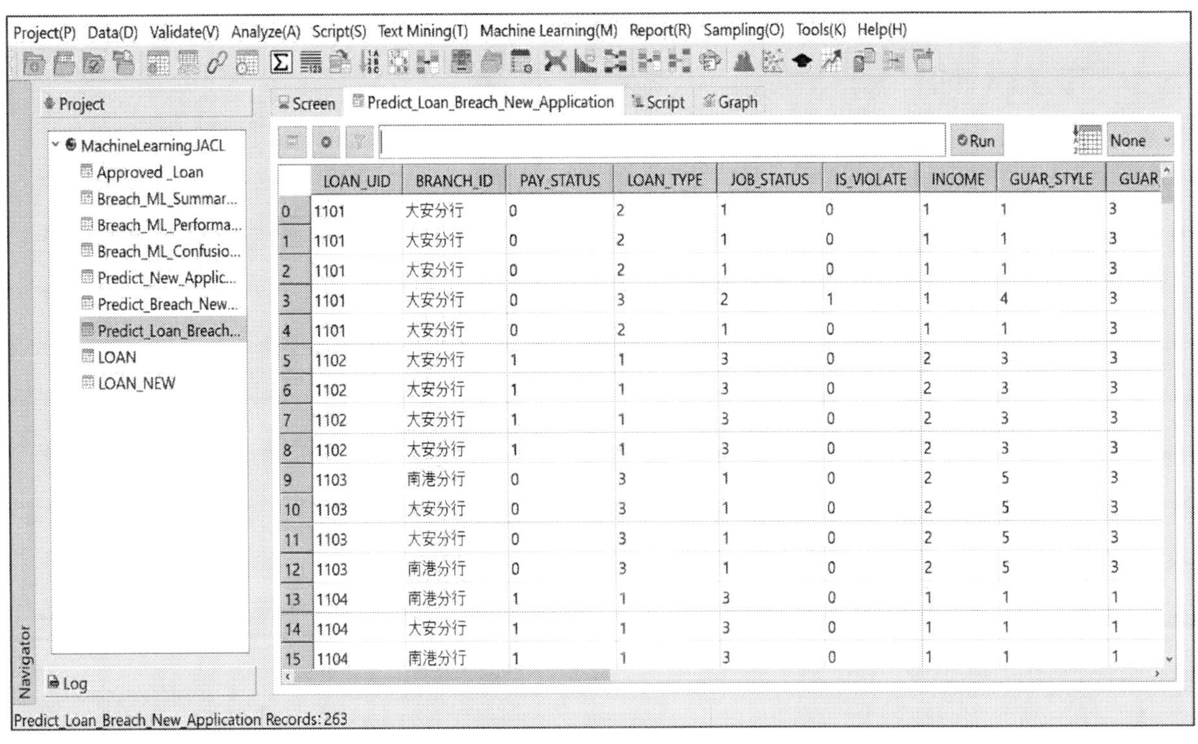

JCAATs-AI Audit Software
Copyright © 2023 JACKSOFT.

Exercise 10.7

Please use a Join command to list the number of cases in the LOAN_NEW file that are predicted to be approved for a loan and predicted to breach contract.

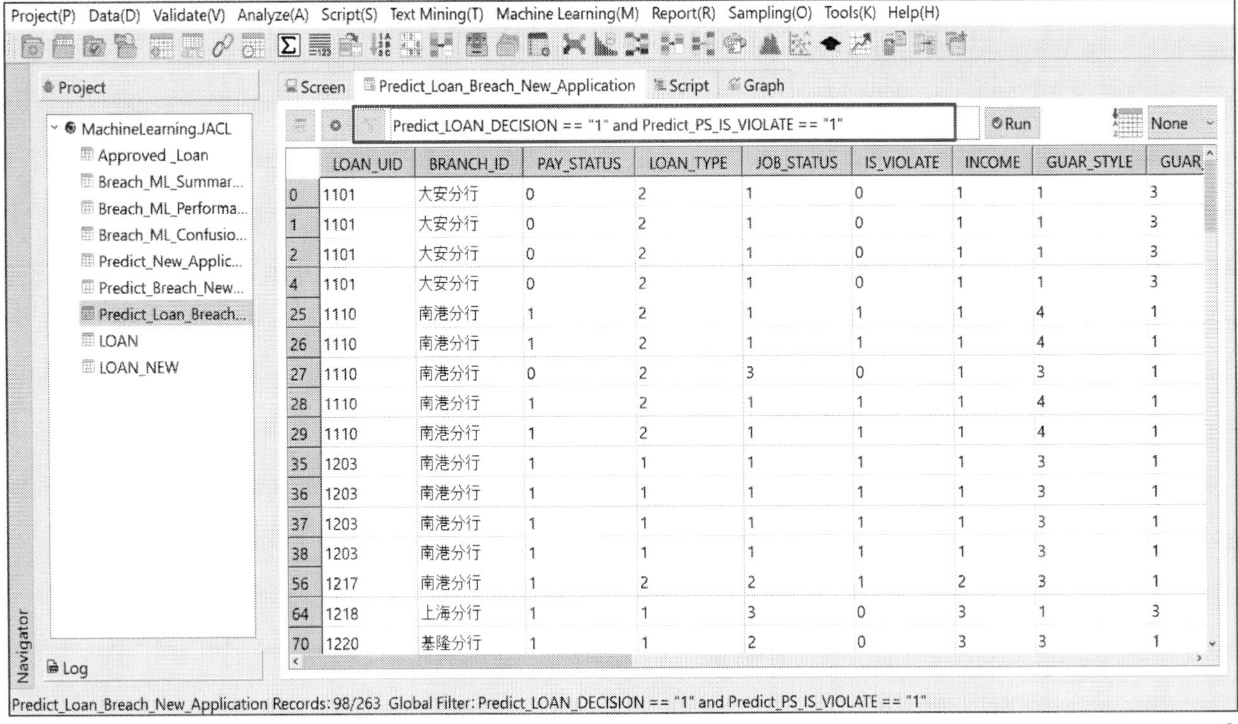

Exercise 10.7

Please use a Join command to list the number of cases in the LOAN_NEW file that are predicted to be approved for a loan and predicted to breach contract.

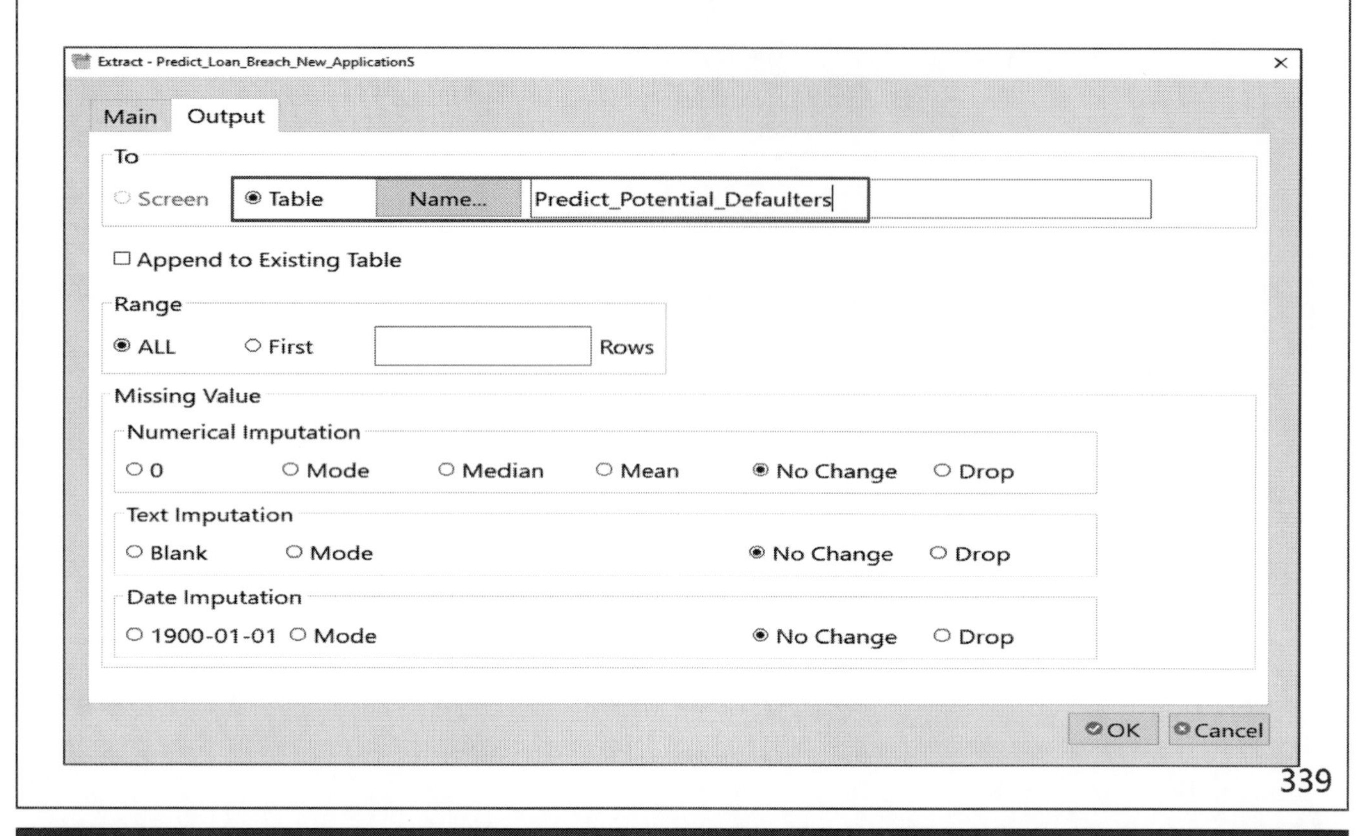

JCAATs-AI Audit Software

Copyright © 2023 JACKSOFT.

Exercise 10.7

Please use a Join command to list the number of cases in the LOAN_NEW file that are predicted to be approved for a loan and predicted to breach contract.

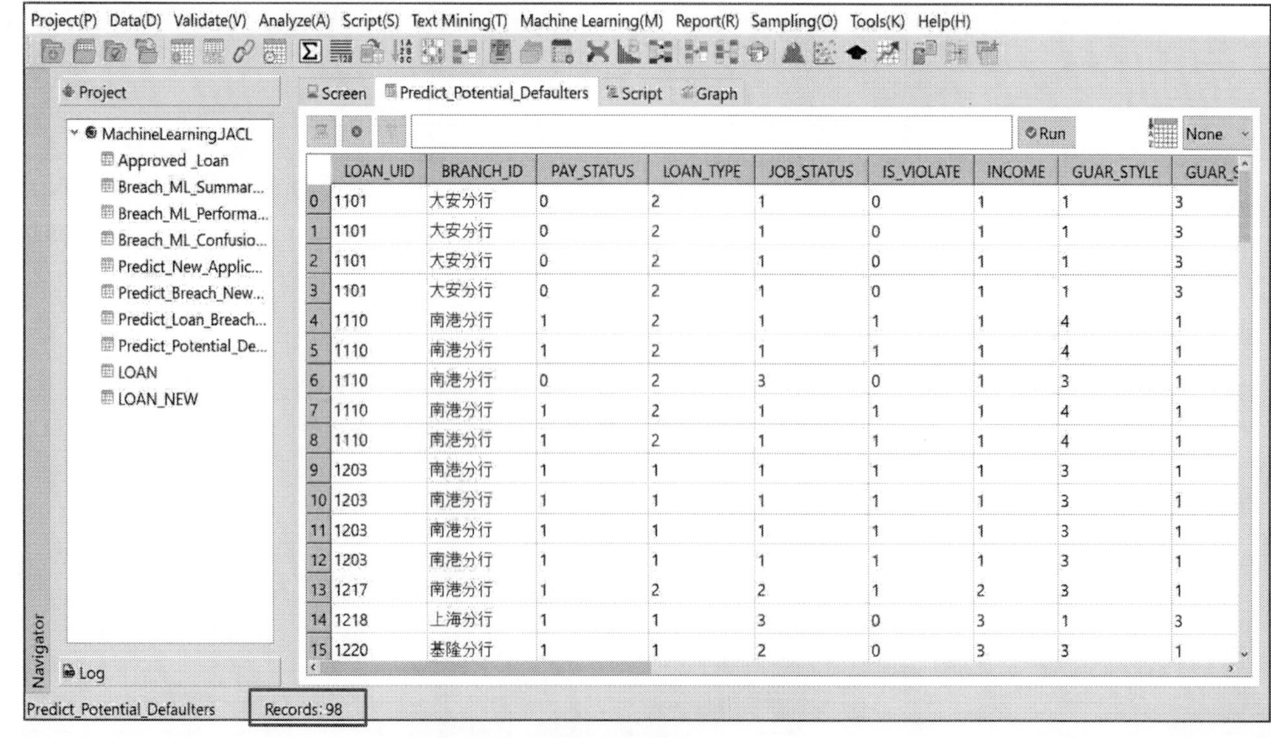

	LOAN_UID	BRANCH_ID	PAY_STATUS	LOAN_TYPE	JOB_STATUS	IS_VIOLATE	INCOME	GUAR_STYLE	GUAR_S
0	1101	大安分行	0	2	1	0	1	1	3
1	1101	大安分行	0	2	1	0	1	1	3
2	1101	大安分行	0	2	1	0	1	1	3
3	1101	大安分行	0	2	1	0	1	1	3
4	1110	南港分行	1	2	1	1	1	4	1
5	1110	南港分行	1	2	1	1	1	4	1
6	1110	南港分行	0	2	3	0	1	3	1
7	1110	南港分行	1	2	1	1	1	4	1
8	1110	南港分行	1	2	1	1	1	4	1
9	1203	南港分行	1	1	1	1	1	3	1
10	1203	南港分行	1	1	1	1	1	3	1
11	1203	南港分行	1	1	1	1	1	3	1
12	1203	南港分行	1	1	1	1	1	3	1
13	1217	南港分行	1	2	2	1	2	3	1
14	1218	上海分行	1	1	3	0	3	1	3
15	1220	基隆分行	1	1	2	0	3	3	1

98 instances detected

JCAATs - AI Audit Software

Chapter 11 - Exercise

- Exercise 11.1: Mike wants to analyze the sales data for the first quarter. He currently has sales data for each month. Please use the **Merge** command to merge these three files into one file.

- Exercise 11.2: Please use the **Export** command to export the first quarter sales data with each amount greater than 10,000 to an Excel file.

Exercise 11.1

Mike wants to analyze the sales data for the first quarter. He currently has sales data for each month. Please use the Merge command to merge these three files into one file.

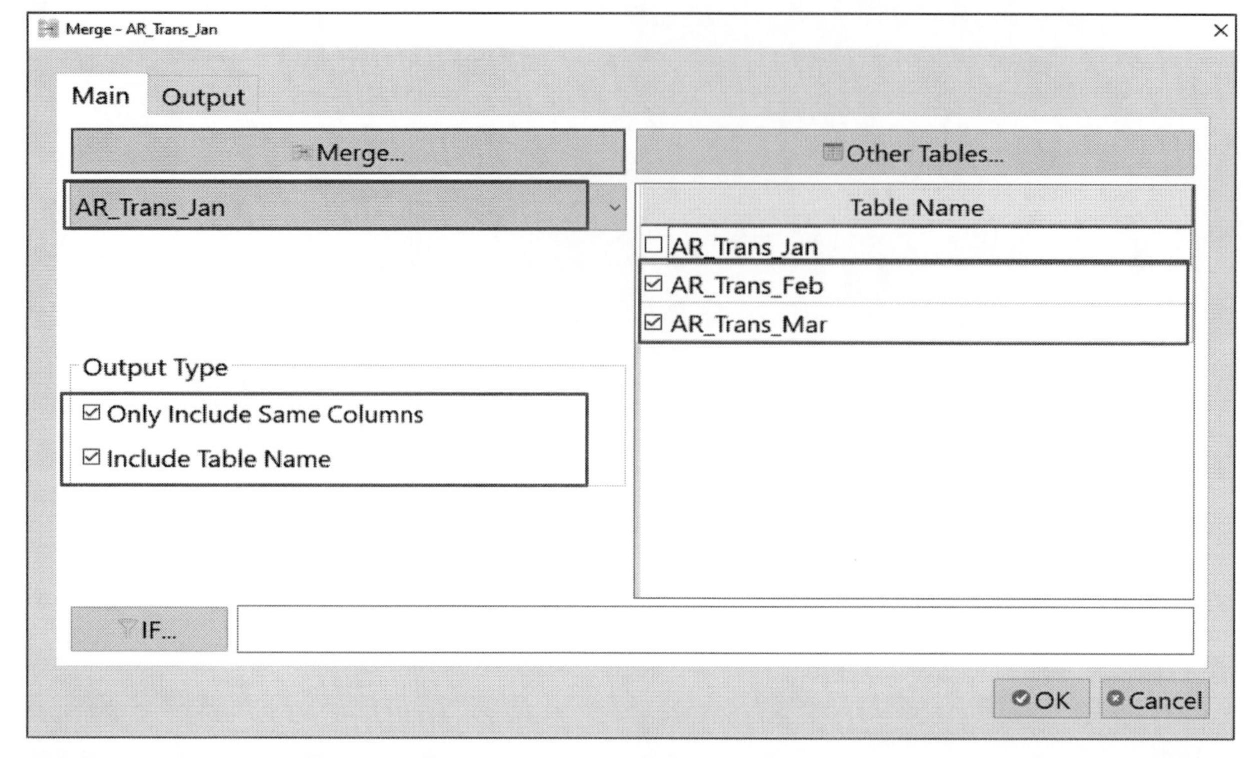

JCAATs-AI Audit Software
Copyright © 2023 JACKSOFT.

Exercise 11.1

Mike wants to analyze the sales data for the first quarter. He currently has sales data for each month. Please use the Merge command to merge these three files into one file.

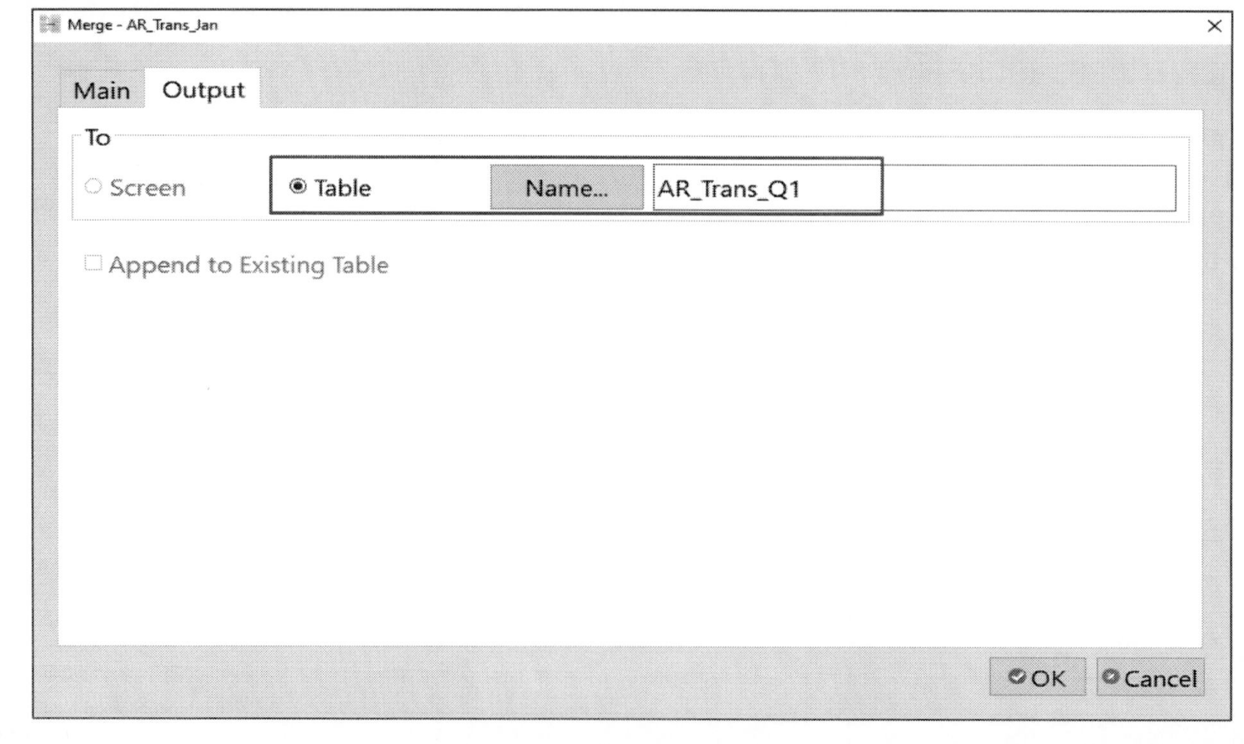

Exercise 11.1

Mike wants to analyze the sales data for the first quarter. He currently has sales data for each month. Please use the Merge command to merge these three files into one file.

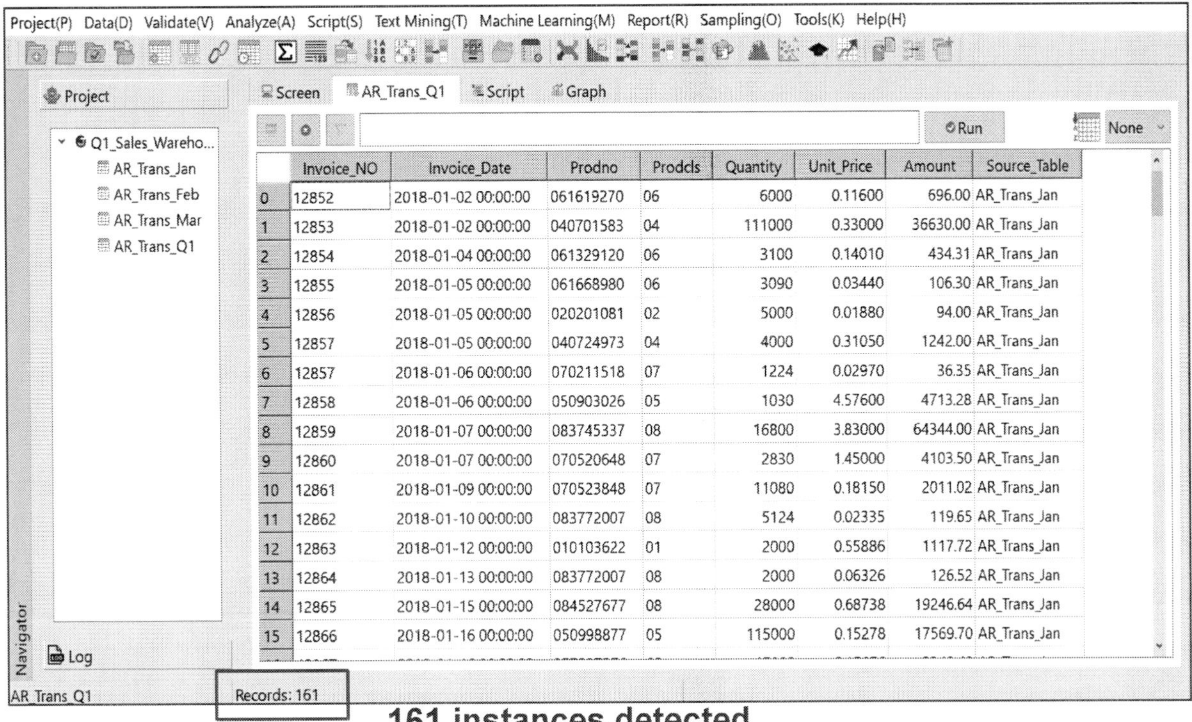

	Invoice_NO	Invoice_Date	Prodno	Prodcls	Quantity	Unit_Price	Amount	Source_Table
0	12852	2018-01-02 00:00:00	061619270	06	6000	0.11600	696.00	AR_Trans_Jan
1	12853	2018-01-02 00:00:00	040701583	04	111000	0.33000	36630.00	AR_Trans_Jan
2	12854	2018-01-04 00:00:00	061329120	06	3100	0.14010	434.31	AR_Trans_Jan
3	12855	2018-01-05 00:00:00	061668980	06	3090	0.03440	106.30	AR_Trans_Jan
4	12856	2018-01-05 00:00:00	020201081	02	5000	0.01880	94.00	AR_Trans_Jan
5	12857	2018-01-05 00:00:00	040724973	04	4000	0.31050	1242.00	AR_Trans_Jan
6	12857	2018-01-06 00:00:00	070211518	07	1224	0.02970	36.35	AR_Trans_Jan
7	12858	2018-01-06 00:00:00	050903026	05	1030	4.57600	4713.28	AR_Trans_Jan
8	12859	2018-01-07 00:00:00	083745337	08	16800	3.83000	64344.00	AR_Trans_Jan
9	12860	2018-01-07 00:00:00	070520648	07	2830	1.45000	4103.50	AR_Trans_Jan
10	12861	2018-01-09 00:00:00	070523848	07	11080	0.18150	2011.02	AR_Trans_Jan
11	12862	2018-01-10 00:00:00	083772007	08	5124	0.02335	119.65	AR_Trans_Jan
12	12863	2018-01-12 00:00:00	010103622	01	2000	0.55886	1117.72	AR_Trans_Jan
13	12864	2018-01-13 00:00:00	083772007	08	2000	0.06326	126.52	AR_Trans_Jan
14	12865	2018-01-15 00:00:00	084527677	08	28000	0.68738	19246.64	AR_Trans_Jan
15	12866	2018-01-16 00:00:00	050998877	05	115000	0.15278	17569.70	AR_Trans_Jan

161 instances detected

Exercise 11.2

Please use the Export command to export the first quarter sales data with each amount greater than 10,000 to an Excel file.

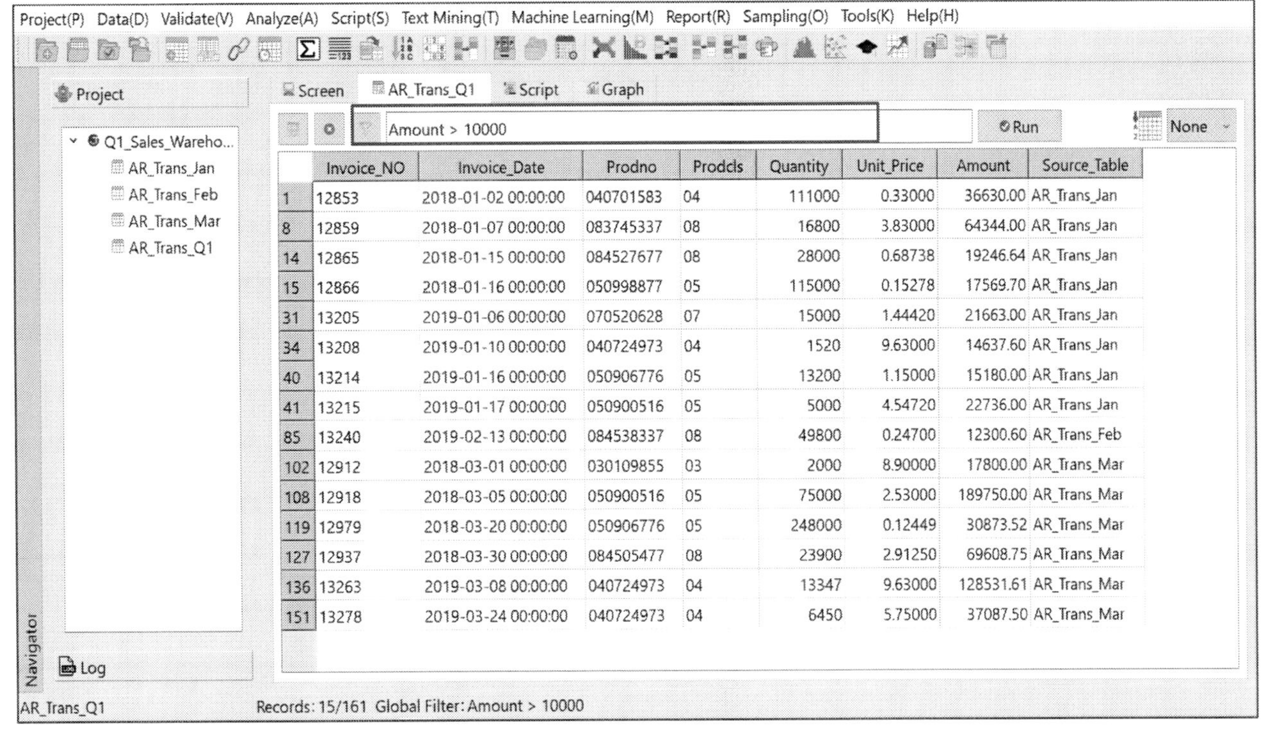

	Invoice_NO	Invoice_Date	Prodno	Prodcls	Quantity	Unit_Price	Amount	Source_Table
1	12853	2018-01-02 00:00:00	040701583	04	111000	0.33000	36630.00	AR_Trans_Jan
8	12859	2018-01-07 00:00:00	083745337	08	16800	3.83000	64344.00	AR_Trans_Jan
14	12865	2018-01-15 00:00:00	084527677	08	28000	0.68738	19246.64	AR_Trans_Jan
15	12866	2018-01-16 00:00:00	050998877	05	115000	0.15278	17569.70	AR_Trans_Jan
31	13205	2019-01-06 00:00:00	070520628	07	15000	1.44420	21663.00	AR_Trans_Jan
34	13208	2019-01-10 00:00:00	040724973	04	1520	9.63000	14637.60	AR_Trans_Jan
40	13214	2019-01-16 00:00:00	050906776	05	13200	1.15000	15180.00	AR_Trans_Jan
41	13215	2019-01-17 00:00:00	050900516	05	5000	4.54720	22736.00	AR_Trans_Jan
85	13240	2019-02-13 00:00:00	084538337	08	49800	0.24700	12300.60	AR_Trans_Feb
102	12912	2018-03-01 00:00:00	030109855	03	2000	8.90000	17800.00	AR_Trans_Mar
108	12918	2018-03-05 00:00:00	050900516	05	75000	2.53000	189750.00	AR_Trans_Mar
119	12979	2018-03-20 00:00:00	050906776	05	248000	0.12449	30873.52	AR_Trans_Mar
127	12937	2018-03-30 00:00:00	084505477	08	23900	2.91250	69608.75	AR_Trans_Mar
136	13263	2019-03-08 00:00:00	040724973	04	13347	9.63000	128531.61	AR_Trans_Mar
151	13278	2019-03-24 00:00:00	040724973	04	6450	5.75000	37087.50	AR_Trans_Mar

Exercise 11.2

Please use the Export command to export the first quarter sales data with each amount greater than 10,000 to an Excel file.

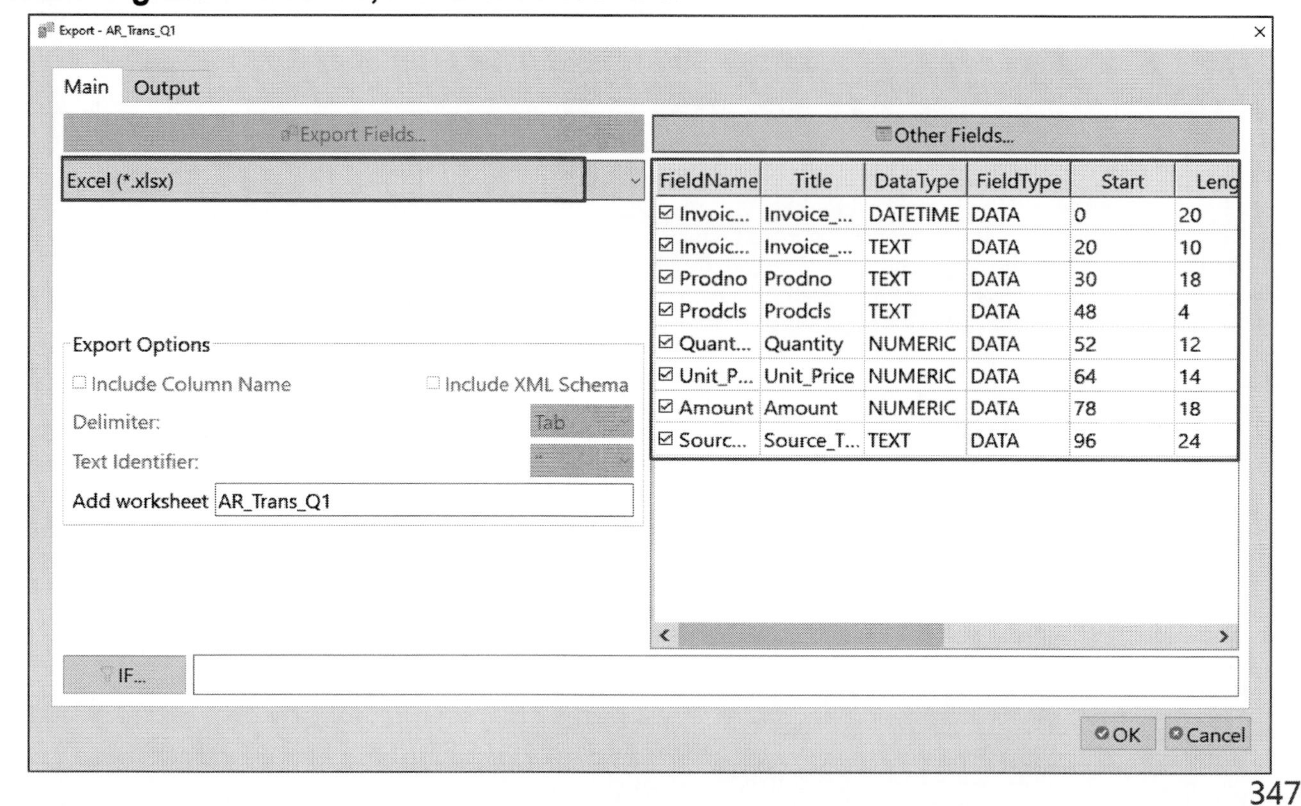

FieldName	Title	DataType	FieldType	Start	Leng
☑ Invoic...	Invoice_...	DATETIME	DATA	0	20
☑ Invoic...	Invoice_...	TEXT	DATA	20	10
☑ Prodno	Prodno	TEXT	DATA	30	18
☑ Prodcls	Prodcls	TEXT	DATA	48	4
☑ Quant...	Quantity	NUMERIC	DATA	52	12
☑ Unit_P...	Unit_Price	NUMERIC	DATA	64	14
☑ Amount	Amount	NUMERIC	DATA	78	18
☑ Sourc...	Source_T...	TEXT	DATA	96	24

Exercise 11.2

Please use the Export command to export the first quarter sales data with each amount greater than 10,000 to an Excel file.

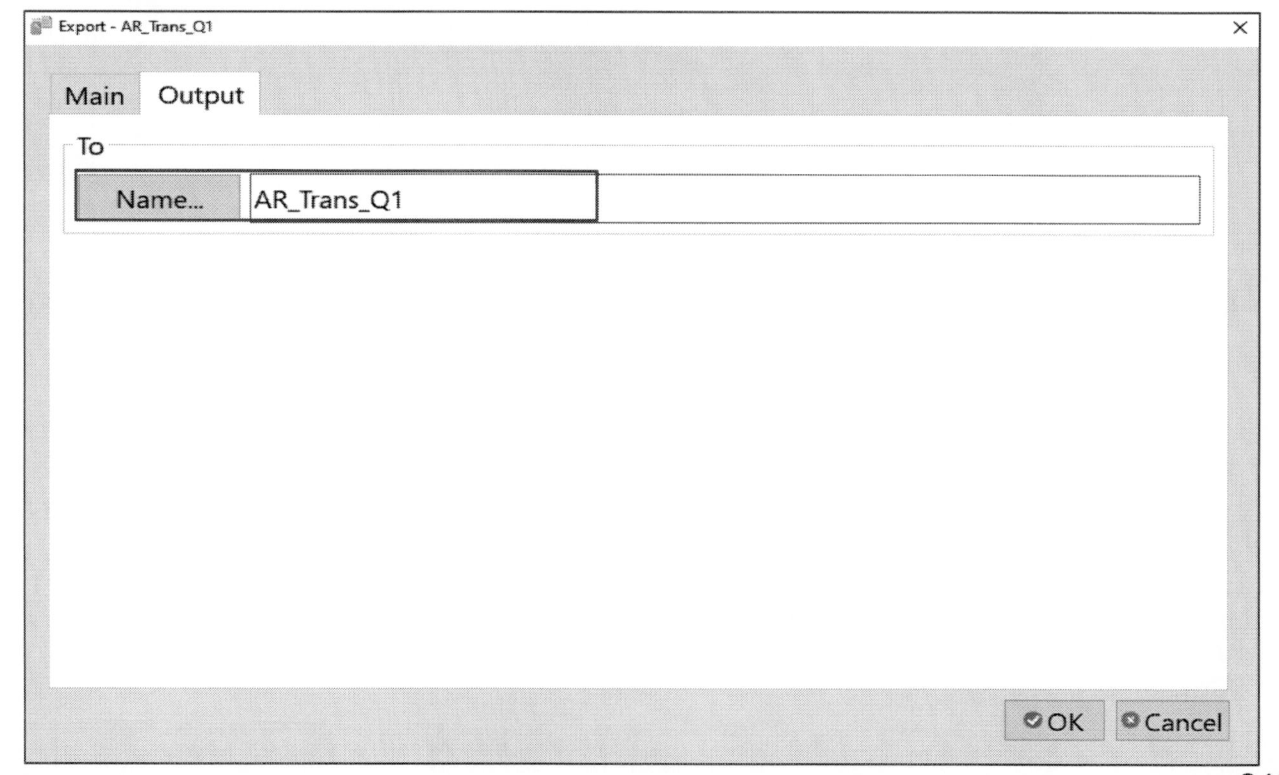

Exercise 11.2

Please use the Export command to export the first quarter sales data with each amount greater than 10,000 to an Excel file.

	A	B	C	D	E	F	G	H	I
1	Invoice_Date	Invoice_NO	Prodno	Prodcls	Quantity	Unit_Price	Amount	Source_Table	
2	2018-01-02 00:00:00	12852	061619270	06	6000	0.116	696	AR_Trans_Jan	
3	2018-01-02 00:00:00	12853	040701583	04	111000	0.33	36630	AR_Trans_Jan	
4	2018-01-04 00:00:00	12854	061329120	06	3100	0.1401	434.31	AR_Trans_Jan	
5	2018-01-05 00:00:00	12855	061668980	06	3090	0.0344	106.3	AR_Trans_Jan	
6	2018-01-05 00:00:00	12856	020201081	02	5000	0.0188	94	AR_Trans_Jan	
7	2018-01-05 00:00:00	12857	040724973	04	4000	0.3105	1242	AR_Trans_Jan	
8	2018-01-06 00:00:00	12857	070211518	07	1224	0.0297	36.35	AR_Trans_Jan	
9	2018-01-06 00:00:00	12858	050903026	05	1030	4.576	4713.28	AR_Trans_Jan	
10	2018-01-07 00:00:00	12859	083745337	08	16800	3.83	64344	AR_Trans_Jan	
11	2018-01-07 00:00:00	12860	070520648	07	2830	1.45	4103.5	AR_Trans_Jan	
12	2018-01-09 00:00:00	12861	070523848	07	11080	0.1815	2011.02	AR_Trans_Jan	
13	2018-01-10 00:00:00	12862	083772007	08	5124	0.02335	119.65	AR_Trans_Jan	
14	2018-01-12 00:00:00	12863	010103622	01	2000	0.55886	1117.72	AR_Trans_Jan	
15	2018-01-13 00:00:00	12864	083772007	08	2000	0.06326	126.52	AR_Trans_Jan	
16	2018-01-15 00:00:00	12865	084527677	08	28000	0.68738	19246.64	AR_Trans_Jan	
17	2018-01-16 00:00:00	12866	050998877	05	115000	0.15278	17569.7	AR_Trans_Jan	
18	2018-01-16 00:00:00	12867	057387376	05	15000	0.15656	2348.4	AR_Trans_Jan	
19	2018-01-18 00:00:00	12869	070563128	07	3736	0.1275	476.34	AR_Trans_Jan	

AR_Trans_Q1

JCAATs-AI Audit Software
Copyright © 2023 JACKSOFT.

JCAATs Learning Note:

JCAATs - AI Audit Software

Chapter 12 - Exercise

- Exercise 12.1: Sabrina is currently conducting an annual audit of accounts receivable. She wants to perform attribute sampling to facilitate the subsequent issuance of confirmation letters. Assuming a 2% tolerable error at a 95% confidence level and using random sampling, how many samples should be selected?

- Exercise 12.2: Mary is currently conducting an annual audit of accounts receivable. She wants to randomly select 5 samples for further analysis. Please provide the data for these 5 samples.

Exercise 12.1

Sabrina is currently conducting an annual audit of accounts receivable. She wants to perform attribute sampling to facilitate the subsequent issuance of confirmation letters. Assuming a 2% tolerable error at a 95% confidence level and using random sampling, how many samples should be selected?

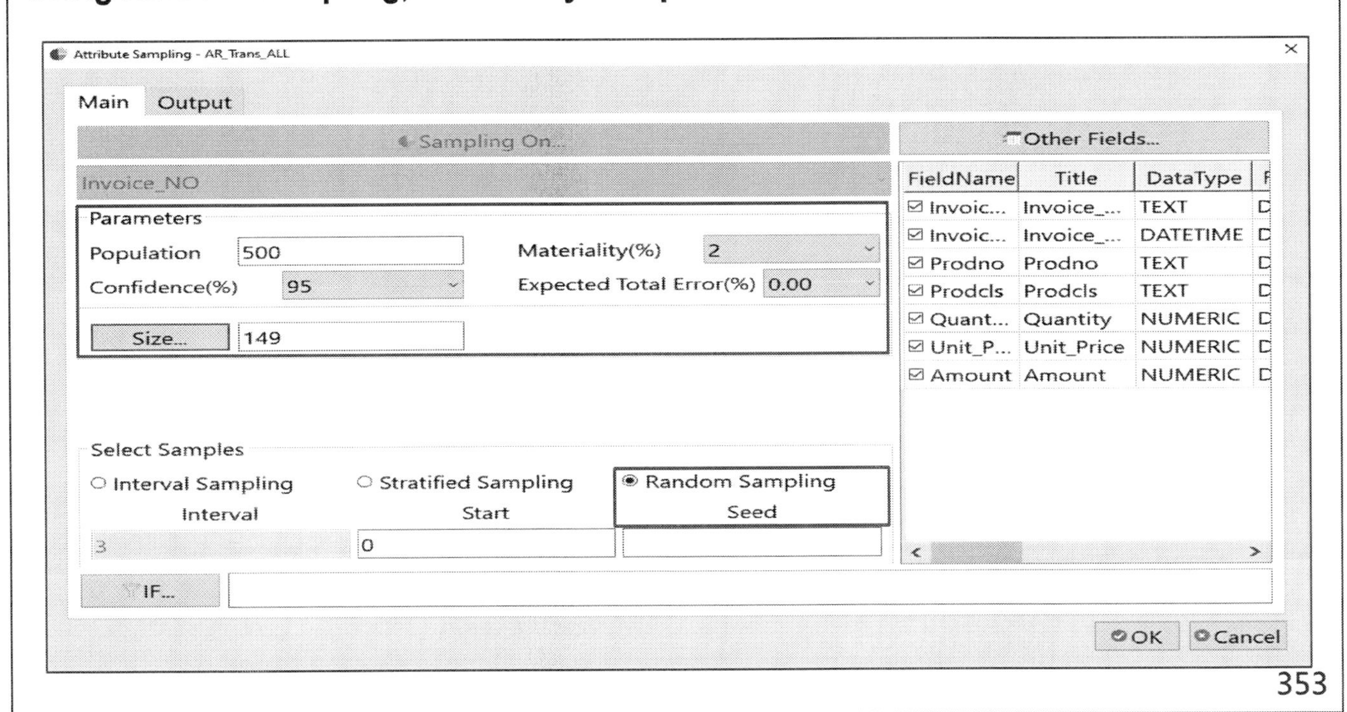

Exercise 12.1

Sabrina is currently conducting an annual audit of accounts receivable. She wants to perform attribute sampling to facilitate the subsequent issuance of confirmation letters. Assuming a 2% tolerable error at a 95% confidence level and using random sampling, how many samples should be selected?

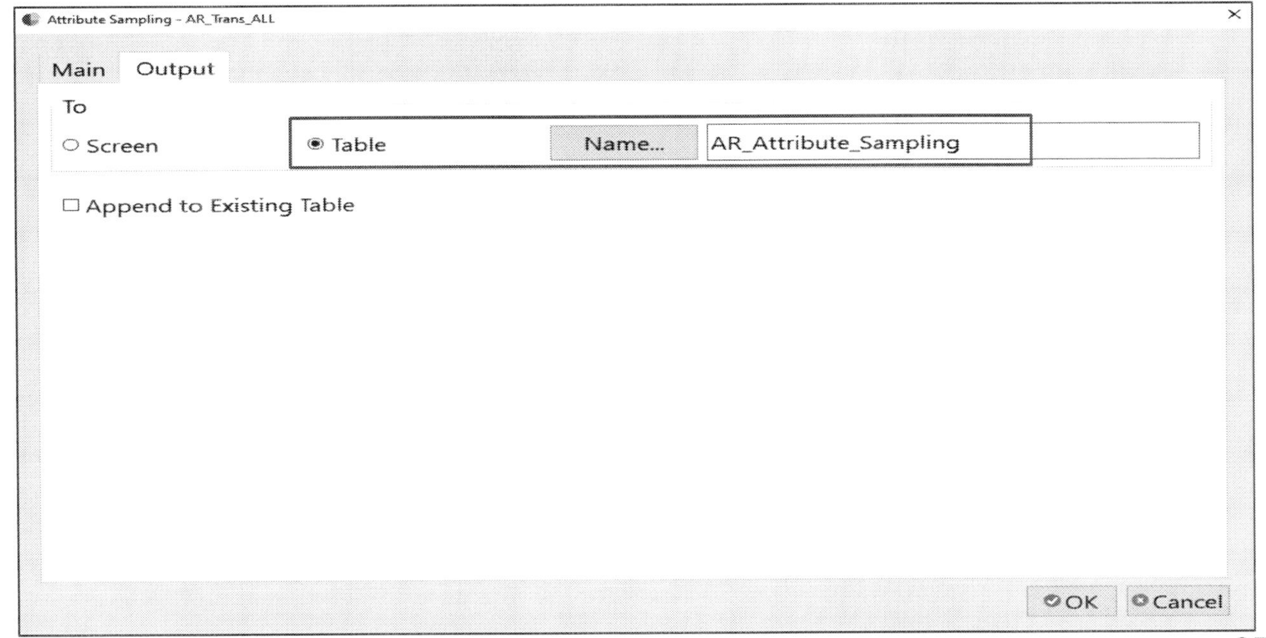

Exercise 12.1

Sabrina is currently conducting an annual audit of accounts receivable. She wants to perform attribute sampling to facilitate the subsequent issuance of confirmation letters. Assuming a 2% tolerable error at a 95% confidence level and using random sampling, how many samples should be selected?

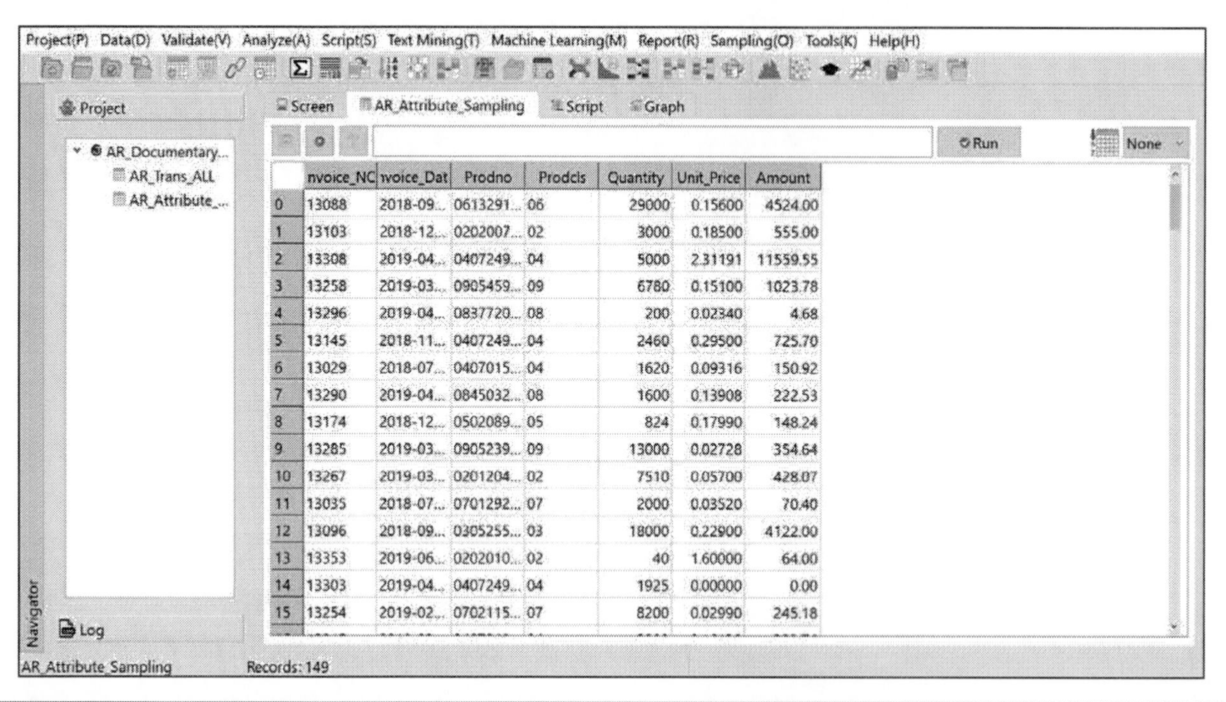

	nvoice_NC	nvoice_Dat	Prodno	Prodcls	Quantity	Unit_Price	Amount
0	13088	2018-09...	0613291...	06	29000	0.15600	4524.00
1	13103	2018-12...	0202007...	02	3000	0.18500	555.00
2	13308	2019-04...	0407249...	04	5000	2.31191	11559.55
3	13258	2019-03...	0905459...	09	6780	0.15100	1023.78
4	13296	2019-04...	0837720...	08	200	0.02340	4.68
5	13145	2018-11...	0407249...	04	2460	0.29500	725.70
6	13029	2018-07...	0407015...	04	1620	0.09316	150.92
7	13290	2019-04...	0845032...	08	1600	0.13908	222.53
8	13174	2018-12...	0502089...	05	824	0.17990	148.24
9	13285	2019-03...	0905239...	09	13000	0.02728	354.64
10	13267	2019-03...	0201204...	02	7510	0.05700	428.07
11	13035	2018-07...	0701292...	07	2000	0.03520	70.40
12	13096	2018-09...	0305255...	03	18000	0.22900	4122.00
13	13353	2019-06...	0202010...	02	40	1.60000	64.00
14	13303	2019-04...	0407249...	04	1925	0.00000	0.00
15	13254	2019-02...	0702115...	07	8200	0.02990	245.18

Exercise 12.1

Mary is currently conducting an annual audit of accounts receivable. She wants to randomly select 5 samples for further analysis. Please provide the data for these 5 samples.

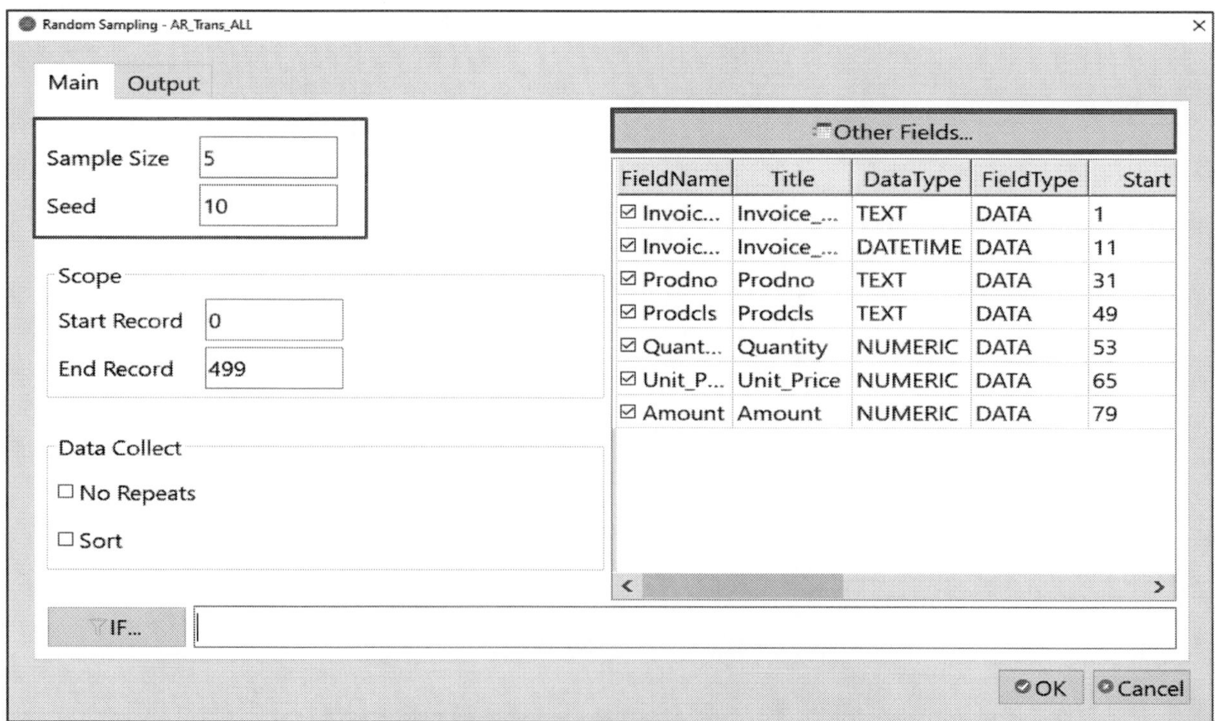

Exercise 12.1

Mary is currently conducting an annual audit of accounts receivable. She wants to randomly select 5 samples for further analysis. Please provide the data for these 5 samples.

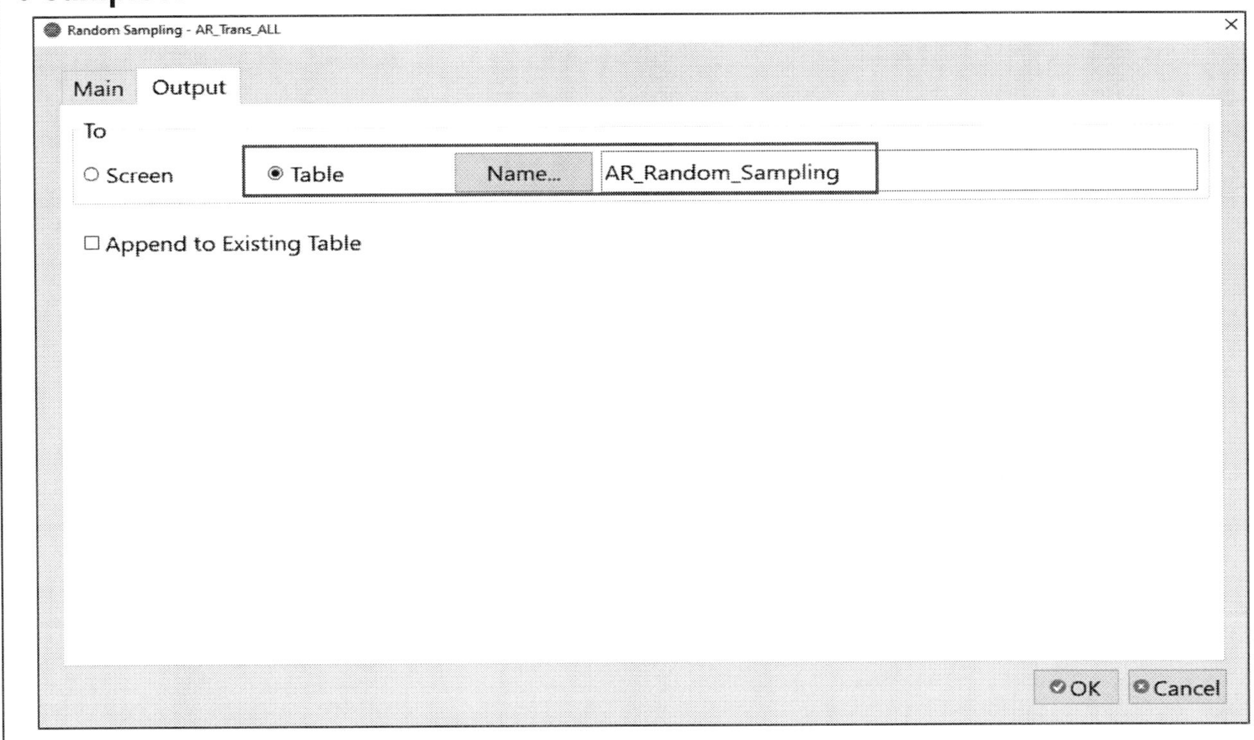

JCAATs-AI Audit Software
Copyright © 2023 JACKSOFT.

Exercise 12.1

Mary is currently conducting an annual audit of accounts receivable. She wants to randomly select 5 samples for further analysis. Please provide the data for these 5 samples.

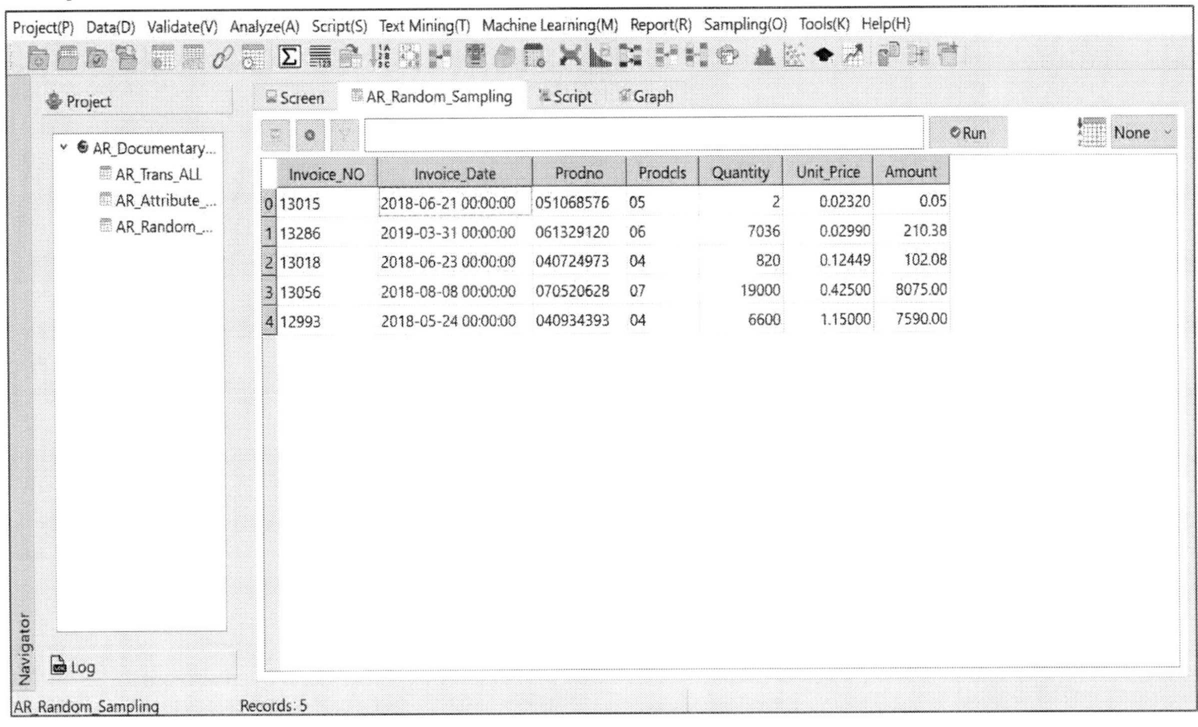

	Invoice_NO	Invoice_Date	Prodno	Prodcls	Quantity	Unit_Price	Amount
0	13015	2018-06-21 00:00:00	051068576	05	2	0.02320	0.05
1	13286	2019-03-31 00:00:00	061329120	06	7036	0.02990	210.38
2	13018	2018-06-23 00:00:00	040724973	04	820	0.12449	102.08
3	13056	2018-08-08 00:00:00	070520628	07	19000	0.42500	8075.00
4	12993	2018-05-24 00:00:00	040934393	04	6600	1.15000	7590.00

JCAATs - AI Audit Software

Chapter 13 - Exercise

Chapter 13 - Exercise

- Practice 13.1: Carol wants to establish a threshold variable for her audit project. The minimum amount is 10,000. Please practice using variable tools to create this variable and store its value.

- Practice 13.2: Please list the data with sales amounts greater than the minimum amount variable and calculate how many records there are.

Exercise 13.1

Carol wants to establish a threshold variable for her audit project. The minimum amount is 10,000. Please practice using variable tools to create this variable and store its value.

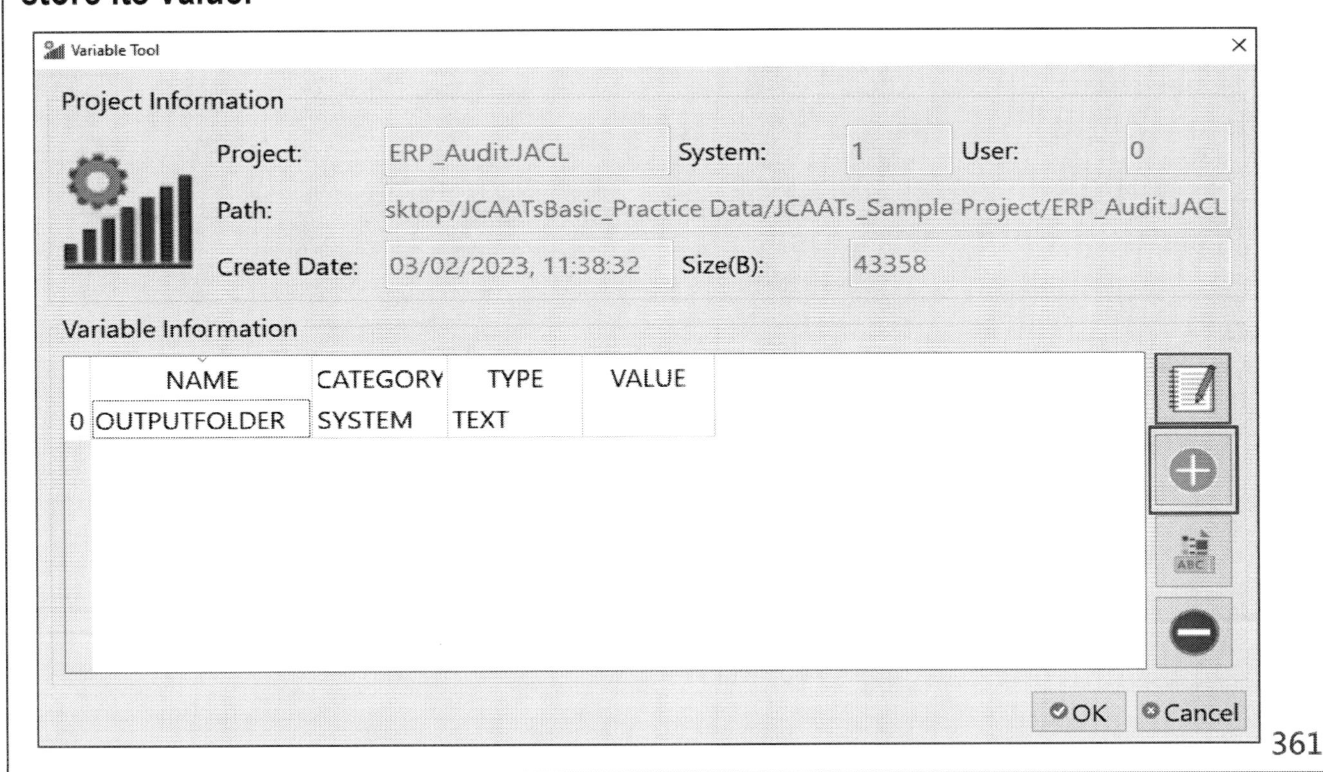

JCAATs-AI Audit Software
Copyright © 2023 JACKSOFT.

Exercise 13.1

Carol wants to establish a threshold variable for her audit project. The minimum amount is 10,000. Please practice using variable tools to create this variable and store its value.

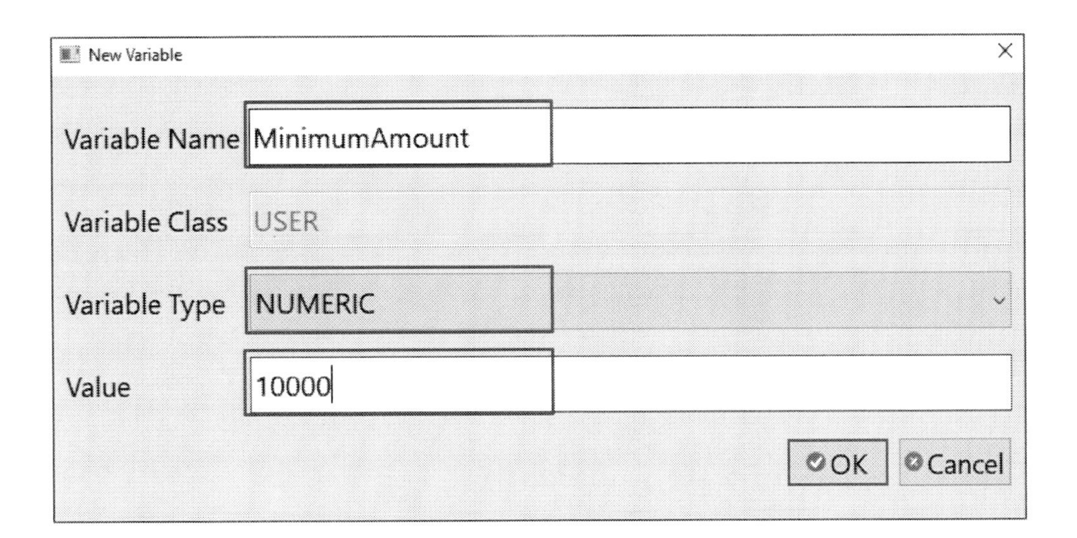

Exercise 13.1

Carol wants to establish a threshold variable for her audit project. The minimum amount is 10,000. Please practice using variable tools to create this variable and store its value.

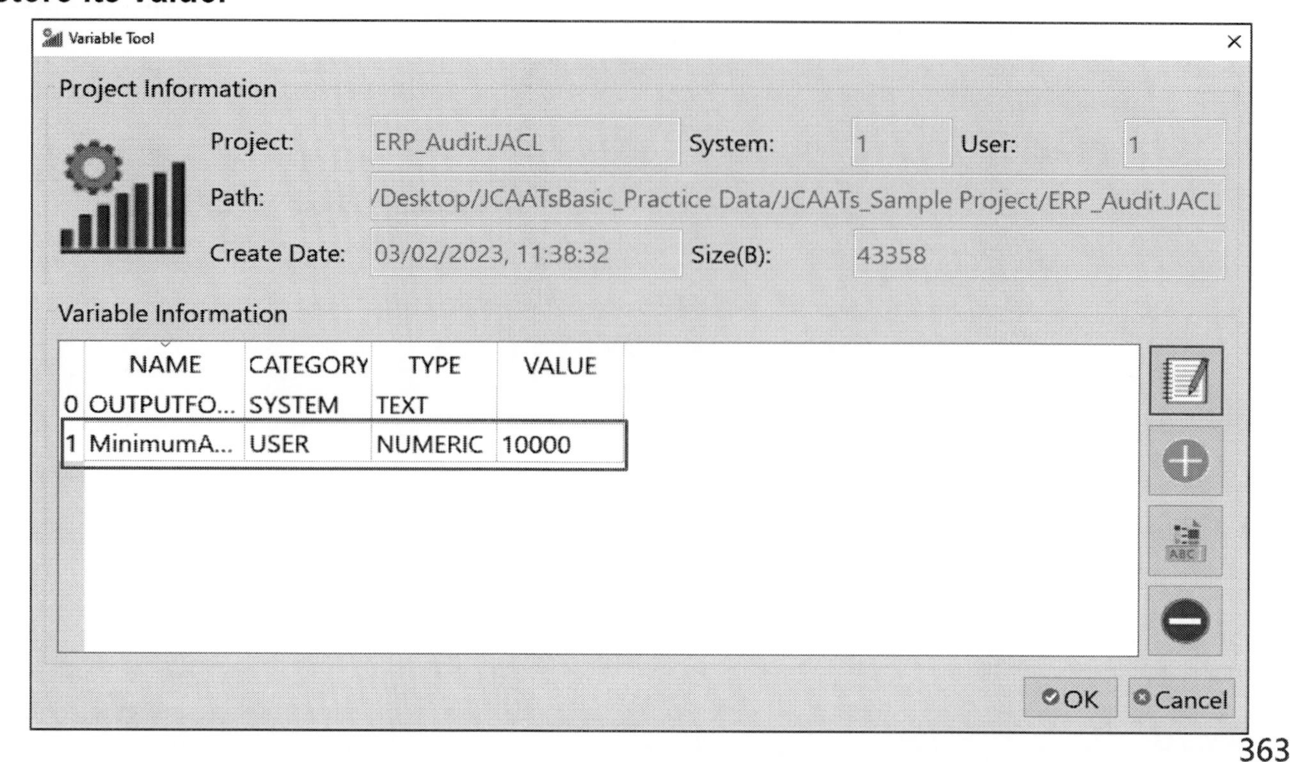

JCAATs-AI Audit Software
Copyright © 2023 JACKSOFT.

Exercise 13.1

Carol wants to establish a threshold variable for her audit project. The minimum amount is 10,000. Please practice using variable tools to create this variable and store its value.

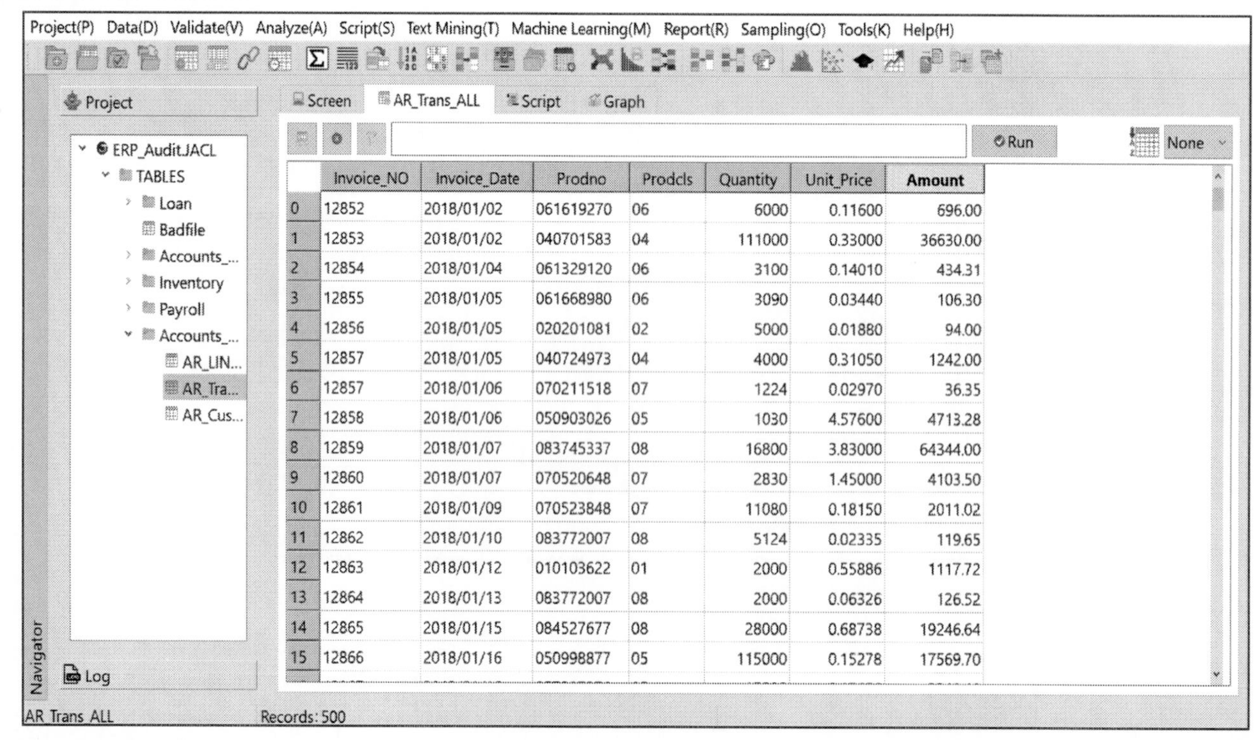

	Invoice_NO	Invoice_Date	Prodno	Prodcls	Quantity	Unit_Price	Amount
0	12852	2018/01/02	061619270	06	6000	0.11600	696.00
1	12853	2018/01/02	040701583	04	111000	0.33000	36630.00
2	12854	2018/01/04	061329120	06	3100	0.14010	434.31
3	12855	2018/01/05	061668980	06	3090	0.03440	106.30
4	12856	2018/01/05	020201081	02	5000	0.01880	94.00
5	12857	2018/01/05	040724973	04	4000	0.31050	1242.00
6	12857	2018/01/06	070211518	07	1224	0.02970	36.35
7	12858	2018/01/06	050903026	05	1030	4.57600	4713.28
8	12859	2018/01/07	083745337	08	16800	3.83000	64344.00
9	12860	2018/01/07	070520648	07	2830	1.45000	4103.50
10	12861	2018/01/09	070523848	07	11080	0.18150	2011.02
11	12862	2018/01/10	083772007	08	5124	0.02335	119.65
12	12863	2018/01/12	010103622	01	2000	0.55886	1117.72
13	12864	2018/01/13	083772007	08	2000	0.06326	126.52
14	12865	2018/01/15	084527677	08	28000	0.68738	19246.64
15	12866	2018/01/16	050998877	05	115000	0.15278	17569.70

Exercise 13.2

Please list the data with sales amounts greater than the minimum amount variable and calculate how many records there are.

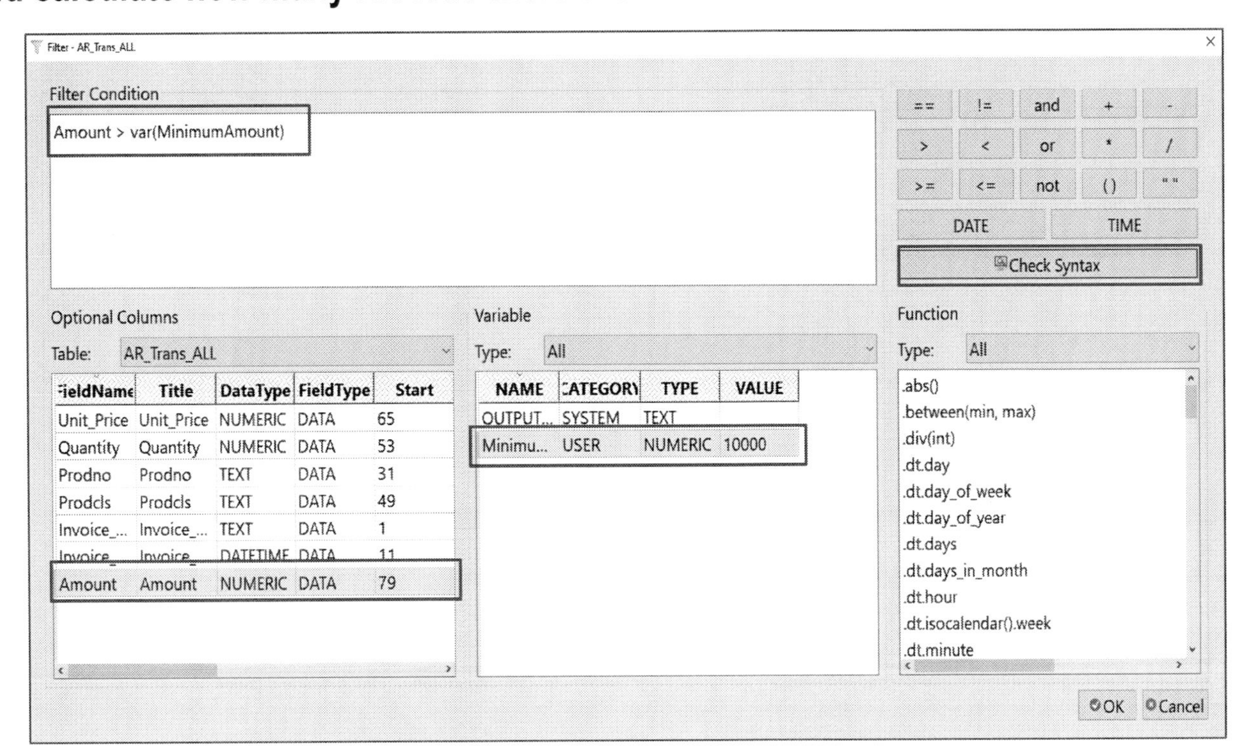

Exercise 13.2

Please list the data with sales amounts greater than the minimum amount variable and calculate how many records there are.

	Invoice_NO	Invoice_Date	Prodno	Prodcls	Quantity	Unit_Price	Amount
1	12853	2018/01/02	040701583	04	111000	0.33000	36630.00
8	12859	2018/01/07	083745337	08	16800	3.83000	64344.00
14	12865	2018/01/15	084527677	08	28000	0.68738	19246.64
15	12866	2018/01/16	050998877	05	115000	0.15278	17569.70
50	12912	2018/03/01	030109855	03	2000	8.90000	17800.00
56	12918	2018/03/05	050900516	05	75000	2.53000	189750.00
67	12979	2018/03/20	050906776	05	248000	0.12449	30873.52
75	12937	2018/03/30	084505477	08	23900	2.91250	69608.75
82	12943	2018/04/05	040725273	04	6450	6.68686	43130.25
84	12945	2018/04/05	073853968	07	18100	2.36000	42716.00
98	12960	2018/04/18	057387376	05	9100	2.45120	22305.92
100	12962	2018/04/22	084511127	08	11600	5.90000	68440.00
111	12974	2018/05/05	090569319	09	1999	12.20000	24387.80
115	12977	2018/05/09	084527677	08	5100	2.50000	12750.00
122	12984	2018/05/12	090576119	09	14800	4.13690	61226.12
160	13024	2018/06/29	030506185	03	2969	3.83000	11371.27

Records: 43/500 Global Filter: Amount > 10000

43 instances detected

JCAATs Learning Note:

Appendix A References

1. AACSB (2014). AACSB International Accounting Accreditation Standard A7: Information Technology Skills and Knowledge for Accounting Graduates: An Interpretation, https://docplayer.net/16013752-Aacsb-international-accounting-accreditation-standard-a7-information-technology-skills-and-knowledge-for-accounting-graduates-an-interpretation.html

2. Adams, J. (2019, March 4). Continuous auditing monitoring architecture. Accounting Today. Retrieved from https://www.accountingtoday.com/magazine/

3. AICPA. (2019). Using Python for Data Analysis in Accounting and Auditing. Retrieved from https://us.aicpa.org/content/dam/aicpa/interestareas/frc/assuranceadvisoryservices/downloadabledocuments/ads-instructional-paper-python.pdf

4. AICPA. (2023). Audit Data Standards. Retrieved March 25, 2023, from https://us.aicpa.org/interestareas/frc/assuranceadvisoryservices/auditdatastandards

5. AICPA. (2021, March 5). AICPA CPA Exam changes. CPA Licensure One Step Closer to Change. Retrieved March 25, 2023, from https://www.aicpa.org/news/article/cpa-licensure-one-step-closer-to-change.html

6. Analytics Vidhya. (2022, August 8). Dealing with outliers using the Z-Score method. Retrieved from https://www.analyticsvidhya.com/blog/2022/08/dealing-with-outliers-using-the-z-score-method/

7. Anuganti Suresh (2020). What is a confusion matrix? https://medium.com/analytics-vidhya/what-is-a-confusion-matrix-d1c0f8feda5

8. APACCIOoutlook (2019), JACKSOFT: RegTech Bots in Action, https://www.apacciooutlook.com/jacksoft

9. Claire Reilly (2018), Robots don't want to take your miserable office job, https://www.cnet.com/science/robots-dont-want-to-take-your-miserable-office-job/

10. Cnyes News. (2020). If Auditors Cannot Go to Mainland China for Audit, FSC Allows Video Audit of Annual Reports. https://news.cnyes.com/news/id/4446011

11. David Denyer. (2017), Organizational Resilience, https://www.cranfield.ac.uk/-/media/images-for-new-website/som-media-room/images/organisational-report-david-denyer.ashx (Source: Cranfield University)

12. Delen, Dursun & Ram, Sudha. (2018). Research challenges and opportunities in business analytics. Journal of Business Analytics. 1. 2-12. 10.1080/2573234X.2018.1507324.

13. Financial News. (2020, February 26). Impact of the Epidemic on Financial Statement Announcements, FSC: Auditors can adopt alternative solutions. TechNews. https://finance.technews.tw/2020/02/26/accountants-can-use-alternatives-for-auditing-financial-statements/

14. Galvanize. (2021). Death of the Tick Mark. Retrieved from https://www.wegalvanize.com/assets/ebook-death-of-tickmark.pdf

15. H. J. Will, 1983, ACL: a language specific for auditors, Communications of the ACM, Volume 26 Issue 5 pp 356–361 https://doi.org/10.1145/69586.358138

16. Huang, S. F. (2011). JOIN Data Comparison Analysis - Analysis Report of Unauthorized False Transaction Audit Activity. Audit Automation, 013, ISSN:2075-0315.

17. Huang, S. M. (2022). ACL Data Analysis and Computer Audit Handbook (8th ed.). Chuan Hwa Book Co.. ISBN 9786263281691.

18. Huang, S. M., Yen, J. C., Ruan, J. S., et al. (2013). Computer Auditing: Theory and Practice (2nd ed.). Chuan Hwa Book Co.

19. Huang, S. M., Huang, S. F., & Chou, L. Y. (2013). Big Data Era: New Challenges for Audit Data Warehouse Construction and Application. Accounting Research Monthly, 337, 124-129.

20. Huang, S. M., Chou, L. Y., & Huang, S. F. (2013). Development Trends in Audit Automation. Accounting Research Monthly, 326.

21. Huang, S. & Huang, S. M. (2017). J-CAATs: a Cloud Data Analytic Platform for Auditors, 2017 International Conference on Computer Auditing, London, UK.

22. ICAEA (2023), ICAEA Code of Ethics and Professional Practice, https://www.icaea.net/English/Certification/Code_of_Ethics.php

23. ICAEA (2023). CAATs (Computer Assisted Audit Techniques) Training Courses. Retrieved from https://www.icaea.net/English/Training/CAATs_Courses_Free_JCAATs.php

24. ICAEA (2023), International Computer Auditing Education Association, https://www.icaea.net.

25. ICAEA (2019, October 2). Audit data analytic case contest [LinkedIn post]. Retrieved from https://www.linkedin.com/groups/10357617/

26. Liu, B. (2011, September 7). Nature language processing. Retrieved from https://www.cnblogs.com/yuxc/archive/2011/09/07/2170385.html

27. Nguyet, T. (2022, February 15). Learn and code confusion matrix with Python. Retrieved from https://www.nbshare.io/notebook/626706996/Learn-And-Code-Confusion-Matrix-With-Python/

28. Oxford Martin Programme on Technology and Employment. (2013). Future of employment. Oxford, UK: Author. Retrieved from https://www.oxfordmartin.ox.ac.uk/downloads/academic/future-of-employment.pdf

29. Phil Leifermann, Shagen Ganason. (2021). Internal Audit Department of Tomorrow, 2021 IIA International Conference, Singapore.

30. Python Software Foundation. (n.d.). Welcome to Python.org. Retrieved from https://www.python.org/

31. The Economist. (2014, January 18). The onrushing wave. Retrieved from https://www.economist.com/briefing/2014/01/18/the-onrushing-wave

32. U.S. Department of the Treasury. (2023). OPEN DATA - the SDN sanctions list from the Office of Foreign Assets Control (OFAC). Retrieved from https://home.treasury.gov/policy-issues/financial-sanctions/specially-designated-nationals-and-blocked-persons-list-sdn-human-readable-lists

33. Wang, T. and Huang, S. (2019), Computer Auditing: The Way Forward, International Journal of Computer Auditing, Vol.1, No.1, pp.1- 3. https://doi.org/10.53106/256299802019120101001

34. Wikipedia (2023, March 11). Tf-idf. In Wikipedia. Retrieved from https://en.wikipedia.org/wiki/Tf-idf

35. Wikipedia (2023, March 11). Benford's law. In Wikipedia. Retrieved from https://en.wikipedia.org/wiki/Benford%27s_law

36. Wikipedia (2023, March 11). Levenshtein distance, In Wikipedia. Retrieved from https://en.wikipedia.org/wiki/Levenshtein_distance

37. Yahoo! News. (2022, February 22). Exposure of False Accounts! Kangyou KY Caused Investors to Lose NT$4.7 Billion, Ernst & Young is ruled by the court for false seizure. https://tw.news.yahoo.com/news/%E5%85%A8%E6%96%87-%E5%81%87%E5%B8%B3%E6%9B%9D%E5%85%89-%E5%BA%B7%E5%8F%8Bky%E5%AE%B3%E6%8A%95%E8%B3%87%E4%BA%BA%E6%90%8D%E5%A4%B147%E5%84%84-%E5%8B%A4%E6%A5%AD%E7%9C%BE%E4%BF%A1%E9%81%AD%E6%B3%95%E9%99%A2%E8%A3%81%E5%AE%9A%E5%81%87%E6%89%A3%E6%8A%BC-215859185.html

Appendix B Additional Learning Resources

About the Author

Sherry Huang

She is a CEO of Jacksoft Ltd. which is one of top audit technology innovative company in Asia. Sherry has more than 20 years of experience in the IT service industry, providing consultancy services to over 500 enterprises in the audit industry. She currently serves as the chair of the ICAEA Taiwan chapter and previously held the position of director of professional development for IIA Taiwan. Sherry also has extensive experience serving as Chief of Internal Auditor and Chief of Accounting for a public company in Taiwan. Before that, she worked as a Senior Auditor at KPMG. Sherry is widely recognized for inventing JCAATs, which were first released in London in 2017. She holds an MBA and various professional certifications, including CIA, CCSA, ICCP, CEAP, CFAP, ISO 14067:2018, ISO27001, ACDA, etc.

JCAATs - Data Analysis and Smart Audit

Author / Sherry Huang

Publisher / Sherry Huang

Publisher / JACKSOFT COMMERCE AUTOMATION LTD.

Address / 3F-2, No. 180, Chang'an West Road, Datong District, Taipei City, Taiwan

Tel / +886-2-2555-7886

Website / www.jacksoft.com.tw

Published Month / March 2023

Edition / 1 edition

ISBN/ 978-626-97151-1-4